SRA Number Worlds

A Prevention/Intervention Math Program

Teacher Edition
Level C

Sharon Griffin

with Building Blocks Douglas H. Clements
Julie Sarama

SRA
Columbus, OH

The McGraw·Hill Companies

Author
Sharon Griffin
*Associate Professor of Education and
Adjunct Associate Professor of Psychology*
Clark University
Worcester, Massachusetts

Building Blocks Authors

Douglas H. Clements
*Professor of Early Childhood
and Mathematics Education*
University at Buffalo
State University of New York, New York

Julie Sarama
Associate Professor of Mathematics Education
University at Buffalo
State University of New York, New York

Contributing Writers
Sherry Booth, *Math Curriculum Developer,* Raleigh, North Carolina
Elizabeth Jimenez, *English Language Learner Consultant,* Pomona, California

Program Reviewers

Jean Delwiche
Almaden Country School
San Jose, California

Cheryl Glorioso
Santa Ana Unified School District
Santa Ana, California

Sharon LaPoint
School District of Indian River County
Vero Beach, Florida

Leigh Lidrbauch
Pasadena Independent School District
Pasadena, Texas

Dave Maresh
Morongo Unified School District
Yucca Valley, California

Mary Mayberry
Mon Valley Education Consortium, AIU 3
Clairton, Pennsylvania

Lauren Parente
Mountain Lakes School District
Mountain Lakes, New Jersey

Juan Regalado
Houston Independent School District
Houston, Texas

M. Kate Thiry
Dublin City School District
Dublin, Ohio

Susan C. Vohrer
Baltimore County Public Schools
Baltimore, Maryland

SRAonline.com

Copyright © 2007 SRA/McGraw-Hill.

All rights reserved. Except as permitted under the United States Copyright Act, no part of this publication may be reproduced or distributed in any form or by any means, or stored in a database or retrieval system, without the prior written permission of the publisher, unless otherwise indicated.

Printed in the United States of America.

Send all inquiries to:
SRA/McGraw-Hill
4400 Easton Commons
Columbus, OH 43219

ISBN 0-07-605337-7

13 14 15 16 17 RMN 15 14 13 12 11

Acknowledgments
Development of the **Number Worlds** program was made possible by generous grants from the James S. McDonnell Foundation. The author gratefully acknowledges this support as well as the contributions of all the teachers and children who used the program in various stages of development and who helped shape its current form.

Building Blocks was supported in part by the National Science Foundation under Grant No. ESI-9730804, "Building Blocks-Foundations for Mathematical Thinking, Pre-Kindergarten to Grade 2: Research-based Materials Development" to Douglas H. Clements and Julie Sarama. The curriculum was also based partly upon work supported in part by the Institute of Educational Sciences (U.S. Dept. of Education, under the Interagency Educational Research Initiative, or IERI, a collaboration of the IES, NSF, and NICHHD) under Grant No. R305K05157, "Scaling Up TRIAD: Teaching Early Mathematics for Understanding with Trajectories and Technologies" and by the IERI through a National Science Foundation NSF Grant No. REC-0228440, "Scaling Up the Implementation of a Pre-Kindergarten Mathematics Curricula: Teaching for Understanding with Trajectories and Technologies." Any opinions, findings, and conclusions or recommendations expressed in this material are those of the authors and do not necessarily reflect the views of the funding agencies.

Photo Credit
C12 ©PhotoDisc/Getty Images, Inc.

Build a solid foundation with

Make a world of difference in your students' **math skills** with a versatile program that has **proven results** in the classroom.

Number Worlds is an intensive intervention program that focuses on students who are one or more grade levels behind in elementary mathematics. It provides all the tools teachers need to assess students' abilities, individualize instruction, build foundational skills and concepts, and make learning fun. And only *Number Worlds* includes a prevention program for Grades Pre-K–1. It's a unique course full of activities that build foundational math skills and prepare younger children to understand more complex concepts later.

☑ Targeted instruction
Through intense identification and development of core concepts, *Number Worlds* creates competency that quickly puts students on-level with their peer groups. The program provides hands-on activities proven effective with even the lowest-level students, including:
- Computer activities
- Discussion activities
- Paper-and-pencil activities

☑ Precise assessment
Number Worlds' easy-to-use assessment component pinpoints the exact unit in which students should begin the curriculum. Weekly and unit tests monitor progress with open response and/or multiple-choice questions to identify when students are ready to return to the main math curriculum.

☑ Flexibility for teachers and students
Number Worlds' lessons are flexible for use in many settings:
- Resource room
- After school
- Summer school

Teacher's aides and parents can use *Number Worlds* after class or at home.

☑ Comprehensive, fully-integrated program
Complete *Number Worlds* program kits are available at every level and include:
- Teacher Edition
- Student workbooks (Levels C–H)
- Student worksheet blackline masters
- Manipulatives (Levels A–F)
- Software for assessment, placement, professional development, and activities

Effective Resources

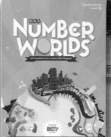

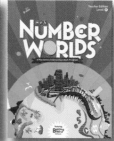

Number Worlds
Teacher Editions

Number Worlds
Manipulatives and Games

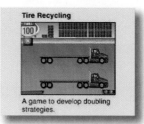

Building Blocks
Screen Capture

Program Components
- Teacher Edition
- Student Workbooks
- Assessment
- Manipulatives
- Games
- Technology featuring **Building Blocks** software

T4   Number Worlds • Resources

Effective Results

Number Worlds Results

Number Worlds has been developed and refined since the mid-1980s and has been the only such program to show proven results through years of rigorous field testing. These tests show how students who began at a disadvantage surpassed the performance of students who began on-level with their peers, simply with the help of the **Number Worlds** program.

One of the tests was a longitudinal study conducted to measure the progress of three groups of children from the beginning of Kindergarten to the end of Grade 2. The treatment and control groups both tested one to two years behind normative measures in mathematical knowledge, while the normative group was on track. The treatment group received the **Number Worlds** program while the other two groups used a variety of other mathematical programs during the entire course of the study.

The chart below shows the progress of each group in mean developmental mathematics-level scores as measured by the **Number Worlds** test. The treatment group using the **Number Worlds** program met and exceeded normative mean developmental-level scores by the end of Grade 2. Meanwhile, the control group continued to fall behind their peers.

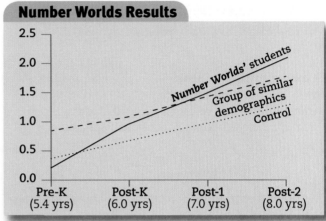

Longitudinal study showing mean developmental scores in mathematical knowledge during Grades K–2

Building Blocks Software Results

The *Building Blocks* software, incorporated into the **Number Worlds** program, is the result of National Science Foundation-funded research. *Building Blocks* includes research-based computer tools with activities and a management system that guides children through research-based learning trajectories.

The program is designed to
- Build upon young children's experiences with mathematics with activities that integrate ways to explore and represent mathematics
- Involve children in "doing mathematics"
- Establish a solid foundation
- Develop a strong conceptual framework
- Emphasize the development of children's mathematical thinking and reasoning abilities
- Develop learning in line with state and national standards

In research studies, *Building Blocks* software was shown to increase young children's knowledge of multiple essential mathematical concepts and skills. One study tested *Building Blocks* against a comparable preschool math program and a no-treatment control group. All classrooms were randomly assigned, the "gold standard" of scientific evaluation. *Building Blocks* children significantly outperformed both the comparison group and control group of children. Results indicate strong positive effects with achievement gains near or exceeding those recorded for individual tutoring.

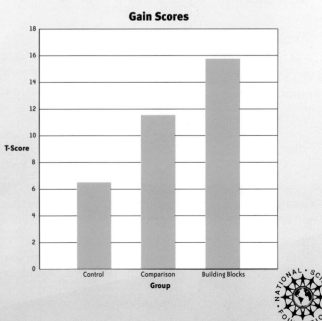

Get students back on track with confidence

Number Worlds is the only program that includes a prevention instruction section for students in **Grades Pre-K–1**. This unique 30-week course of daily instruction improves students' grasp of the world of math so they can move forward with the head start they need.

Prevention

Fundamental Concepts Levels A–C

Level A	Level B	Level C
Children acquire well-developed counting and quality schemas.	Children develop a well-consolidated central conceptual structure for single-digit numbers.	Children link their central conceptual structure of number to the formal symbol system.

For students in **Grades 2–6** who are one or more grade levels behind in math, **Number Worlds** intervention program is an invaluable tool. It builds on students' current level of understanding with six 4-week intensive units per grade.

With each daily lesson, teachers have the opportunity to reach the program's knowledge objectives through problem-solving activities, small-group interaction, and discussion. By asking good questions in the classroom, teachers can encourage learning while defining areas that require extra work, ultimately helping to bring students to their appropriate grade level.

Intervention

Core Content Topics Levels D–H

Level D	Level E	Level F	Level G	Level H
Math Intervention for Grades				
1–3	2–4	3–5	4–6	5–7
Develops Concepts Covered in Grades				
K–1	1–2	2–3	3–4	4–5
Number Sense	Number Sense	Number Sense	Number Sense	Number Sense
Number Patterns and Relationships (Algebra)	Number Patterns and Relationships (Algebra)	Number Patterns and Relationships (Algebra)	Number Patterns and Relationships (Algebra)	Number Patterns and Relationships (Algebra)
Addition	Addition	Addition & Subtraction	Multiplication	Fractions, Decimals & Percents
Subtraction	Subtraction	Multiplication and Beginning Division	Division	Multiplication & Division
Geometry & Measurement	Geometry & Measurement	Geometry & Measurement	Geometry & Measurement	Geometry & Measurement
Data Analysis & Applications	Data Analysis & Applications	Data Analysis & Applications	Data Analysis & Applications	Data Analysis & Applications

Engage your students with step-by-step lessons

Number Worlds Lesson Planner provides a wide array of helpful information before lessons even begin. Background overviews, activity ideas, and tips prepare teachers to the fullest extent.

Weekly Planners map out an entire week of lessons, complete with pacing options, goals, and the resources necessary to get the most out of every class period.

Manipulative-rich lessons are proven to help students turn **abstract concepts into concrete understanding**.

Math Background gives teachers math **context for the lesson**.

Get **insight into your students' capabilities** and how their minds work.

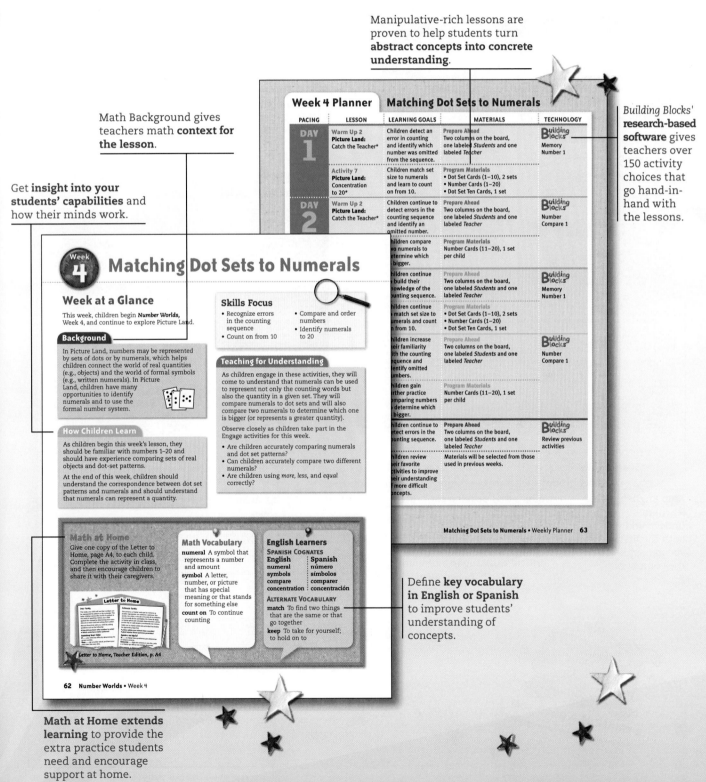

Building Blocks' research-based software gives teachers over 150 activity choices that go hand-in-hand with the lessons.

Define **key vocabulary in English or Spanish** to improve students' understanding of concepts.

Math at Home extends learning to provide the extra practice students need and encourage support at home.

Number Worlds • Lesson Plan   **T7**

Thorough lesson plans guide you every step of the way

Every comprehensive *Number Worlds* lesson is divided into four distinct sections for simplified time management in the classroom. Whether it's time for concept building or skill building, in-depth discussion or assessment, *Number Worlds* always helps you keep learning objectives within reach.

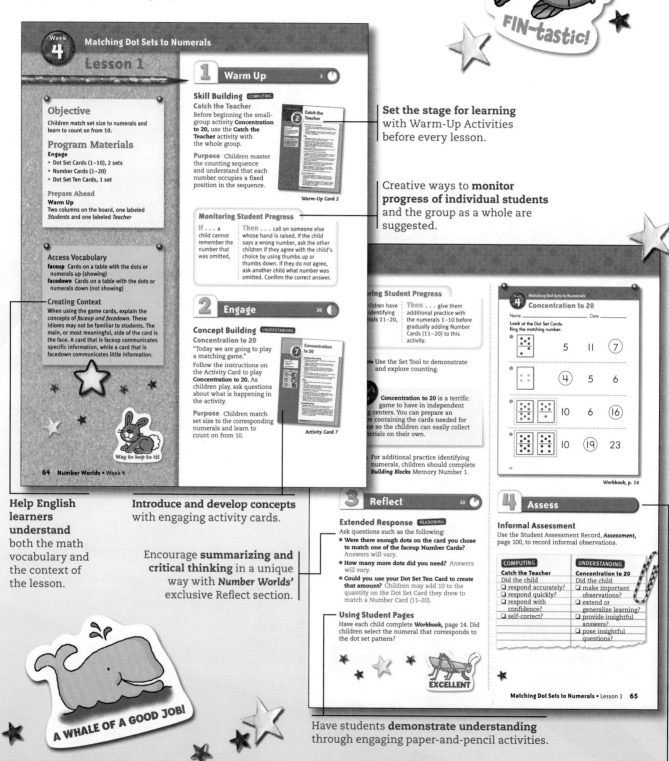

Set the stage for learning with Warm-Up Activities before every lesson.

Creative ways to **monitor progress of individual students** and the group as a whole are suggested.

Help English learners understand both the math vocabulary and the context of the lesson.

Introduce and develop concepts with engaging activity cards.

Encourage **summarizing and critical thinking** in a unique way with *Number Worlds'* exclusive Reflect section.

Have students **demonstrate understanding** through engaging paper-and-pencil activities.

Assess student progress after each lesson.

T8   **Number Worlds** • Lesson Plan

Program Authors

Sharon Griffin, Ph.D., is a Professor of Education and Psychology at Clark University. She specializes in child development and mathematics education and has been studying how playing games that involve numbers helps children structure and understand the world. She conducted research on the development of math competence in the preschool and early school years and used this theoretical work as the basis to create the *Number Worlds* curriculum. Dr. Griffin has worked closely with teachers as they have introduced this curriculum into their classrooms. She has also worked collaboratively with schools around the country and in Canada as they have sought to systematically reform their mathematics programs.

Dr. Griffin's work has been widely published, and she is the author of the chapter "Fostering the Development of Whole-Number Sense: Teaching Mathematics in the Primary Grades" that appeared in a recent book published by the National Research Council. She is a member of the Mathematical Sciences Education Board in the National Academies of Science. She is also involved in an Organization of Economic Cooperation and Development project that brings together leading researchers in neuroscience and cognitive science from several countries to allow each to inform and advance the others' work.

Dr. Griffin holds a B.A. in Psychology from McGill University, a M.Ed. in Learning and Instruction from the University of New Hampshire, and a Ph.D. in Cognitive Science from the University of Toronto.

Douglas H. Clements, Professor of Early Childhood, Mathematics, and Computer Education at the University at Buffalo, State University of New York, has conducted research and published widely on the learning and teaching of geometry, computer applications in mathematics education, the early development of mathematical ideas, and the effects of social interactions on learning. Along with Julie Sarama, Dr. Clements has directed several research projects funded by the National Science Foundation and the U.S. Department of Education's Institute of Educational Sciences, one of which resulted in the mathematics software and activities included in *Building Blocks*.

Julie Sarama is an Associate Professor of Mathematics Education at the University at Buffalo, State University of New York. She conducts research on the implementation and effects of software and curricula in mathematics classrooms, young children's development of mathematical concepts and competencies, implementation and scale-up of educational reform, and professional development. Dr. Sarama has taught secondary mathematics and computer science, gifted mathematics at the middle school level, and preschool and Kindergarten mathematics methods and content courses for elementary to secondary teachers.

Sherry Booth is a mathematics curriculum specialist. Her past projects include the JASON web-based mathematics courses, the ATLAS Project, and the Math Partners project funded by the National Science Foundation. She has collaborated with researchers and designers to develop mathematics curricula that includes software, video, teacher guides, and student materials.

Contents

Getting Started T14

| Week 1 | *Overview* .... | **26** |
| Week 1 | Counting .... | **28** |

| Week 2 | *Overview* .... | **38** |
| Week 2 | Counting and Comparing .... | **40** |

| Week 3 | *Overview* .... | **50** |
| Week 3 | More Counting and Comparing .... | **52** |

| Week 4 | *Overview* .... | **62** |
| Week 4 | Matching Dot Sets to Numerals .... | **64** |

| Week 5 | *Overview* .... | **74** |
| Week 5 | Number Sequence and Number Lines .... | **76** |

| Week 6 | *Overview* .... | **86** |
| Week 6 | More Number Sequence and Number Lines .... | **88** |

| Week 7 | *Overview* .... | **98** |
| Week 7 | Number Neighborhoods .... | **100** |

| Week 8 | *Overview* .... | **110** |
| Week 8 | More Number Neighborhoods .... | **112** |

| Week 9 | *Overview* .... | **122** |
| Week 9 | Adding Numbers .... | **124** |

| Week 10 | *Overview* .... | **134** |
| Week 10 | More Adding .... | **136** |

Contents

Week 11	*Overview*	**146**
Week 11	Sequencing Numbers	**148**
Week 12	*Overview*	**158**
Week 12	Writing Equations	**160**
Week 13	*Overview*	**170**
Week 13	Counting and Adding	**172**
Week 14	*Overview*	**182**
Week 14	Making Equations	**184**
Week 15	*Overview*	**194**
Week 15	Graphing and Comparing Numbers	**196**
Week 16	*Overview*	**206**
Week 16	More Counting and Adding	**208**
Week 17	*Overview*	**218**
Week 17	Solving Equations	**220**
Week 18	*Overview*	**230**
Week 18	Adding and Subtracting	**232**
Week 19	*Overview*	**242**
Week 19	Subtracting	**244**
Week 20	*Overview*	**254**
Week 20	Subtracting and Predicting	**256**

Contents

Week 21	*Overview*	**266**
Week 21	Adding and Comparing	**268**
Week 22	*Overview*	**278**
Week 22	Subtracting to Zero	**280**
Week 23	*Overview*	**290**
Week 23	More Adding and Subtracting	**292**
Week 24	*Overview*	**302**
Week 24	Numbers to 100	**304**
Week 25	*Overview*	**314**
Week 25	More Numbers to 100	**316**
Week 26	*Overview*	**326**
Week 26	Addition Stories	**328**
Week 27	*Overview*	**338**
Week 27	Tens and Ones	**340**
Week 28	*Overview*	**350**
Week 28	Adding and Subtracting Length	**352**
Week 29	*Overview*	**362**
Week 29	Addition and Subtraction Stories	**364**
Week 30	*Overview*	**374**
Week 30	Making a Map	**376**

Contents

Appendix

Letters to Home **A1**

Activity Sheets **B1**

About Math Intervention **C1**

Building Number Sense with *Number Worlds* **C2**

Math Proficiencies **C7**

Content Strands of Mathematics **C9**

Technology Overview **C12**

Building Blocks **C13**

eMathTools **C15**

English Learners **C16**

Learning Trajectories **C17**

Trajectory Progress Chart **C31**

Glossary **C33**

Scope and Sequence **C37**

Getting Started

Preparing to Use *Number Worlds*

This section provides an overview of classroom management issues and explanations of the Number Worlds program elements and how to use them.

Program Goal

Number Worlds was designed to foster the development of good intuitions about number and number environments. It also was designed to give children a desire to explore such environments and a sense of confidence in moving around within them. Children who develop a solid base of Number Sense have developed the core foundation on which all higher-order understandings are built.

Levels A–C (PreKindergarten, Kindergarten, and First Grade) are intended to serve as **Prevention** for later problems in mathematics. The thoroughly tested, engaging activities bring all students up to level so that they are ready for elementary mathematics. The activities are specifically targeted to address foundational understandings, but are engaging for all students.

Levels D–H (Grades 2–6) provide **Intervention** for students who are 1–2 years behind their peers in mathematics. Targeted units focus on computing, understanding, reasoning, applying, and engaging—the specific concepts and skills that build mathematics proficiency.

Program Objectives

- Teach concepts and skills that build a strong foundation for later learning
- Expose children to the major ways numbers are represented in the world—as objects, symbols, horizontal and vertical lines, and on dials
- Ensure that children acquire the interconnected knowledge that underlies number sense
- Include hands-on and computer activities that provide concrete representations of concepts
- Encourage communication using the language and vocabulary of formal mathematics
- Develop concept understanding with activities that are engaging and appropriate for children from all social and cultural backgrounds

Building Blocks

In addition to hands-on and workbook activities, **Number Worlds** includes **Building Blocks** software to provide additional exposure to and practice with foundational math concepts. **Building Blocks** software is an essential element of the **Number Worlds** curriculum.

Building Blocks software has these advantages

- It combines visual displays, animated graphics, and speech.
- It links "concrete" (graphical) and symbolic (e.g., numerals or spoken words) representations, which build understanding.
- It provides feedback.
- It provides opportunities to explore.
- It focuses children's attention and increases their motivation.
- It individualizes—gives children tasks at children's own ability levels.
- It provides undivided attention, proceeding at the child's pace.
- It keeps a variety of records.
- It provides more manageable manipulatives (e.g., manipulatives "snap" into position).
- It offers more flexible and extensible manipulatives (e.g., manipulatives can be cut apart).
- It provides more manipulatives (you never run out!).
- It stores and retrieves children's work so they can work on it again and again, which facilitates reflection and long-term projects.
- It records and replays children's actions.
- It presents clearer mathematics (e.g., using tools such as a "turn tool," helps children become aware of mathematical processes).

Building Blocks software stores records of how children are doing on every activity. It assigns them to just the right difficulty level. You can also view records of how the whole group or any individual is doing at any time.

Usage Models

The **Number Worlds** program is designed for flexible use. Each lesson is designed to take from 45–60 minutes. It is highly recommended that students spend at least one hour daily using the **Number Worlds** program.

Intervention Models

The program can be effectively used in each of the following environments:

• During Class Time

The Prevention program is designed for whole-class implementation at PreK–1. It can also be used in the same way as the Intervention program with a teacher or teacher's aide working with small groups apart from the rest of the class.

• Math Resource Room

Number Worlds is effective used in a math resource room with a teacher or teacher's aide working with a small group of students.

• After School

Because the activities are engaging, after school programs have used **Number Worlds** effectively for both intervention and regular education students. Both benefit from the intensive math activities.

• Intervention Classrooms

Number Worlds is very effective at every grade level for classes that need intervention in mathematics.

• Summer School

Number Worlds is ideal for summer school programs with an intensive focus on mathematics. Depending on the length of the summer school session, students can work through more than one lesson a day and make substantial progress.

• Tutoring

Number Worlds is also very effective when used in a one-on-one tutoring situation in which a teacher or aide participates in activities with the child.

Many teachers who have used the **Number Worlds** program have found that a teaching assistant—a student teacher, a teacher aide, or a parent volunteer—can help students become extremely familiar with the participatory structures that are required to teach the program so students are able to assume more control over their own learning and become less reliant on the help of a guide or coach. In fact, many **Number Worlds** teachers find it useful to institute "mini-teachers," students who have experience with an activity and understand its concepts, to help their peers.

Although **Number Worlds** activities can be effective with on-level students, they are truly beneficial for students who are not making adequate progress in their core program.

Supplemental Intervention

Number Worlds is geared for students identified with math deficiencies and who have not responded to reteaching efforts. It provides scientifically-based math instruction emphasizing the five critical elements of mathematics proficiency: understanding, computing, applying, reasoning, and engaging. This can be accomplished in a variety of environments for 45–90 minutes per day with a teacher or teacher's aide.

Intensive Intervention

Number Worlds is also effective for students with low skills and a sustained lack of adequate progress in mathematics. The program provides intensive focus on developing mathematical understanding and skills, and includes explicit instruction designed to meet the individual needs of struggling students. This can be accomplished when **Number Worlds** is used as replacement of core lessons for 60–90 minutes a day. The program must be implemented by a teacher or specialized math teacher or teacher's aide.

Understanding Students

For successful math intervention, it is imperative that teachers understand what students know and don't know about math. Teachers also need to know where students' current knowledge fits within the expected developmental sequence, what knowledge students have available to build upon, and what knowledge—the next steps in the sequence—students have yet to master. Finally, teachers need to become familiar with the problem-solving strategies students are using and the range of strategies they need to acquire to become efficient mathematics thinkers and problem-solvers. Using the Number Worlds program and the assessment tools it provides will help teachers acquire a rich understanding of their students along all of these dimensions.

Grades PreK–1 (Levels A–C)

Children with impoverished math backgrounds may come to preschool already 1–2 years behind their peers. Adults frequently overestimate children's understanding of number. These children may not have had experience playing board games, singing counting songs, or participating in number experiences at home. These children may not realize that the same number on a line and on a clock face represent the same quantity.

In the Levels A–C *Number Worlds* program, children explore five different ways number and quantity is represented. Each of the five "lands" of *Number Worlds* that children encounter in the lower grades exposes children to a different representation of number and helps children learn the language used to talk about number in that context.

In **Object Land** students explore the world of counting numbers by counting and comparing sets of objects or pictures of objects. In Object Land you might ask:
- How many or few do you have?
- Which is bigger or smaller?

In **Picture Land** numbers are represented as sets of stylized, semi-abstract dot-set patterns such as in a die and also as tally marks and numerals. In Picture Land you might ask:
- What did you roll/pick?
- Which has more or less?

In **Line Land** number is represented as a position on a path or a line. The language used for numbers in Line Land refers to a particular place on a line and also to the moves along a line. These types of questions are asked in Line Land:
- Where are you now? How far did you go?
- Who is farther or less far along the line?
- Do you go forward or backward?

In **Sky Land** number is represented as a position on a vertical scale such as on a thermometer or a bar graph. Sky Land inspires these questions:
- How high or low are you now?
- What number or amount is higher or lower?

In **Circle Land** number is represented as a point on a dial, such as a clock face or a sundial. In Circle Land you might ask:
- How many times did you go around the dial?
- Which number is farther or less far around?

Grades 2–6 (Levels D–H)

At grades 2–6 (Levels D–H) students may have difficulty with one, two, or many different math concepts. The goal of the upper grades is to develop foundational understandings in each concept so that students develop mathematical proficiency. Every lesson involves activities that actively engage students in understanding, computing, applying, reasoning, and engagement with the concept. At the end of each week, an assessment will help teachers determine whether a student has reached proficiency. The units are carefully sequenced to develop concepts across grade levels and can be used flexibly to meet student needs.

Program Resources

A variety of program materials are designed to help teachers provide a quality mathematics curriculum.

Teacher Edition
The **Teacher Edition** is the heart of the **Number Worlds** curriculum. It provides background for teachers and complete lesson plans with thorough instructions on how to develop math concepts. It explains when and how to use the program resources.

Activity Cards
These cards are packaged so that each can be carried around and used to direct or teach an activity. The activities, particularly at grades PreK–1 are employed again and again. The Activity Cards provide detailed descriptions of each activity, with suggested questions to ask during the activity.

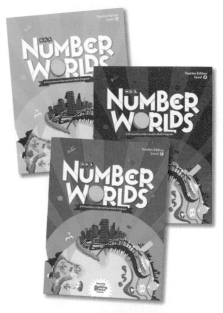

Activity Sheets
In the **Teacher Edition,** you will find numbered Activity Sheets in blackline master form that accompany activities developed for **Number Worlds.** These sheets can be copied in sufficient quantity for the number of children in your class who will be using that activity on a particular day.

Student Workbook
The **Student Workbook** includes developmental activities to help students develop higher-order thinking skills and practice basic skills.

Assessment
Number World's flexible assessment component allows for prescriptive placement and reliable curriculum and criterion assessment.

Each level of the program includes the following:

- **Placement Tests** to identify where students should begin the **Number Worlds** curriculum.
- **Weekly Tests** to measure student comprehension of the week's five daily lessons.
- **Unit Tests** to evaluate concept acquisition for the entire unit. These tests are available in both open-response and multiple-choice formats.
- **Rubrics** to informally evaluate student understanding of each lesson.
- The **Number Knowledge Test,** designed to measure students' intuitive knowledge of number—the knowledge that helps children make sense of quantitative problems.

Manipulatives and Technology

Number Worlds provides a wealth of manipulative resources. Establishing procedures for use of materials will simplify management issues and allow students to spend more time developing mathematical understanding.

Intervention Packages

A *Number Worlds Intervention Package* is available at every level and includes everything small groups need to use the program.
- Teacher Guide
- Assessment
- Student Workbooks
- Activity Manipulatives and Props
- Software

The package provides storage space for the program so it can be stored and moved as convenient. Activity game mats, manipulatives, demonstration props, playing cards, and pieces are all included. Students will benefit from a demonstration on how you want students to remove materials and return materials to the kit.

Manipulative Modules

Manipulatives are also available for specific topics: Counting, Base-Ten, Fractions, Geometry, Measurement, Money, and Time. Every unit of *Number Worlds* uses manipulative material. If these manipulatives are not already an integral part of your math curriculum, they are available from SRA.

Manipulative Topic Modules Grades K–6

Base-Ten: Cubes, Flats, Rods, Units
ISBN 0-07-605418-7

Geometry I: Pattern Blocks, Attribute Blocks, Solids, Mirror Cards
ISBN 0-07-605414-4

Geometry II: Geoboards, Protractors, Safety Compass, Gummed Tape
ISBN 0-07-605415-2

Measurement I: Rulers, Tape Measures, Measuring Cups, Liter Pitcher, Double-Pan Balance, Thermometer
ISBN 0-07-605416-0

Measurement II: Platform Scale, Metric Weight Set, Customary Weight Set
ISBN 0-07-605417-9

Money: Pennies, Nickels, Dimes, Quarters, Half Dollars, Money Packet
ISBN 0-07-605419-5

Counting: Craft Sticks, Rubber Bands, Panda Bear Counters, Math-Link Cubes
ISBN 0-07-605412-8

Probability: Spinners, Counters
ISBN 0-07-605420-9

Time: Clock Faces, Stopwatches
ISBN 0-07-605421-7

Fractions: Fraction Tiles, Fraction Circles
ISBN 0-07-605413-6

Technology Resources

For Students

Building Blocks activities are engaging, research-based software activities designed to reinforce levels of mathematical development in different strands of mathematics.

eMathTools are electronic tools to help students solve problems and explore and demonstrate concepts.

For Teachers

eAssess An assessment tool to grade, track, and report student progress

For more information on *Number Worlds* manipulatives and software see Appendix C.

Program Organization

Levels A–C
(Grades PreK–1)
Includes 30 weeks of daily lessons.

Levels D–H (Grades 2–6)
Each level includes 6 4-week units that address specific concepts and skills for a total of 24 weeks of instruction. Units are carefully sequenced to develop concepts for students who are one to two grade levels behind their peers. Placement Tests in the **Assessment** Book help teachers place students at the appropriate level and concept based on their demonstrated understanding.

Weekly Overview
- **Teaching for Understanding** provides the big ideas of the chapter.
- **Background** provides a refresher of the mathematics principles relevant to the chapter.
- **How Children Learn** offers insight into how children learn and gives research-based teaching strategies.
- **Weekly Planner** includes objectives that explain how the key concepts are developed lesson by lesson and which resources can be used with each lesson.
- **Lessons** provide overview, ideas for differentiating instruction, complete lesson plans, teaching strategies, and assessments that inform instruction.
- **Cumulative Tests** are provided incrementally to allow you to evaluate whether students are retaining previously developed concepts and skills.

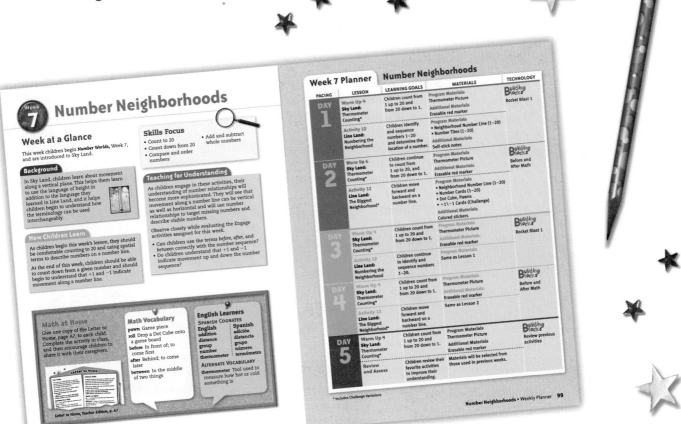

Teacher Edition, Level C

Routines

Lesson Plans

Every lesson throughout *Number Worlds* is structured the same way.

1. Warm Up
2. Engage
3. Reflect
4. Assess

Routines for each part of the lesson are explained in the following discussions:

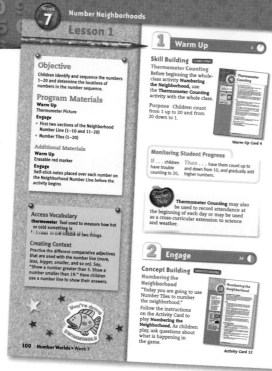

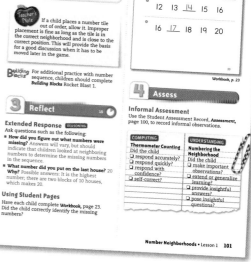

Teacher Edition, Level C

Warm Up exercises provide cumulative review and computation practice for students and give you opportunities to assess students' skills quickly. Warm Up is an essential component of *Number Worlds* because it helps students review concepts they will need in the Engage activities. It also gives students daily opportunities to sharpen their counting and mental math skills.

Begin every day with the lesson Warm Up with the entire group. These short activities are used in a whole-group format at the start of each day's lesson.

2 Engage

Engage is the heart of the lesson instruction. Here are suggestions for how to introduce lesson concepts, ideas for Guided Discussion, Skill Building, Game Demonstrations, and Strategy Building activities to develop student understanding. Before beginning the Engage activity, familiarize yourself with the lesson objective, and make sure materials are available for student use. Introduce the activity, and use math talk to help students explore the lesson concepts.

Math Talk is expected in every *Number Worlds* lesson. When students speak the language of mathematics, they communicate mathematically, explain their thinking, and demonstrate understanding. Routines or rules for Math Talk established at the beginning of the year can make discussions more productive and promote listening and speaking skills.

1. **Pay attention to others.** Give your full attention to the person who is speaking. This includes looking at the speaker and nodding to show that you understand.
2. **Wait for speakers to answer and complete their thoughts.** Sometimes teachers and other students get impatient and move on and ask someone else or give the answer before someone has a chance to think and speak. Giving students time to answer is a vital part of teaching for understanding.
3. **Listen.** Let yourself finish listening before you begin to speak. You can't listen if you are busy thinking about what you want to say next.
4. **Respect speakers.** Take turns and make sure that everyone gets a chance to speak and that no one dominates the conversation.
5. **Build on others' ideas.** Make connections, draw analogies, or expand on the idea.
6. **Ask questions.** Asking questions of another speaker shows that you were listening. Ask if you are not sure you understand what the speaker has said, or ask for clarification or explanation. It is a good idea to repeat in your own words what the speaker said so you can be sure your understanding is correct.

Routines

The **Activity Cards** include suggestions for some of the many questions that can be asked during an activity. The questions were selected because they are the kinds of questions children will be able to learn "by heart" and ask themselves. Eventually, the children can and should assume the role of asking the questions during a game. This will help them to become independent learners, responsible for their own learning.

The kinds of questions you ask will depend upon the way in which number is represented, the stage during the activity that you are asking the questions, and the level of difficulty the children are ready to explore. Whatever the combination of these three elements, you should ask the children as often as applicable, "How do you know?" and "How did you figure it out?"

Familiarizing children with these different ways of talking about number and quantity is a major goal of the **Number Worlds** program. It will enable children to realize, for example, that words such as *bigger, more, farther, higher,* and *further around* can all refer to an increase in quantity of the same magnitude, even though this change is expressed in different words, and very likely is represented in a different form. This understanding lies at the heart of number sense.

Using Student Pages

In every week of levels C–H, you will assign student pages to be completed during class. Students will finish at different times and should know what they can do to use their time productively until the Reflect part of the lesson. Students should not feel penalized for finishing early but should do something that is mathematically rewarding.

Student Workbook exercises in Number Worlds are primarily non-mechanical. Students cannot do all the problems on a page in a mechanical, non-thinking manner. Student workbook pages help the students learn to think about the problems.

Because student workbook exercises are non-mechanical, they sometimes require your active participation.

1. Make sure students know what pages to work on and any special requirements of those pages.
2. Tell students whether they should work independently or in small groups as they complete the pages.
3. Tell students how long they have to work on the student pages before you plan to begin the Reflect part of the lesson.

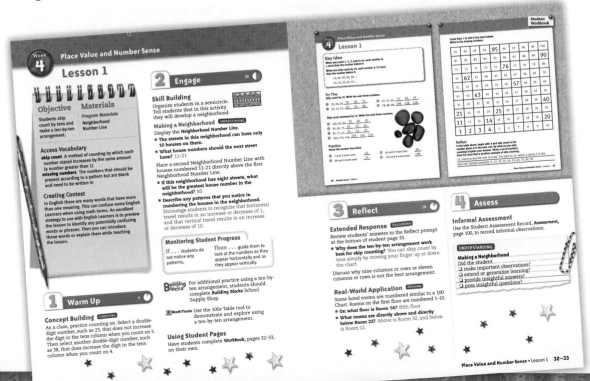

Teacher Edition, Level F

Routines

4. Tell students what their options are if they finish early. Technology resources are listed under the Assign Student Pages heading in each lesson. These activities further explore and develop lesson concepts. They include
   a. suggested **eMathTools** to use.
   b. **Building Blocks** activities to do.
5. As students work on the student pages, circulate around the room to monitor their progress. Use the Monitoring Student Progress suggestions for ideas on what to look for. Comment positively on student work, and stop to ask exploring, synthesizing, clarifying, or refocusing questions.
6. You may also use this time to work with English Learners or students who need intervention.
7. Because activities are an important and integral part of the program that provide necessary practice in traditional basic skills as well as higher-order thinking skills, when activities are included in a lesson, be sure to stop work on student pages early enough to leave enough time to play the game. Students may have to complete the student pages outside school in that case.
8. Complete the Informal Assessment Checklists on the last page of each lesson.

3 Reflect

Reflect is a vital part of the lesson that offers ways to help students summarize and reflect on their understanding of lesson concepts. Engaging children in reflection is as important as assessing—and a good way to assess. When children talk about their thinking, using their own words, they engage in mathematical generalizing and communicating. Allowing children to discuss what they did during an activity helps build mathematical reasoning but also develops social skills such as turn taking, listening, and speaking. At the designated time, have students stop working and direct their attention to reflecting on the lesson.

Use the suggested questions in Reflect or ask students to consider these ideas:

A. Summarize their ideas about the lesson concepts
B. Compare how the lesson concept or skill is like or different from other skills
C. Ask how students have seen or can apply the lesson in other curricular areas, other strands of mathematics, or in the world outside of school
D. Think about related matters that go beyond the scope of the lesson

Discuss student solutions to Extended Response questions.

Reflect Questions

A powerful reflection question is, "How do you know?" or "How did you figure that out?" Children may or may not answer you and often cannot provide reasons for their answers. They may shrug their shoulders and say, "I don't know," "Because," or, "Because I'm smart." As the year progresses, children become accustomed to explaining their ideas and their answers give more insight into their mathematical thinking. Young children who have such discussions with teachers and with each other begin to question and correct each other. Incorporate time for reflection into your classroom mathematics activities to develop deep understanding. The following are good questions and challenges.

- How do you know?
- How did you figure that out?
- Why?
- Show me how you did that.
- Tell me about . . .
- How is that the same?
- How is that different?

Routines

4 Assess

Assess helps you use informal and formal assessments to summarize and analyze evidence of student understanding and plan for differentiating instruction.

Goals of Assessment

1. **Improve instruction** by informing teachers about the effectiveness of their lessons
2. **Promote growth** of students by identifying where they need additional instruction and support
3. **Recognize accomplishments**

Phases of Assessment

Planning As you develop lesson plans, you can consider how you might assess the instruction, determining how you will tell if students have grasped the material.

Gather Evidence Throughout the instructional phase, you can informally and formally gather evidence of student understanding. The Informal Assessment Checklist and **Student Assessment Record** are provided to help you record data. The end of every lesson is designed to help in conducting meaningful assessments.

Summarize Findings Taking time to reflect on the assessments to summarize findings and make plans for follow-up is a critical part of any lesson.

Use Results Use the results of your findings to differentiate instruction or to adjust or confirm future lessons.

Number Worlds is rich in opportunities to monitor student progress to accomplish these goals.

Informal Daily Assessment

Informal Daily assessments evaluate students' math proficiencies in computational fluency, reasoning, understanding, applying, and engagement.

Warm-Up exercises, activities, and *Student Workbook* pages can be used for day-to-day observation and assessment of how well each student is learning skills and grasping concepts. Because of their special nature, these activities are an effective and convenient means of monitoring students.

Activities, for example, allow you to watch students practice particular skills under conditions more natural to them than most classroom activities. Warm-Up exercises allow you to see individual responses, give immediate feedback, and involve the entire class.

Simple rubrics enable teachers to record and track their observations. These can later be recorded by hand on the Student Assessment Record or in *eAssess* to help provide a more complete view of student proficiency.

Formal Assessment

The *Student Workbook* and the *Assessment Book* provide formal assessments for each chapter. Included are Entry Tests, Weekly Tests, and Cumulative Reviews to evaluate students' understanding of chapter concepts.

Routines

Classroom Management

The majority of the *Number Worlds* activities involve small groups and materials that will be new and intriguing to your students. Children as young as five years can learn to function well in semi-autonomous small learning groups. A clear-cut set of classroom expectations must be in place to support and encourage independent learning behaviors. Creating this set of expectations and preparing children to work effectively in small groups will be a task you may wish to address right from the start, as early as the first week of school.

Hints for Starting Small-Group Activities

The following suggestions will help you handle several small groups at once. You might decide to use alternative suggestions with different activities.

- Arrange the class into groups, and have the materials ready and divided ahead of time. Lead the activity with all the groups following along at the same time.
- Have the whole group sit in a semicircle around you. Have volunteers demonstrate the activity, giving many children an opportunity to participate. Once the children are familiar with the activity (perhaps on another day), organize them into small groups and have each group do the activity.
- Have a class helper lead a whole-group activity while you take one group at a time.
- Concentrate on one group while the other groups are involved with another, independent activity.

Because managing several small learning groups at once can be a challenge, several aids and devices have been included in the *Number Worlds* program to make this easier.

Activity Cards are available for each activity in the program. They can be carried around by the teacher or teacher's helper and used, on the spot, to direct the activity. The ***Activity Cards*** have a built-in categorization system that will tell you how difficult an activity is, how many children can play, what the children will be doing, what they can learn, and your role in the activity.

Activity Card 48, Level C

The Activity Cards in Levels A–C are also sequenced by difficulty level.

- At Level 1, children generally count and identify quantities.
- At Level 2, children generally match, change, or compare quantities.
- At Level 3, children generally solve more complex problems and predict the results of operations.
- Activities that show more than one difficulty level offer substantial learning opportunities at more than one level.

Activity Sheets are available in this *Teacher Edition*. These activity sheets accompany activities developed for each *Number Worlds* land. These sheets are intended to be copied in sufficient quantities for the number of children in your class who will be using that activity on a particular day.

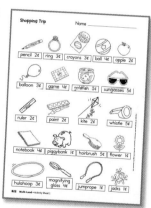

page B22, Level C

Manipulatives are included for all the activities in the five lands. Once the children are familiar with the materials, they can be responsible for gathering them for activities in small groups.

The materials themselves should be organized within the *Number Worlds Intervention Package* so the children can gather the supplies themselves. It is important that you establish a procedure for the management of the materials so that your students will be able to collect and return the materials efficiently.

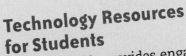

Technology Resources for Teachers

Number Worlds includes several pieces of integrated technology for teachers designed to increase efficiency and effectiveness of instruction and assessment.

eAssess
- **Daily Records** Use **eAssess** to record daily formal and informal assessments.
- **Report Cards** Use **eAssess** to print student and class reports to determine grades.
- **Parent-Teacher Conferences** Use **eAssess** to print student reports to discuss with parents.

Technology Resources for Students

Number Worlds provides engaging technology resources to enrich, apply, and extend learning.

eMathTools are electronic tools that students can use to solve problems, test solutions, explore concepts, or demonstrate understanding.

Building Blocks activities are designed to reinforce key concepts and develop mathematics understanding.

Using Technology

A. Determine rules for computer use and communicate them to students. Rules should include
- sharing available computers. Some teachers have a computer sign-up chart for each computer. Some teachers have the students track these themselves.
- computer time. You might limit the amount of time students can be at the computer or allow students to work in pairs. Some teachers have students work until they complete an activity. Others allow students to continue on with additional activities.

B. Train students on your rules for proper use of computers, including how to turn computers on, load programs, and shut down the computer. Some teachers manage computers themselves; others have an aide or student in charge of computer management.

C. Using the suggestions for eMathTools and Building Blocks, make sure the computers are on and the programs are loaded and that students know how to access the software.

D. At the beginning of each week, demonstrate and discuss any new Building Blocks or eMath Tools computer activities.

E. Make sure all children work on the assigned computer activities individually at least twice per week for about 15 minutes each time.

F. When the children have finished the assigned activities, they should always get the chance to play and learn with the "free explore" activity. This might be individual but is also an excellent opportunity for children to explore cooperatively, posing problems for each other, solving problems together, or just learning through play.

G. Remember that preparation and followup are as necessary for computer activities as they are for any other activities. Do not omit critical whole group discussion sessions following computer work. Help children communicate their solution strategies and reflect on what they've learned.

H. Make sure children make sense of the mathematics.

For more information about Technology, see Appendix C.

Week 1: Counting

Week at a Glance

This week children begin **Number Worlds,** Week 1, and are introduced to Object Land.

Background

In Object Land, numbers are represented as groups of objects. This is the first way numbers were represented historically, and this is the first way children naturally learn about numbers. In Object Land, children work with real, tangible objects, such as Counters.

How Children Learn

Children may enter school having already learned to count and perhaps even to add small groups of objects. They may also understand terms such as *more* and *less*.

This week, children should become as familiar with numbers from 11–100 as they are with numbers from 1–10. Children should also learn to count down just as easily as they count up, which paves the way for subtraction at higher grade levels.

Skills Focus
- Count to 100
- Compare and order numbers
- Predict the next number

Teaching for Understanding

As children play the Object Land activities, they will learn to move back and forth between the world of objects and the world of numbers without counting. For example, children will be able to say which is bigger, seven cents or nine cents, without counting out two sets of objects and comparing them.

Observe closely while evaluating the Engage activities assigned for this week.

- Are children counting to 20?
- Can children order numbers from smallest to biggest?
- Can children predict the next number in a sequence?

Math at Home

Give one copy of the Letter to Home, page A1, to each child. Complete the activity in class, and then encourage children to share it with their caregivers.

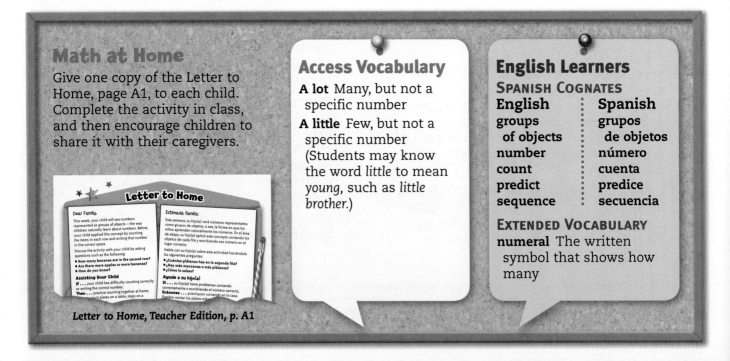

Letter to Home, Teacher Edition, p. A1

Access Vocabulary

A lot Many, but not a specific number

A little Few, but not a specific number (Students may know the word *little* to mean young, such as *little brother*.)

English Learners

SPANISH COGNATES

English	Spanish
groups of objects	grupos de objetos
number	número
count	cuenta
predict	predice
sequence	secuencia

EXTENDED VOCABULARY

numeral The written symbol that shows how many

Week 1 Planner Counting

PACING	LESSON	LEARNING GOALS	MATERIALS	TECHNOLOGY
DAY 1	**Warm Up 1** **Object Land:** Pointing and Winking*	Children count and stop at a specified number, then say the next number up in sequence.	No additional materials required	**Building Blocks** School Supply Shop
	Activity 1 **Object Land:** Drop and Count	Children count to 100 by ones and tens.	**Program Materials** 100 Counters **Additional Materials** • Coffee can or box • Chart paper and marker	
DAY 2	**Warm Up 1** **Object Land:** Pointing and Winking*	Children continue to count and stop at a specified number, then say the next number up.	No additional materials required	**Building Blocks** Dino Shop 1
	Activity 2 **Object Land:** Feed the Animals	Children count a set of objects and arrange quantities from smallest to biggest.	**Program Materials** • Counters, 4 bags with varying quantities from 11 to 20 • Number Cards (11–20) • Zoo Pictures	
DAY 3	**Warm Up 1** **Object Land:** Pointing and Winking*	Children continue to count higher as they master the counting sequence.	No additional materials required	**Building Blocks** School Supply Shop
	Activity 1 **Object Land:** Drop and Count	Children understand that when 10 is added, the number in the tens column is increased by 1.	**Program Materials** 100 Counters **Additional Materials** • Coffee can or box • Chart paper and marker	
DAY 4	**Warm Up 1** **Object Land:** Pointing and Winking*	Children continue to master the counting sequence.	No additional materials required	**Building Blocks** Dino Shop 1
	Activity 2 **Object Land:** Feed the Animals	Children continue to match numerals to increasingly bigger set sizes.	**Program Materials** • Counters, 4 bags with varying quantities from 11 to 20 • Number Cards (11–20) • Zoo Pictures	
DAY 5	**Warm Up 1** **Object Land:** Pointing and Winking*	Children continue to count objects.	No additional materials required	**Building Blocks** Review previous activities
	Review and Assess	Children review their favorite activities to improve understanding.	Materials will be selected from those used in previous lessons.	

* Includes Challenge Variations

Week 1 Counting
Lesson 1

Objective
Children learn to count sequentially and to count to 100 by ones and tens.

Program Materials
Engage
100 Counters

Additional Materials
Engage
- Coffee can
- Chart paper and marker

Access Vocabulary
count forward To begin with a number and say a bigger number each time; for example, 5, 6, 7, 8

count backward To begin with a number, then say a smaller number each time: 8, 7, 6, 5

counting up Counting up is the same as counting forward.

counting down Counting down is the same as counting backward.

figure it out To think about a problem and solve it; discover the answer

Creating Context
In this lesson you will introduce an activity called **Pointing and Winking.** These gestures may not be universally equivalent. Make sure students understand that these gestures are meant for quiet prompting while playing the game.

1 Warm Up   5

Skill Building COMPUTING
Pointing and Winking
Before beginning the whole-class activity **Drop and Count,** use the **Pointing and Winking** activity with the whole group.

Purpose Children will learn to count and stop at a specified number and to say the next number up in a sequence.

Warm-Up Card 1

Monitoring Student Progress
If . . . children are not ready to count to 20,
Then . . . start with a lower number.

2 Engage   30

Concept Building UNDERSTANDING
Drop and Count
"Today you will predict the number of counters dropped into a can."

Follow the instructions on the Activity Card to play **Drop and Count.** As children play, ask questions about what is happening in the game.

Purpose Children will understand that when a group of 10 is added, the number in the ones column does not change and the number in the tens column is increased by 1.

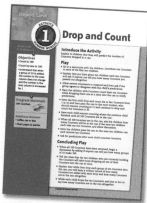

Activity Card 1

Teacher's Note Make sure the can does not have any sharp edges. You may choose to use a box or other opaque container instead.

Monitoring Student Progress

If . . . children have trouble remembering where the previous child stopped counting,

Then . . . write on the board the number at which each child stops counting or use tally marks to keep a record of how many Counters are dropped.

 MathTools Use the 100 Table to demonstrate and explore place value.

 Teacher's Note This game works best with a group of ten students.

Building Blocks For additional practice with number sequence and skip counting, students should complete **Building Blocks** School Supply Shop.

3 Reflect

Extended Response — REASONING
Ask questions such as the following:

- **When it was your turn to drop 10 Counters into the can, how did you know there would be 30 (or 40, etc.) altogether?** Possible answer: I noticed that when I added 10 Counters the numbers in the tens column got one bigger.
- **How did you figure it out?** Possible answer: I watched the other students add 10, and saw that the ones column did not change.

Using Student Pages
Have each child complete **Workbook,** page 5. Did the child write the correct number story?

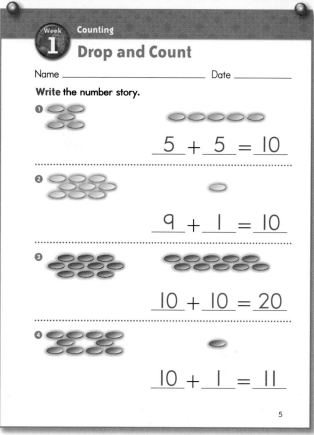

Workbook, p. 5

4 Assess

Informal Assessment
Use the Student Assessment Record, **Assessment,** page 100, to record informal observations.

COMPUTING	UNDERSTANDING
Pointing and Winking	**Drop and Count**
Did the child	Did the child
❏ respond accurately?	❏ make important observations?
❏ respond quickly?	❏ extend or generalize learning?
❏ respond with confidence?	❏ provide insightful answers?
❏ self-correct?	❏ pose insightful questions?

Counting • Lesson 1   29

Week 1 Counting
Lesson 2

Objective
Children will count a set of objects and also arrange quantities and numerals from smallest to biggest.

Program Materials
Engage
- Counters, four bags with varying quantities from 11 to 20; use four colors so that each bag has a different color
- Number Cards (11–20)
- Zoo Pictures

Access Vocabulary
wink To open and shut one eyelid
smallest, biggest These adjectives are superlatives, used in comparing. Introduce the English pattern *big–bigger–biggest*. Make a comparison chart that includes the adjectives *small, high, low,* and *fast*.
higher number Here the word *higher* does not mean location but a number with a greater value when counting "up."

Creating Context
Some English Learners may not have enough English proficiency to describe their understanding of the word *more*. Use comparison checks that do not rely heavily on language by having children use manipulatives or by using prompts such as "Show me thumbs up if you need more."

1 Warm Up 5

Concept Building COMPUTING
Pointing and Winking
Before beginning the small-group activity **Feed the Animals**, use the **Pointing and Winking** activity with the whole group.

Purpose Children will learn to count and stop at a specified number, and will also learn to say the next number up in a sequence.

Warm-Up Card 1

Monitoring Student Progress
If . . . children can count fluently to 20, **Then . . .** gradually have them count to higher numbers.

2 Engage 30

Concept Building ENGAGING
Feed the Animals
"Today you are going to pretend to work at a zoo and feed the cheetahs, tigers, zebras, and monkeys."

Follow the instructions on the Activity Card to play **Feed the Animals**. As children play, ask questions about what is happening in the game.

Purpose Children arrange quantities and numerals from smallest to biggest.

Activity Card 2

30 Number Worlds • Week 1

Monitoring Student Progress

| If . . . a child makes a mistake, | Then . . . help the child find the correct numeral and say its name. Recount to confirm that the numeral matches the set size. |

eMathTools Use the Set Tool to demonstrate and explore counting and comparing quantities.

 For additional practice matching numerals to set size, children should complete **Building Blocks** Dino Shop 1.

3 Reflect — 10

Extended Response REASONING

Ask questions such as the following:

- **How did you know which bag had more Counters?** Possible answer: Because this bag had 17 pieces and this bag had only 12, and 17 is more than 12. Encourage children to use numbers in their responses.

- **How do you know 17 is more than 12?** Possible answers: Because it has more Counters; because it comes after 12 when you're counting. You have to count 13, 14, 15, 16, 17—five more numbers after 12.

Teacher's Note Allow ample time for discussion so children can justify their answers and come to understand that 17 not only has 5 more chips than 12 but 17 also comes 5 numbers after 12 when counting. Give children time to discuss all the comparisons they make in a similar fashion.

Using Student Pages

Have each child complete **Workbook,** page 6. Did the child match the Counters to the correct number?

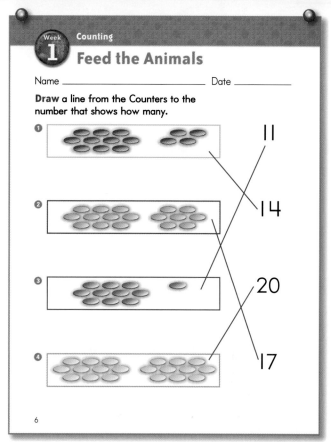

Workbook, p. 6

4 Assess

Informal Assessment

Use the Student Assessment Record, **Assessment,** page 100, to record informal observations.

COMPUTING	ENGAGING
Pointing and Winking	**Feed the Animals**
Did the child	Did the child
❏ respond accurately?	❏ pay attention to the contributions of others?
❏ respond quickly?	
❏ respond with confidence?	❏ contribute information and ideas?
❏ self-correct?	
	❏ improve on a strategy?
	❏ reflect on and check accuracy of work?

Counting • Lesson 2 31

Week 1 Counting

Lesson 3

Objective
Children continue to count sequentially and understand adding 10.

Program Materials
Engage
100 Counters

Additional Materials
Engage
- Coffee can
- Chart paper and marker

Access Vocabulary
predict To say what you think will happen
know To understand or figure out (e.g. "How do you know?"). For students with limited English proficiency who are familiar with the word *no*, hearing this question may be puzzling. Write the question on the board and point out the difference in spelling. Using gestures, give a few examples of the use of *no* vs. *know*. For example, "I know your name. It is _____."

Creating Context
Introduce children to the concept of predicting. Ask, "How many of you watch the weather report on television? The reporters *know* what the weather is like today and *predict* what the weather will be like for the next few days. They are not always correct, but their predictions are based on information they have. What do you predict the weather will be tomorrow? Will it rain? Will it be sunny?" (Write down the prediction and save it until tomorrow to check the outcome.) "In this activity we will predict how many Counters are dropped in the can."

1 Warm Up 5

Skill Building COMPUTING
Pointing and Winking
Before beginning the whole-class activity, **Drop and Count,** use the **Pointing and Winking** activity with the whole group.

Purpose Children will learn to count and stop at a specified number, and will learn to say the next number up in a sequence.

Warm-Up Card 1

Monitoring Student Progress
If . . . children are not ready to count to 20,
Then . . . start with a lower number.

2 Engage 30

Concept Building REASONING
Drop and Count
"Today you will predict the number of counters dropped into a can."

Follow the instructions on the Activity Card to play **Drop and Count.** As children play, ask questions about what is happening in the game.

Purpose Children will understand that when a group of 10 is added, the number in the ones column does not change and the number in the tens column is increased by one.

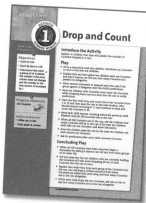

Activity Card 1

32 Number Worlds • Week 1

Monitoring Student Progress

If . . . children are not sure about the resulting quantities after any drop into the can,

Then . . . have them empty the can and count the Counters to verify.

 Teacher's Note Encourage children to use this week's vocabulary words as they engage in activities, discuss math concepts, and make predictions.

Building Blocks For additional practice with number sequence and skip counting, students should complete **Building Blocks** School Supply Shop.

3 Reflect

Extended Response REASONING
Ask questions such as the following:
- **How could you check your answers?** Possible answer: Count the Counters in the can.
- **What patterns do you see?** Possible answers: The numbers in the tens column keep getting bigger. They go 0, 1, 2, 3, 4, 5. The numbers in the ones column stay the same. They are all 0.

4 Assess

Informal Assessment
Use the Student Assessment Record, **Assessment,** page 100, to record informal observations.

COMPUTING	REASONING
Pointing and Winking	**Drop and Count**
Did the child	Did the child
❏ respond accurately?	❏ provide a clear explanation?
❏ respond quickly?	❏ communicate reasons and strategies?
❏ respond with confidence?	❏ choose appropriate strategies?
❏ self-correct?	❏ argue logically?

Counting • Lesson 3 33

Week 1 Counting
Lesson 4

Objective
Children will continue to count sets of objects and arrange quantities and numerals from smallest to biggest.

Program Materials

Engage
- Counters, four bags with varying quantities from 11 to 20; use four colors so that each bag has a different color
- Number Cards (11–20)
- Zoo Pictures

Access Vocabulary

pretend To make believe. *Pretend that you are at the zoo* means to act like you are at the zoo even though we all know that you are really in class.

lower number/higher number The words *lower* and *higher* are used as comparing words, not location words. A lower number has less. A higher number has more.

Creating Context
Show children pictures of various zoo animals to help them think about relative size. Have volunteers say whether each animal is bigger or smaller than they are. Be sure to include animals of various sizes, such as an elephant, a snake, a hippopotamus, a penguin, a tiger, and a parrot.

1 Warm Up 5

Skill Building COMPUTING
Pointing and Winking
Before beginning the small-group activity **Feed the Animals**, use the **Pointing and Winking** activity with the whole group.

Purpose Children will learn to count and stop at a specified number and will also learn to say the next number up in a sequence.

Warm-Up Card 1

Monitoring Student Progress
If . . . children are not ready to count to 20, **Then . . .** start with a lower number.

2 Engage 30

Concept Building APPLYING
Feed the Animals
"Today you are going to pretend to work at a zoo and feed the cheetahs, tigers, zebras, and monkeys."

Follow the instructions on the Activity Card to play **Feed the Animals**. As children play, ask questions about what is happening in the game.

Purpose Children arrange quantities and numerals from smallest to biggest.

Activity Card 2

Monitoring Student Progress

| If . . . a child makes a mistake, | Then . . . help the child find the correct numeral and say its name. Recount to confirm that the numeral matches the set size. |

 For additional practice matching numerals to set size, children should complete **Building Blocks** Dino Shop 1.

3 Reflect 10

Extended Response REASONING

Ask questions such as the following:
- **Which animal gets which bag of food?** Possible answer: The tigers get the green bag.
- **Why?** Possible answer: Because the tiger gets the most, and the green bag has the most food.

4 Assess

Informal Assessment

Use the Student Assessment Record, **Assessment**, page 100, to record informal observations.

COMPUTING	APPLYING
Pointing and Winking	**Feed the Animals**
Did the child	Did the child
❏ respond accurately?	❏ apply learning to new situations?
❏ respond quickly?	❏ contribute concepts?
❏ respond with confidence?	❏ contribute answers?
❏ self-correct?	❏ connect mathematics to the real world?

Counting • Lesson 4 35

Week 1 Counting

Lesson 5

Review

Objective
Children review the material presented in Week 1 of *Number Worlds*.

Program Materials
none

Access Vocabulary
free choice To choose any activity. (Students may know *free* as "no cost.")
like To find fun or interesting. (Beginning English Learners may know *like* as a comparison word.)

Creating Context
Many math concepts rely on comparison. In English, comparatives and superlatives are formed by adding the endings *-er* and *-est* to monosyllabic adjectives (e.g., *She is taller than her brother*). For adjectives of three or more syllables, *more* and *most* are used (e.g., *He is more intelligent than his dog*). Post a chart with the following words in class and add to it throughout the weeks. Review the definition of each word as you post it.

little	less	least
many	more	most
far	farther	farthest
near	nearer	nearest

1 Warm Up

Skill Practice COMPUTING
Pointing and Winking
Before beginning the **Free-Choice** activity, use the **Pointing and Winking** activity with the whole group.

Purpose Children continue to count and stop at a specified number.

Warm-Up Card 1

Monitoring Student Progress
If . . . children can count fluently to 20,
Then . . . gradually have them count to higher numbers.

2 Engage

Concept Building APPLYING
Free-Choice Activity
For the last day of the week, allow children to choose any activity from Week 1.

Make a note of the activity that students select. Do they prefer easy or challenging activities? Continue to provide Challenge opportunities for children who have mastered the basic activities.

Monitoring Student Progress
If . . . children would benefit from extra practice on specific skills,
Then . . . choose an activity for them.

Reflect — 10

Extended Response REASONING
Ask questions such as the following:
- What did you like about playing Drop and Count?
- Was there anything about playing this game you didn't like?
- Did this game help you do something you couldn't do before? What did it help you do?
- What was easy when you were playing Feed the Animals?
- What was hard when you were playing Feed the Animals?
- What do you remember most about this game?
- What was your favorite part?

Using Student Pages
Have each child complete **Workbook,** page 7. Is the child correctly applying the skills learned in Week 1?

Workbook, p. 7

Assess — 10

A Gather Evidence
Formal Assessment
Have students complete the weekly test on **Assessment,** page 25. Record formal assessment scores on the Student Assessment Record, **Assessment,** page 100.

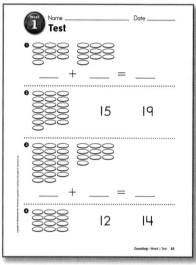

Assessment, p. 25

B Summarize Findings
Review the Student Assessment Records. Determine whether students have Minimal, Basic, or Secure understanding of the concepts presented in Week 1.

C Differentiate Instruction
Based on your observations, use these teaching strategies next week to follow up.

Minimal Understanding
- Repeat the Warm-Up and Engage activities in subsequent weeks to develop concepts of counting objects.
- Repeat the Object Land activities from this week. Observe student understanding of counting concepts and reinforce as necessary.
- Use **Building Blocks** computer activities beginning with Count to develop and reinforce counting concepts.

Basic Understanding
- Repeat Engage activities in subsequent weeks to reinforce basic counting concepts.
- Use **Building Blocks** computer activities beginning with School Supply Shop to reinforce counting concepts.

Secure Understanding
- Use Challenge variations of **Pointing and Winking.**
- Use **Building Blocks** computer activities beginning with Dino Shop 1.

Counting • Lesson 5 Review 37

Week 2: Counting and Comparing

Week at a Glance

This week children begin **Number Worlds**, Week 2, and continue to explore Object Land.

Background

In Object Land, numbers are represented as groups of objects. This is the first way numbers were represented historically, and this is the first way children naturally learn about numbers. In Object Land, children work with real, tangible objects and with pictures of objects.

How Children Learn

As children begin this week's lesson, they should be comfortable counting objects to determine how many are in a set. At the end of this week, children should understand the concepts of *more* and *equal* and will be able to use written numerals to tell which sets have more, less, or the same amount.

Skills Focus

- Count to 20
- Add tens
- Understand *more* and *equality*
- Solve missing addend problems

Teaching for Understanding

As children engage in these activities, they should come to understand that written numerals can not only be used to show how many objects are in a set, but can also be used to compare set size without recounting.

Observe closely while evaluating the Engage activities assigned for this week.

- Are students counting to 20?
- Are students using *more* and *equal* correctly?
- Are students solving missing addend problems?

Math at Home

Give one copy of the Letter to Home, page A2, to each child. Complete the activity in class, and then encourage children to share it with their caregivers.

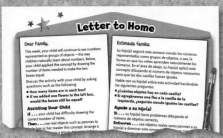

Letter to Home, Teacher Edition, p. A2

Math Vocabulary

set A group of objects that go together

equal Having the same amount

numeral A symbol that represents a number

compare To think about how things are alike and different

English Learners

SPANISH COGNATES

English	Spanish
activity	actividad
compare	compare
equal	igual
equality	igualdad
sequence	secuencia

ALTERNATE VOCABULARY

pretend To make believe

38 Number Worlds • Week 2

Week 2 Planner: Counting and Comparing

PACING	LESSON	LEARNING GOALS	MATERIALS	TECHNOLOGY
DAY 1	Warm Up 1 **Object Land:** Pointing and Winking*	Children count and stop at a specified number, then say the next number.	No additional materials required	Building Blocks Dino Shop 1
DAY 1	Activity 3 **Object Land:** Count and Compare*	Children use written numerals to tell which sets have more, less, or the same amount.	**Program Materials** Object Land Activity Sheet 1 **Additional Materials** Pencils, one for each child	
DAY 2	Warm Up 1 **Object Land:** Pointing and Winking*	Children continue to count and stop at a specified number, then say the next number up in sequence.	No additional materials required	Building Blocks Pizza Pizzazz 1
DAY 2	Activity 4 **Object Land:** Food Fun	Children acquire understanding of the concepts of *more* and *equality*.	**Program Materials** Counters	
DAY 3	Warm Up 1 **Object Land:** Pointing and Winking*	Children continue to count higher as they master counting sequences.	No additional materials required	Building Blocks Dino Shop 1
DAY 3	Activity 3 **Object Land:** Count and Compare*	Children continue to use written numerals to compare set size.	**Program Materials** Object Land Activity Sheet 1 for each child **Additional Materials** Pencils, one for each child	
DAY 4	Warm Up 1 **Object Land:** Pointing and Winking*	Children keep counting higher as they master counting sequences.	No additional materials required	Building Blocks Pizza Pizzazz 1
DAY 4	Activity 4 **Object Land:** Food Fun	Children continue to solve missing addend problems.	**Program Materials** Counters	
DAY 5	Warm Up 1 **Object Land:** Pointing and Winking*	Children continue to count objects.	No additional materials required	Building Blocks Review previous activities
DAY 5	Review and Assess	Children review their favorite activities to improve their understanding of more difficult concepts.	Materials will be selected from those used in previous activities.	

* Includes Challenge Variations

Week 2 — Counting and Comparing

Lesson 1

Objective
Children continue to count sequentially and learn that numerals can be used to compare set size.

Program Materials
Engage
Object Land Activity Sheet 1, 1 for each child

Additional Materials
Engage
- Pencils, one for each child
- Pictures of small objects (optional)
- Glue (optional)

Access Vocabulary
next The one after this one
compare To think about how things are the same and different

Creating Context
In this lesson children are asked which set has the most and which set has the least. To determine whether children understand the concepts of *most* and *least*, ask them to point to the part of the classroom with the most desks, the least number of students, and so on. This allows students at beginning levels of proficiency to participate before launching the math instruction on comparing sets.

1 Warm Up   5

Skill Building COMPUTING
Pointing and Winking
Before beginning the whole-class activity **Count and Compare**, use the **Pointing and Winking** activity with the whole class.

Purpose **Pointing and Winking** will help children learn to count and stop at a specified number. They will also learn to say the next number up in a sequence.

Warm-Up Card 1

Monitoring Student Progress
If . . . children are not ready to count to 20, **Then . . .** have them count to a lower number instead.

2 Engage   30

Concept Building UNDERSTANDING
Count and Compare
"Today you will use numerals to tell how many objects are in a set."

Follow the instructions on the Activity Card to play **Count and Compare.** As children play, ask questions about what is happening in the game.

Purpose Children will understand that written numerals tell how many objects are in a set and that numerals can be used to tell which sets have more, less, or the same amount without seeing the actual objects.

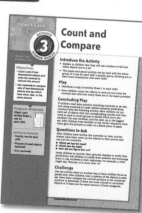
Activity Card 3

Monitoring Student Progress

If . . . children are having trouble remembering what the numerals look like,

Then . . . refer them to the number line at the bottom of the page.

eMathTools Use the Set Tool to demonstrate and explore counting.

Teacher's Note Encourage children to use this week's vocabulary words as they engage in the activities, discuss math concepts, and make predictions.

 For additional practice with counting and number sequence, children should complete **Building Blocks** Dino Shop 1.

3 Reflect   10

Extended Response REASONING

Ask questions such as the following:

- **Which set had the most?** The set with the highest numeral.
- **How did you know?** The numerals tell how many are in each set.

Using Student Pages

Have each child complete **Workbook,** page 8. Did the child correctly identify the quantity in each set?

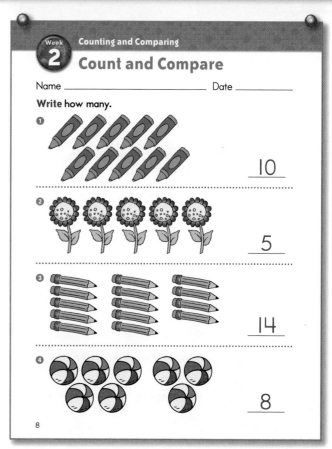

Workbook, p. 8

4 Assess

Informal Assessment

Use the Student Assessment Record, **Assessment,** page 100, to record informal observations.

COMPUTING	UNDERSTANDING
Pointing and Winking	**Count and Compare**
Did the child	Did the child
❏ respond accurately?	❏ make important observations?
❏ respond quickly?	❏ extend or generalize learning?
❏ respond with confidence?	❏ provide insightful answers?
❏ self-correct?	❏ pose insightful questions?

Counting and Comparing • Lesson 1   **41**

Week 2 — Counting and Comparing

Lesson 2

Objective
Children count sequentially and understand the concepts of *more* and *equality* while gaining initial experience with missing addend problems.

Program Materials
Engage
Counters

Access Vocabulary
pretend To make believe
equal Having the same amount

Creating Context
Some English learners may not have enough English proficiency to describe their understanding of the word *more*. Use comprehension checks that do not rely heavily on language by having children use manipulatives or by using prompts such as "Show me thumbs up if you need more."

1 Warm Up 5

Concept Building COMPUTING
Pointing and Winking
Before beginning the small-group activity **Food Fun,** use the **Pointing and Winking** activity with the whole group.

Purpose Children will learn to count and stop at a specified number and to say the next number up in a sequence.

Warm-Up Card 1

Monitoring Student Progress
If . . . children are not ready to count to 20, **Then . . .** have them count to a lower number.

2 Engage 30

Concept Building ENGAGING
Food Fun
"Today you are going to make sure that everyone gets the same amount of food."

Follow the instructions on the Activity Card to play **Food Fun.** As children play, ask questions about what is happening in the game.

Purpose Children will understand the concepts of *more* and *equality* while gaining initial experience with missing addend problems.

Activity Card 4

42 Number Worlds • Week 2

Monitoring Student Progress

If ... children have difficulty figuring out what the missing addend is,

Then ... demonstrate counting on with your fingers or some other concrete object.

Teacher's Note A common mistake with missing addend problems is for the child to say that he or she needs the same amount as what the total will be. When asked how many more are needed, a child may not be able to shift his or her focus off the goal and onto the partial set and the missing set. If children make this common mistake, do one of the following:

- Agree that the amount they requested is the goal, remind them that they already have some, and then help them by counting on the missing addend.
- Give the children the amount that they ask for, and allow them to discover their mistakes and correct them on their own.

Building Blocks For additional practice with counting and missing-addend problems, children should complete **Building Blocks** Pizza Pizzazz 1.

Extended Response REASONING

Ask questions such as the following:

- **How many did you already have?** Answers will vary.
- **How many did you want to have altogether?** Accept all reasonable answers.
- **How many more did you need? How did you figure that out?** Possible answer: I counted on from how many I had to how many I wanted.

Using Student Pages

Have each child complete **Workbook,** page 9. For Questions 3 and 4 ask: "Do you need more apples if each child wants one? How many more do you need?" Did the child identify the missing addend correctly?

Workbook, p. 9

Informal Assessment

Use the Student Assessment Record, **Assessment,** page 100, to record informal observations.

COMPUTING	ENGAGING
Pointing and Winking	**Food Fun**
Did the child	Did the child
❏ respond accurately?	❏ pay attention to the contributions of others?
❏ respond quickly?	
❏ respond with confidence?	❏ contribute information and ideas?
❏ self-correct?	
	❏ improve on a strategy?
	❏ reflect on and check accuracy of work?

Counting and Comparing • Lesson 2   **43**

Week 2 — Counting and Comparing

Lesson 3

Objective

Children continue to count sequentially and learn that numerals can be used to represent and compare set size.

Program Materials

Engage
Object Land Activity Sheet 1, 1 for each child

Additional Materials

Engage
- Pencils, one for each child
- Pictures of small objects (optional)
- Glue (optional)

Access Vocabulary

numeral A symbol that represents a number
set Group of objects

Creating Context

Discuss the concepts of *most* and *least* with children. Review the chart of comparatives and superlatives from Week 1 to be sure children understand the words *most* and *least*. Ask children how many brothers and sisters they have, then determine which student has the most and which student has the least. Discuss which child has the most pencils in his or her desk, who has the least number of crayons, and so on.

1 Warm Up

Skill Building — COMPUTING

Pointing and Winking
Before beginning the whole-class activity, **Count and Compare**, use the **Pointing and Winking** activity with the whole group.

Purpose **Pointing and Winking** will help children learn to count and stop at a specified number. They will also learn to say the next number up in a sequence.

Warm-Up Card 1

Monitoring Student Progress

If . . . children can count fluently to 20,

Then . . . challenge them to count to a higher number.

2 Engage

Concept Building — REASONING

Count and Compare

"Today you will use numerals to tell how many objects are in a set."

Follow the instructions on the Activity Card to play **Count and Compare.** As children play, ask questions about what is happening in the game.

Purpose Children will understand that numerals tell how many objects are in a set and can be used to compare set size without seeing the actual objects that make up the set.

Activity Card 3

Monitoring Student Progress

| If . . . children have trouble remembering what the numerals look like, | Then . . . refer them to the number line at the bottom of the page. |

Teacher's Note You can also play **Count and Compare** backward. Give each child a numeral between 11 and 20. Have children compare numerals with a classmate, predict which numeral will make a bigger set, and then count out Counters to check their predictions.

Building Blocks For additional practice with counting and number sequence, children should complete **Building Blocks** Dino Shop 1.

3 Reflect 10

Extended Response REASONING
Ask questions such as the following:
- **Which set had the least?** The hats.
- **How can you check your answer?** Possible answers: Count the sets; look at the numerals to see that 11 is the smallest.

Assess

Informal Assessment
Use the Student Assessment Record, **Assessment,** page 100, to record informal observations.

COMPUTING	REASONING
Pointing and Winking Did the child ❑ respond accurately? ❑ respond quickly? ❑ respond with confidence? ❑ self-correct?	**Count and Compare** Did the child ❑ provide a clear explanation? ❑ communicate reasons and strategies? ❑ choose appropriate strategies? ❑ argue logically?

Counting and Comparing • Lesson 3 45

Week 2 — Counting and Comparing
Lesson 4

Objective
Children understand the concepts of *more* and *equality* while gaining initial experience with missing addend problems and counting sequentially.

Program Materials
Engage
Counters

Access Vocabulary
more A bigger number or amount
altogether In all; total

Creating Context
Discuss the concept of *equality* with children. Divide groups of classroom objects (pencils, books, and so on) into sets. Make some of the sets equal and some unequal. Discuss with children what makes two sets equal. *(The sets have the same number of objects in them.)* Can the children think of times when it is important for things to be equal? *(Making sure there are equal numbers of books and students, making sure children have equal amounts of snacks, and so on.)*

---

1 Warm Up — 5

Skill Building COMPUTING
Pointing and Winking
Before beginning the small-group activity **Food Fun**, use the **Pointing and Winking** activity with the whole group.

Purpose Children will learn to count and stop at a specified number.

Warm-Up Card 1

Monitoring Student Progress
If . . . children can count fluently to 20,
Then . . . challenge them to count to a higher number.

---

2 Engage — 30

Concept Building APPLYING
Food Fun
"Today you are going to make sure that everyone gets the same amount of food."

Follow the instructions on the Activity Card to play **Food Fun**. As children play, ask questions about what is happening in the game.

Purpose Children will understand the concepts of *more* and *equality* while gaining experience with missing addend problems.

Activity Card 4

46 Number Worlds • Week 2

Monitoring Student Progress

| If . . . children have difficulty figuring out what the missing addend is, | Then . . . demonstrate counting on with your fingers or some other concrete object. |

 For additional practice with counting and missing-addend problems, children should complete **Building Blocks** Pizza Pizzazz 1.

Reflect 10

Extended Response REASONING

Ask questions such as the following:

- **What is this problem about?** Making sure each person has the same amount of food.
- **Can you think of other times at school or at home when you need to make sure everyone has the same amount?** Possible answers include game pieces, art supplies, and so on.

Assess

Informal Assessment

Use the Student Assessment Record, **Assessment,** page 100, to record informal observations.

COMPUTING	APPLYING
Pointing and Winking	**Food Fun**
Did the child	Did the child
❏ respond accurately?	❏ apply learning to new situations?
❏ respond quickly?	❏ contribute concepts?
❏ respond with confidence?	❏ contribute answers?
❏ self-correct?	❏ connect mathematics to the real world?

Counting and Comparing • Lesson 4   **47**

Week 2: Counting and Comparing

Lesson 5
Review

Objective
Children review the material presented in Week 2 of **Number Worlds**.

Program Materials
Engage
See game chosen for materials.

Additional Materials
Engage
See game chosen for materials.

Access Vocabulary
wink To quickly close and open one eye

Creating Context
Discuss with children multiple meanings of the word *free*. Children may know that *free* means "no cost," but may not know that "free choice" means that they can choose which activity they would like to do.

---

1 Warm Up 5

Skill Practice COMPUTING
Pointing and Winking
Before beginning the **Free-Choice** activity, use the **Pointing and Winking** activity with the whole group.

Purpose This activity helps children count on from a specified number, which lays the foundation for later work with addition.

Warm-Up Card 1

Monitoring Student Progress
| If . . . children can fluently count on, | Then . . . challenge them to count on from higher numbers. |

---

2 Engage 20

Concept Building APPLYING
Free-Choice Activity

For the last day of the week, allow children to choose an activity from the previous week. Some activities they may choose include the following:
- Object Land: **Drop and Count**
- Object Land: **Feed the Animals**

Make a note of the activities children select. Do they prefer easy or challenging activities? Continue to provide Challenge opportunities for children who have mastered the basic activities.

Monitoring Student Progress
| If . . . children would benefit from extra practice on specific skills, | Then . . . choose an activity for them. |

3 Reflect — 10

Extended Response REASONING
Ask questions such as the following:
- What did you like about playing Count and Compare?
- Was there anything about playing this game you didn't like?
- Did this game help you do something you couldn't do before? What did it help you do?
- What was easy when you were playing Food Fun?
- What was hard when you were playing Food Fun?
- What do you remember most about this game?
- What was your favorite part?

Using Student Pages
Have each child complete **Workbook,** page 10. Is the child correctly applying the skills they learned in Week 2?

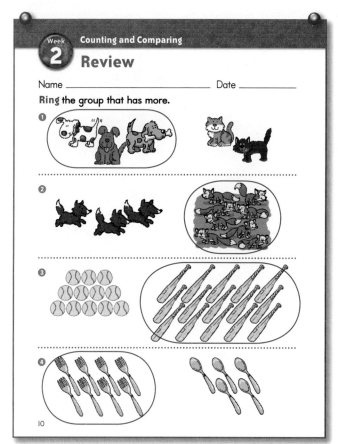

Workbook, p. 10

4 Assess — 10

A Gather Evidence
Formal Assessment
Have students complete the weekly test on *Assessment,* page 27. Record formal assessment scores on the Student Assessment Record, *Assessment,* page 100.

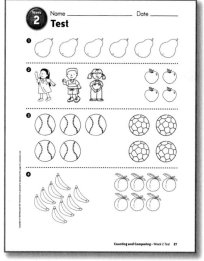

Assessment, p. 27

B Summarize Findings
Review the Student Assessment Records. Determine whether students have Minimal, Basic, or Secure understanding of the concepts presented in Week 2.

C Differentiate Instruction
Based on your observations, use these teaching strategies next week to follow up.

Minimal Understanding
- Repeat the Warm-Up and Engage activities in subsequent weeks to develop concepts of counting objects.
- Repeat the Object Land activities from this week. Observe children's understanding of these concepts and reinforce as necessary.
- Use **Building Blocks** computer activities beginning with Count to develop and reinforce numeration concepts.

Basic Understanding
- Repeat Engage activities in subsequent weeks to reinforce basic counting concepts.
- Use **Building Blocks** computer activities beginning with Dino Shop 1 to reinforce numeration concepts.

Secure Understanding
- Use Challenge variations of **Pointing and Winking.**
- Use **Building Blocks** computer activities beginning with Pizza Pizzazz 1.

Counting and Comparing • Lesson 5   **49**

Week 3: More Counting and Comparing

Week at a Glance

This week children begin **Number Worlds,** Week 3, and are introduced to Picture Land.

Background

In Picture Land, numbers are represented as a set of dots. When a child is asked, "What number do you have?" he or she might count each dot to determine the size of the set or identify the amount by recognizing the pattern that marks that amount.

How Children Learn

As children begin this week's lesson, they should be comfortable counting objects.

At the end of the week, each student should understand the correspondence between two sets. For example, they should notice that the pattern for five is the same as four with one extra dot in the center.

Skills Focus

- Count to 20
- Compare and order numbers
- Describe and extend patterns

Teaching for Understanding

As children engage in these activities, they will gradually come to think of these patterns as forming the same sort of ordered series as do the number words themselves. Numerals, another way of representing numbers, are also part of Picture Land.

Observe closely while evaluating the Engage activities assigned for this week.

- Are students counting to 20?
- Are students using *many* and *few* correctly?
- Are students matching numbers to picture patterns?

Math at Home

Give one copy of the Letter to Home, page A3, to each child. Complete the activity in class, and then encourage children to share it with their caregivers.

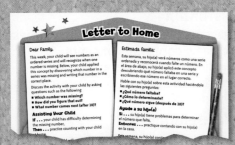

Letter to Home, Teacher Edition, p. A3

Math Vocabulary

compare To think about how things are alike and different

many Made up of a large number

few Made up of a small number

numeral A symbol that represents a number

English Learners

SPANISH COGNATES

English	Spanish
compare	compare
many	muchos
numeral	número

ALTERNATE VOCABULARY

concentrate To think carefully about

left over Not used

Week 3 Planner — More Counting and Comparing

PACING	LESSON	LEARNING GOALS	MATERIALS	TECHNOLOGY
DAY 1	Warm Up 2 **Picture Land:** Catch the Teacher*	Children detect an error in counting and identify the number omitted from the sequence.	**Prepare Ahead** Two columns on the board, one labeled *Students* and one labeled *Teacher*	**Building Blocks** Memory Number 1
	Activity 5 **Picture Land:** Concentration*	Children match patterns and quantities and verify equivalence by counting.	**Program Materials** • Dot Set Cards (1–10), 2 sets • Number Cards (1–10), for a Challenge	
DAY 2	Warm Up 2 **Picture Land:** Catch the Teacher*	Children keep detecting errors in counting and identifying the omitted number.	**Prepare Ahead** Two columns on the board, one labeled *Students* and one labeled *Teacher*	**Building Blocks** Party Time 1
	Activity 6 **Object Land:** Party!	Children compare objects and say whether there are enough, too many, or too few.	**Program Materials** Object Land Activity Sheet 2, for each child **Additional Materials** Pencils or crayons	
DAY 3	Warm Up 2 **Picture Land:** Catch the Teacher*	Children continue to count higher as they master counting sequences.	**Prepare Ahead** Two columns on the board, one labeled *Students* and one labeled *Teacher*	**Building Blocks** Memory Number 1
	Activity 5 **Picture Land:** Concentration*	Children continue to match patterns and quantities and verify equivalence by counting.	**Program Materials** • Dot Set Cards (1–10), 2 sets • Number Cards (1–10), for a Challenge	
DAY 4	Warm Up 2 **Picture Land:** Catch the Teacher*	Children keep detecting errors in counting and identifying omitted numbers.	**Prepare Ahead** Two columns on the board, one marked *Students* and one marked *Teacher*	**Building Blocks** Party Time 1
	Activity 6 **Object Land:** Party!	Children continue to compare objects and say whether there are enough, too many, or too few.	**Program Materials** Object Land Activity Sheet 2, for each child **Additional Materials** Pencils or crayons	
DAY 5	Warm Up 2 **Picture Land:** Catch the Teacher*	Children continue to detect errors in counting and identify omitted numbers.	**Prepare Ahead** Two columns on the board, one labeled *Students* and one labeled *Teacher*	**Building Blocks** Review previous activities
	Review and Assess	Children review their favorite activities to improve their understanding of concepts.	Materials will be selected from those used in days 1 through 4.	

* Includes Challenge Variations

Week 3 Lesson 1: More Counting and Comparing

Objective
Children will continue to explore the counting sequence and gain experience matching patterns and quantities.

Program Materials
Engage
- Dot Set Cards (1–10), 2 sets
- Number Cards (1–10), for the Challenge

Prepare Ahead
Warm Up
Two columns on the board, one labeled *Students* and one labeled *Teacher,* to keep tally mark scores.

Access Vocabulary
compare To think about how things are alike and different
concentrate To think carefully about

Creating Context
When using the game cards, explain the concept of *faceup* and *facedown*. These idioms may not be familiar to children. The main, or most meaningful, side of the card is the face. A card that is faceup communicates specific information, while a facedown card gives little information.

1 Warm Up

Skill Building — COMPUTING
Catch the Teacher
Before beginning the small-group activity **Concentration**, use the **Catch the Teacher** activity with the whole group.

Purpose Children will learn to detect an error in counting and identify which number was omitted from the sequence.

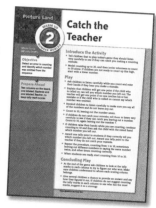

Warm-Up Card 2

Monitoring Student Progress
If . . . a child who is called on cannot remember the number that was omitted,

Then . . . call on someone else whose hand is raised. If the child says a wrong number, ask the other children if they agree with the child's choice by using thumbs up or thumbs down. If they do not agree, ask another child what number was left out. Then confirm this child's answer by asking, "Do you all agree that I forgot to say the number 7?"

2 Engage

Concept Building — UNDERSTANDING
Concentration
"Today you will use the patterns you see as well as the number of dots in each pattern to tell you how much is in each set."

Follow the instructions on the Activity Card to play **Concentration**. As children play, ask questions about what is happening in the game.

Activity Card 5

Purpose Children will learn to identify a pattern or quantity on a Dot Set Card and find a card that matches.

Monitoring Student Progress

If . . . children have trouble keeping track of the placement of the facedown cards,

Then . . . allow them to play on a piece of paper marked with placement lines.

Teacher's Note The most important part of this activity is for children to gain experience identifying the patterns associated with different quantities and comparing the values of these patterns. Use this activity until children can recognize these patterns easily without having to count the dots.

 For additional practice with number and quantity, children should complete **Building Blocks** Memory Number 1.

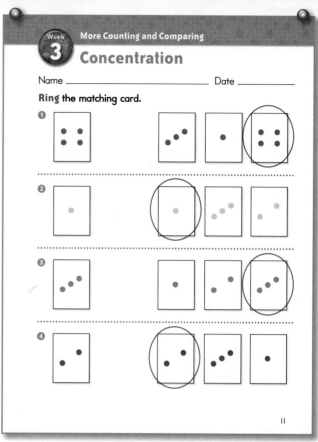

Workbook, p. 11

3 Reflect 10

Extended Response REASONING

Ask questions such as the following:
- **How could you tell whether a faceup card matched the card you chose?** Possible answers: count the dots; recognize the dot pattern.
- **How could you tell that one pattern (e.g., 8) has more than another pattern (e.g., 7)?** Possible answer: the pattern has one more dot. 7 + 1 = 8.

Using Student Pages

Have each child complete **Workbook,** page 11. Tell children, "Look at the Dot Set Card on the left. Ring the matching card." Did the child correctly match the dot set patterns?

4 Assess

Informal Assessment

Use the Student Assessment Record on **Assessment,** page 100, to record informal observations.

COMPUTING	UNDERSTANDING
Catch the Teacher	**Concentration**
Did the child	Did the child
❑ respond accurately?	❑ make important observations?
❑ respond quickly?	
❑ respond with confidence?	❑ extend or generalize learning?
❑ self-correct?	❑ provide insightful answers?
	❑ pose insightful questions?

More Counting and Comparing • Lesson 1 53

Week 3 — More Counting and Comparing
Lesson 2

Objective
Children continue to identify missing numbers in the counting sequence and also compare objects to determine whether there are enough, too many, or too few.

Program Materials
Engage
Object Land Activity Sheet 2, for each child

Additional Materials
Warm Up
Two columns on the board, one marked *Students* and one marked *Teacher,* to keep tally mark scores

Engage
Pencils or crayons for each student

Access Vocabulary
equal Having the same amount
left over Not used

Creating Context
Keep in mind that the word *party* may bring different experiences to mind for children. Party games, traditions, and foods may vary widely. With the children, create a word web using the word *party* as the central theme. Have children work in groups to draw pictures of party activities and foods, then have them share their drawings with the whole class.

---

1. Warm Up (5)

Skill Building — COMPUTING
Catch the Teacher
Before beginning the small-group activity **Party!**, use the **Catch the Teacher** activity with the whole group.

Purpose Children will master the counting sequence and understand that each number occupies a fixed position in the sequence.

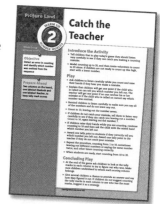
Warm-Up Card 2

Monitoring Student Progress
If . . . children have trouble identifying the missing number,

Then . . . practice counting to 10 with them throughout the day.

2. Engage (30)

Concept Building — ENGAGING
Party!
"Today you will pretend to have a party and make sure that every guest has enough of everything they need."

Follow the instructions on the Activity Card to play **Party!** As children play, ask questions about what is happening in the game.

Activity Card 6

Purpose Children will learn to use numbers to compare sets of objects and say whether there are enough, too many, or too few in each set.

54 Number Worlds • Week 3

Monitoring Student Progress

If . . . children have trouble counting the party items,

Then . . . make sure they are counting only one item at a time or give them tangible items to count.

Teacher's Note Encourage children to use this week's vocabulary words as they engage in the activities, discuss math concepts, and make predictions.

 Building Blocks For additional practice with one-to-one correspondence, children should complete **Building Blocks** Party Time 1.

3 Reflect 10

Extended Response REASONING

Ask questions such as the following:
- **How many guests were at the party?** 16 **How did you know?** Counted them.
- **How many hats were there?** 17 **Were there enough hats for each guest?** Yes. **Were there any left over?** Yes, there was 1 hat left over.
- **How did you figure that out?** 17 is one more than 16; when you give 16 hats to the 16 guests, you have 1 left over.

Using Student Pages

Have each child complete **Workbook,** page 12. For Question 3 ask, "Are there too few or too many horns if each child wants one horn?" For Question 4 ask, "Are there enough hats for each child to have one hat?"
- Did children count correctly?
- Do children know if there are exactly enough items or extra items?

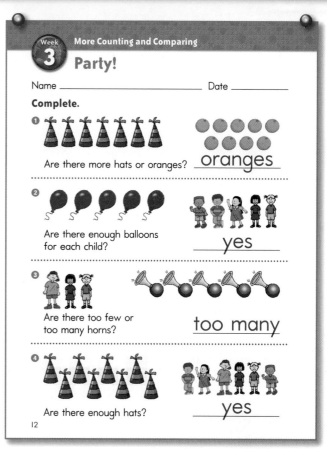

Workbook, p. 12

4 Assess

Informal Assessment

Use the Student Assessment Record on **Assessment,** page 100, to record informal observations.

COMPUTING	ENGAGING
Catch the Teacher	**Party!**
Did the child	Did the child
❏ respond accurately?	❏ pay attention to the contributions of others?
❏ respond quickly?	
❏ respond with confidence?	❏ contribute information and ideas?
❏ self-correct?	
	❏ improve on a strategy?
	❏ reflect on and check accuracy of work?

More Counting and Comparing • Lesson 2 55

Week 3 — More Counting and Comparing
Lesson 3

Objective
Children will continue to explore the counting sequence and gain experience matching patterns and quantities.

Program Materials
Engage
- Dot Set Cards (1–10), 2 sets
- Number Cards (1–10), for the Challenge

Prepare Ahead
Warm Up
Two columns on the board, one labeled *Students* and one labeled *Teacher,* to keep tally mark scores

Access Vocabulary
pattern Dots arranged a certain way
pick Choose

Creating Context
Prepositions can be confusing for English learners; consider the difference in meaning between *pick up, pick out,* and *pick on.* As a class, create a chart of prepositions. Have the children draw pictures that illustrate the meaning of each preposition or find pictures in magazines. Post the chart in the classroom for children to refer to throughout the year.

1 Warm Up — 5

Skill Building COMPUTING
Catch the Teacher
Before beginning the small-group activity **Concentration,** use the **Catch the Teacher** activity with the whole group.

Purpose Children will practice the counting sequence and understand that each number occupies a fixed position in the sequence.

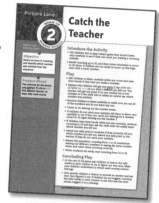
Warm-Up Card 2

Monitoring Student Progress
If . . . children are ready for a challenge,
Then . . . play **Catch the Teacher** counting backward.

2 Engage — 30

Concept Building REASONING
Concentration
"Today you will use the patterns you see, as well as the number of dots in each pattern, to tell you how much is in each set."

Follow the instructions on the Activity Card to play **Concentration.** As children play, ask questions about what is happening in the game.

Activity Card 5

Purpose Children learn to identify a pattern or quantity on a Dot Set Card, and find a card that matches.

56 Number Worlds • Week 3

Monitoring Student Progress

| If . . . children are ready for a challenge, | Then . . . have them try the Challenge 1 or Challenge 2 variation of **Concentration.** |

eMathTools Use the Set Tool to demonstrate and explore counting.

> **Teacher's Note** Once children are familiar with this activity, it is a great game to keep at independent learning centers. You can prepare an envelope containing the game cards so the children can easily collect the materials on their own.

Building Blocks For additional practice with number and quantity, children should complete **Building Blocks** Memory Number 1.

3 Reflect 10

Extended Response REASONING

Ask questions such as the following:

- **How did you solve this problem?** If children say they counted the dots to match the cards, ask them if they see any patterns in the dot sets. Help them discover that the pattern for 3 is the same as the pattern for 2 with one more dot, and so on.

- **Could you have solved this another way?** Help children understand that counting the dots and recognizing the dot patterns are two ways to reach the same answer (the quantity in each dot set).

4 Assess

Informal Assessment

Use the Student Assessment Record, **Assessment,** page 100, to record informal observations.

COMPUTING	REASONING
Catch the Teacher	**Concentration**
Did the child	Did the child
❏ respond accurately?	❏ provide a clear explanation?
❏ respond quickly?	❏ communicate reasons and strategies?
❏ respond with confidence?	
❏ self-correct?	❏ choose appropriate strategies?
	❏ argue logically?

More Counting and Comparing • Lesson 3

Week 3 — More Counting and Comparing

Lesson 4

Objective
Children compare objects and say whether there are enough, too many or too few.

Program Materials

Engage
Object Land Activity Sheet 2, for each child

Additional Materials

Warm Up
Two columns on the board, one labeled *Students* and one labeled *Teacher*, to keep tally mark scores

Engage
Pencils or crayons for each student

Access Vocabulary
tally To count; keep score
tally marks Straight lines grouped in sets of five

Creating Context
To extend children's vocabulary, pair children and have them draw and label pictures of six different items they would see at a party or celebration. Have groups share their pictures with the class. You may choose to copy the pictures to use when playing **Party!**

---

1. Warm Up — 5

Skill Building COMPUTING
Catch the Teacher

Before beginning the whole-class activity **Party!**, use the **Catch the Teacher** activity with the whole group.

Purpose Children practice the counting sequence and understand that each number occupies a fixed position in the sequence.

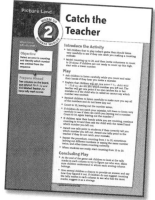

Warm-Up Card 2

Monitoring Student Progress

If . . . children are comfortable using tally marks,

Then . . . challenge them to tally other items throughout the day. They can tally the number of children wearing blue, the books on a bookshelf, and so on.

2. Engage — 30

Concept Building APPLYING
Party!

"Today you will pretend to have a party and make sure that each guest has enough of what they need."

Follow the instructions on the Activity Card to play **Party!** As children play, ask questions about what is happening in the game.

Purpose Children learn to use numbers to compare sets of objects and say whether there are enough, too many, or too few in each set.

Activity Card 6

58 Number Worlds • Week 3

Monitoring Student Progress

If . . . children have trouble counting the party items,

Then . . . make sure they are counting only one item at a time.

 For additional practice with one-to-one correspondence, children should complete **Building Blocks** Party Time 1.

3 Reflect

Extended Response REASONING

Ask questions such as the following:

- **Could you have solved this problem another way?** Accept all reasonable answers.
- **What would happen if there were only 15 hats?** 15 is less than 16; there would not be enough hats for all the guests.

4 Assess

Informal Assessment

Use the Student Assessment Record on **Assessment,** page 100, to record informal observations.

COMPUTING	APPLYING
Catch the Teacher	**Party!**
Did the child	Did the child
❏ respond accurately?	❏ apply learning to new situations?
❏ respond quickly?	❏ contribute concepts?
❏ respond with confidence?	❏ contribute answers?
❏ self-correct?	❏ connect mathematics to the real world?

More Counting and Comparing • Lesson 4 59

Week 3: More Counting and Comparing

Lesson 5
Review

Objective
Children review the material presented in Week 3 of **Number Worlds**.

Prepare Ahead
Two columns on a board, one marked *Students,* and one marked *Teacher,* to keep tally mark scores.

Access Vocabulary
like To enjoy

Creating Context
Discuss with children different meanings of the word *like*. Make a two-column chart on the board. Label one column *I Like* and the other column *Because*. Have children give reasons why they like something as you complete the chart on the board. This will help them describe why they like certain **Number Worlds** activities.

---

1 Warm Up 5

Skill Practice — COMPUTING
Catch the Teacher

Before beginning the **Free-Choice** activity, use the **Catch the Teacher** activity with the whole group.

Purpose Children practice the counting sequence and understand that each number occupies a fixed position in the sequence.

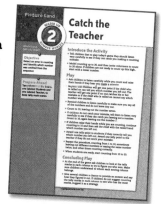

Warm-Up Card 2

Monitoring Student Progress

| If . . . children struggle with **Catch the Teacher** backward, | Then . . . give them more opportunities to practice counting down from 5 or 10 before playing this Challenge activity again. |

---

2 Engage 20

Concept Building — APPLYING
Free-Choice Activity

Allow children to choose an activity from the previous weeks, such as:
- Object Land: **Pointing and Winking**
- Object Land: **Drop and Count**
- Object Land: **Feed the Animals**

Make a note of the activities children select. Do they prefer easy or challenging activities? Continue to provide Challenge opportunities for children who have mastered the basic activities.

Monitoring Student Progress

| If . . . children need practice on specific skills, | Then . . . choose an activity for them. |

Reflect 10

Extended Response REASONING

Ask questions such as the following:
- What did you like about playing Concentration?
- Was there anything about playing this game you didn't like?
- Did this game help you do something you couldn't do before? What did it help you do?
- What was easy when you were playing Party?
- What was hard when you were playing Party?
- What do you remember most about this game?
- What was your favorite part?

Using Student Pages

Have each child complete **Workbook,** page 13. For Question 4 ask, "How many hats do you need? Ring the number that shows how many hats these children need if each child wants one hat." Is the child correctly applying the skills he or she learned in Week 3?

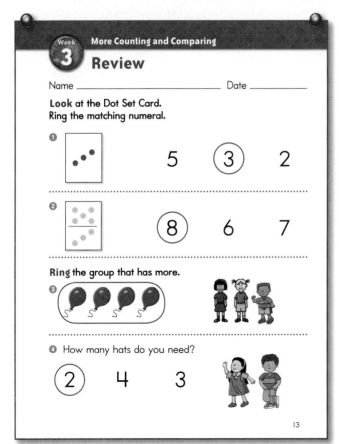

Workbook, p. 13

 Assess 10

A Gather Evidence

Formal Assessment

Have students complete the weekly test on *Assessment,* page 29. Record formal assessment scores on the Student Assessment Record, *Assessment,* page 100.

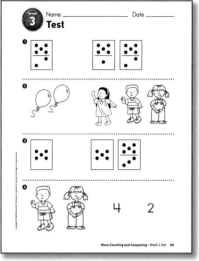

Assessment, p. 29

B Summarize Findings

Review the Student Assessment Records. Determine whether students have Minimal, Basic, or Secure understanding of the concepts presented in Week 3.

C Differentiate Instruction

Based on your observations, use these teaching strategies next week to follow up.

Minimal Understanding
- Repeat the Warm-Up and Engage activities in subsequent weeks to develop concepts of counting objects.
- Repeat the activities from this week. Observe children's understanding of these concepts and reinforce as necessary.
- Use **Building Blocks** computer activities beginning with Count to develop and reinforce numeration concepts.

Basic Understanding
- Repeat Engage activities in subsequent weeks to reinforce basic counting concepts.
- Use **Building Blocks** computer activities beginning with Party Time 1 to reinforce numeration and addition concepts.

Secure Understanding
- Use Challenge variations of **Catch the Teacher.**
- Use **Building Blocks** computer activities beginning with Number Compare.

More Counting and Comparing • Lesson 5 Review **61**

Week 4: Matching Dot Sets to Numerals

Week at a Glance

This week, children begin **Number Worlds**, Week 4, and continue to explore Picture Land.

Background

In Picture Land, numbers may be represented by sets of dots or by numerals, which helps children connect the world of real quantities (e.g., objects) and the world of formal symbols (e.g., written numerals). In Picture Land, children have many opportunities to identify numerals and to use the formal number system.

How Children Learn

As children begin this week's lesson, they should be familiar with numbers 1–20 and should have experience comparing sets of real objects and dot-set patterns.

At the end of this week, children should understand the correspondence between dot set patterns and numerals and should understand that numerals can represent a quantity.

Skills Focus

- Recognize errors in the counting sequence
- Count on from 10
- Compare and order numbers
- Identify numerals to 20

Teaching for Understanding

As children engage in these activities, they will come to understand that numerals can be used to represent not only the counting words but also the quantity in a given set. They will compare numerals to dot sets and will also compare two numerals to determine which one is bigger (or represents a greater quantity).

Observe closely as children take part in the Engage activities for this week.

- Are children accurately comparing numerals and dot set patterns?
- Can children accurately compare two different numerals?
- Are children using *more, less,* and *equal* correctly?

Math at Home

Give one copy of the Letter to Home, page A4, to each child. Complete the activity in class, and then encourage children to share it with their caregivers.

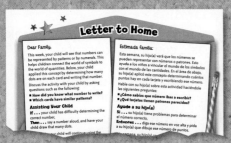

Letter to Home, Teacher Edition, p. A4

Math Vocabulary

numeral A symbol that represents a number and amount

symbol A letter, number, or picture that has special meaning or that stands for something else

count on To continue counting

English Learners

SPANISH COGNATES

English	Spanish
numeral	número
symbols	símbolos
compare	comparer
concentration	concentración

ALTERNATE VOCABULARY

match To find two things that are the same or that go together

keep To take for yourself; to hold on to

Week 4 Planner — Matching Dot Sets to Numerals

PACING	LESSON	LEARNING GOALS	MATERIALS	TECHNOLOGY
DAY 1	**Warm Up 2** Picture Land: Catch the Teacher*	Children detect an error in counting and identify which number was omitted from the sequence.	**Prepare Ahead** Two columns on the board, one labeled *Students* and one labeled *Teacher*	**Building Blocks** Memory Number 1
	Activity 7 Picture Land: Concentration to 20*	Children match set size to numerals and learn to count on from 10.	**Program Materials** • Dot Set Cards (1–10), 2 sets • Number Cards (1–20) • Dot Set Ten Cards, 1 set	
DAY 2	**Warm Up 2** Picture Land: Catch the Teacher*	Children continue to detect errors in the counting sequence and identify an omitted number.	**Prepare Ahead** Two columns on the board, one labeled *Students* and one labeled *Teacher*	**Building Blocks** Number Compare 1
	Activity 8 Picture Land: Bravo!*	Children compare two numerals to determine which is bigger.	**Program Materials** Number Cards (11–20), 1 set per child	
DAY 3	**Warm Up 2** Picture Land: Catch the Teacher*	Children continue to build their knowledge of the counting sequence.	**Prepare Ahead** Two columns on the board, one labeled *Students* and one labeled *Teacher*	**Building Blocks** Memory Number 1
	Activity 7 Picture Land: Concentration to 20*	Children continue to match set size to numerals and count on from 10.	**Program Materials** • Dot Set Cards (1–10), 2 sets • Number Cards (1–20) • Dot Set Ten Cards, 1 set	
DAY 4	**Warm Up 2** Picture Land: Catch the Teacher*	Children increase their familiarity with the counting sequence and identify omitted numbers.	**Prepare Ahead** Two columns on the board, one labeled *Students* and one labeled *Teacher*	**Building Blocks** Number Compare 1
	Activity 8 Picture Land: Bravo!*	Children gain further practice comparing numbers to determine which is bigger.	**Program Materials** Number Cards (11–20), 1 set per child	
DAY 5	**Warm Up 2** Picture Land: Catch the Teacher*	Children continue to detect errors in the counting sequence.	**Prepare Ahead** Two columns on the board, one labeled *Students* and one labeled *Teacher*	**Building Blocks** Review previous activities
	Review and Assess	Children review their favorite activities to improve their understanding of more difficult concepts.	Materials will be selected from those used in previous weeks.	

* Includes Challenge Variations

Week 4 — Matching Dot Sets to Numerals

Lesson 1

Objective
Children match set size to numerals and learn to count on from 10.

Program Materials
Engage
- Dot Set Cards (1–10), 2 sets
- Number Cards (1–20)
- Dot Set Ten Cards, 1 set

Prepare Ahead
Warm Up
Two columns on the board, one labeled *Students* and one labeled *Teacher*

Access Vocabulary
faceup Cards on a table with the dots or numerals up (showing)
facedown Cards on a table with the dots or numerals down (not showing)

Creating Context
When using the game cards, explain the concepts of *faceup and facedown*. These idioms may not be familiar to students. The main, or most meaningful, side of the card is the face. A card that is faceup communicates specific information, while a card that is facedown communicates little information.

1 Warm Up — 5

Skill Building COMPUTING
Catch the Teacher
Before beginning the small-group activity **Concentration to 20**, use the **Catch the Teacher** activity with the whole group.

Purpose Children master the counting sequence and understand that each number occupies a fixed position in the sequence.

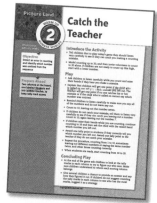

Warm-Up Card 2

Monitoring Student Progress

| If . . . a child cannot remember the number that was omitted, | Then . . . call on someone else whose hand is raised. If the child says a wrong number, ask the other children if they agree with the child's choice by using thumbs up or thumbs down. If they do not agree, ask another child what number was omitted. Confirm the correct answer. |

2 Engage — 30

Concept Building UNDERSTANDING
Concentration to 20
"Today we are going to play a matching game."

Follow the instructions on the Activity Card to play **Concentration to 20**. As children play, ask questions about what is happening in the activity.

Purpose Children match set size to the corresponding numerals and learn to count on from 10.

Activity Card 7

64 Number Worlds • Week 4

Monitoring Student Progress

| If . . . children have difficulty identifying the numerals 11–20, | Then . . . give them additional practice with the numerals 1–10 before gradually adding Number Cards (11–20) to this activity. |

eMathTools Use the Set Tool to demonstrate and explore counting.

Teacher's Note **Concentration to 20** is a terrific game to have in independent learning centers. You can prepare an envelope containing the cards needed for the game so the children can easily collect the materials on their own.

Building Blocks For additional practice identifying numerals, children should complete **Building Blocks** Memory Number 1.

Reflect

Extended Response REASONING

Ask questions such as the following:

- **Were there enough dots on the card you chose to match one of the faceup Number Cards?** Answers will vary.
- **How many more dots did you need?** Answers will vary.
- **Could you use your Dot Set Ten Card to create that amount?** Children may add 10 to the quantity on the Dot Set Card they drew to match a Number Card (11–20).

Using Student Pages

Have each child complete **Workbook,** page 14. Did children select the numeral that corresponds to the dot set pattern?

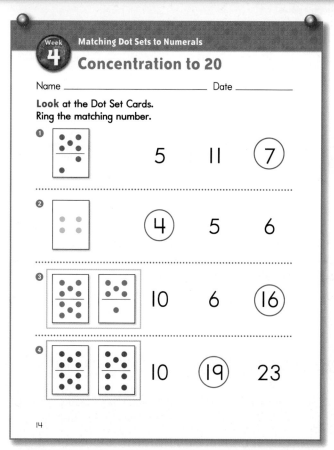

Workbook, p. 14

Assess

Informal Assessment

Use the Student Assessment Record, **Assessment,** page 100, to record informal observations.

COMPUTING	UNDERSTANDING
Catch the Teacher	**Concentration to 20**
Did the child	Did the child
❏ respond accurately?	❏ make important observations?
❏ respond quickly?	❏ extend or generalize learning?
❏ respond with confidence?	❏ provide insightful answers?
❏ self-correct?	❏ pose insightful questions?

Matching Dot Sets to Numerals • Lesson 1 **65**

Week 4 — Lesson 2: Matching Dot Sets to Numerals

Objective
Children compare numerals and identify errors in the counting sequence.

Program Materials

Engage
Number Cards (11–20), 1 set per child

Prepare Ahead

Warm Up
Two columns on the board, one labeled *Students* and one labeled *Teacher*

Engage
Shuffle together two sets of Number Cards (11–20) for each pair of children.

Access Vocabulary
shuffle To mix together
bravo A Spanish word that means "hooray!"

Creating Context
Some students who have attended school outside the United States may find it puzzling to see the teacher make a mistake or ask questions when the teacher clearly knows the answer. Help children and their families realize that games such as **Catch the Teacher** are designed to enhance learning.

1. Warm Up — 5

Concept Building — COMPUTING
Catch the Teacher
Before beginning the small-group activity **Bravo!**, use the **Catch the Teacher** activity with the whole group.

Purpose Children master the counting sequence and understand that each number occupies a fixed position in the sequence.

Warm-Up Card 2

Monitoring Student Progress
If . . . children have trouble identifying the missing number,

Then . . . practice counting to 20 with them throughout the day.

2. Engage — 30

Concept Building — ENGAGING
Bravo!
"Today we are going to play a fast-paced comparing game with numerals."

Follow the instructions on the Activity Card to play **Bravo!** As children play, ask questions about what is happening in the activity.

Purpose Children compare two numerals and determine which is bigger.

Activity Card 8

Monitoring Student Progress

If . . . children are having trouble comparing correctly,

Then . . . include a third child in the group, and have the children alternate between playing the game and serving as a group leader who makes sure the players are comparing correctly.

 Teacher's Note Encourage children to use this week's vocabulary words as they engage in the activities, discuss math concepts, and make predictions.

Building Blocks For additional practice comparing numerals, children should complete **Building Blocks** Number Compare 1.

3 Reflect 10

Extended Response REASONING

Ask questions such as the following:

- **What was the number on each of your cards? Did you both agree that you named the numerals correctly?** Answers will vary.
- **How could we figure out which number is bigger?** Students may suggest counting sets of objects that correspond to the numerals, or may say that one numeral comes after the other in the counting sequence.
- **Which number is higher when we count up?** Answers will vary.

Using Student Pages

Have each child complete **Workbook,** page 15. Did children successfully ring the bigger number?

Workbook, p. 15

4 Assess

Informal Assessment

Use the Student Assessment Record, **Assessment,** page 100, to record informal observations.

COMPUTING	ENGAGING
Catch the Teacher	**Bravo!**
Did the child	Did the child
❏ respond accurately?	❏ pay attention to the contributions of others?
❏ respond quickly?	
❏ respond with confidence?	❏ contribute information and ideas?
❏ self-correct?	
	❏ improve on a strategy?
	❏ reflect on and check accuracy of work?

Matching Dot Sets to Numerals • Lesson 2   67

Week 4: Matching Dot Sets to Numerals

Objective
Children match set size to numerals and count on from 10.

Program Materials

Engage
- Dot Set Cards (1–10), 2 sets
- Number Cards (1–20)
- Dot Set Ten Cards, 1 set

Prepare Ahead
Warm Up
Two columns on the board, one labeled *Students* and one labeled *Teacher*

Access Vocabulary
set A group of things that go together
match To put together things that are alike

Creating Context
Primary language can be a very helpful tool for ensuring comprehension of new concepts. Pair English learners of beginning English proficiency with students of the same primary language background who have greater English proficiency. Have them work through the activity using their primary language to support their understanding of the concept.

1 Warm Up

Skill Building COMPUTING
Catch the Teacher
Before beginning the small-group activity **Concentration to 20,** use the **Catch the Teacher** activity with the whole group.

Purpose Children master the counting sequence and understand that each number occupies a fixed position in the sequence.

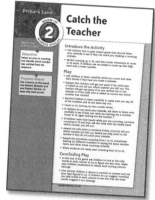

Warm-Up Card 2

Monitoring Student Progress
If . . . children are ready for a challenge,
Then . . . play **Catch the Teacher** backward.

2 Engage

Concept Building REASONING
Concentration to 20
"Today we are going to play a matching game with dot patterns and numerals."

Follow the instructions on the Activity Card to play **Concentration to 20.** As children play, ask questions about what is happening in the activity.

Purpose Children match set size to the corresponding numerals and learn to count on from 10.

Activity Card 7

Monitoring Student Progress

If . . . children can fluently match numerals to the corresponding dot set patterns,

Then . . . have them try the Challenge variation of **Concentration to 20.**

 Building Blocks For additional practice identifying numerals, children should complete **Building Blocks** Memory Number 1.

Teacher's Note The most important outcomes of this activity are for children to gain experience identifying numerals, and to identify or create dot set combinations that match the numerals in magnitude.

3 Reflect

Extended Response REASONING

For the Challenge variation, ask questions such as the following:

- **What number did you pick?** Answers will vary.
- **How many dots did you need? How did you know?** Children should indicate that they needed the number of dots represented by the numeral on their Number Card.
- **Did any of the faceup cards have that many dots? Could you use your Dot Set Ten Card to create the right amount?** For numbers 11–20, children could combine a Dot Set Ten Card and another Dot Set Card to match the Number Card they drew.

4 Assess

Informal Assessment

Use the Student Assessment Record, **Assessment**, page 100, to record informal observations.

COMPUTING	REASONING
Catch the Teacher	**Concentration to 20**
Did the child	Did the child
❑ respond accurately?	❑ provide a clear explanation?
❑ respond quickly?	❑ communicate reasons and strategies?
❑ respond with confidence?	
❑ self-correct?	❑ choose appropriate strategies?
	❑ argue logically?

Matching Dot Sets to Numerals • Lesson 3 69

Week 4 — Matching Dot Sets to Numerals

Objective
Children compare numerals and identify errors in the counting sequence.

Program Materials

Engage
Number Cards (11–20), 1 set per child

Prepare Ahead
Warm Up
Two columns on the board, one labeled *Students* and one labeled *Teacher*

Engage
Shuffle together two decks of Number Cards (11–20) for each pair of children.

Access Vocabulary
numeral A symbol that represents a number or amount
compare To think about how two things are alike and different

Creating Context
Some English learners might be confused by numbers described as *higher* or *lower* and *counting up* and *counting down* because these words commonly refer to position. Discuss with children that *higher* and *up* refer to numbers that come later in the counting sequence, while *lower* and *down* refer to numbers that come earlier in the sequence.

1 Warm Up 5

Skill Building — COMPUTING
Catch the Teacher
Before beginning the small-group activity **Bravo!,** use the **Catch the Teacher** activity with the whole group.

Purpose Children master the counting sequence and understand that each number occupies a fixed position in the sequence.

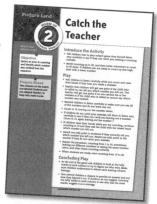

Warm-Up Card 2

Monitoring Student Progress
If . . . children are ready for a challenge, **Then . . .** play **Catch the Teacher** backward.

2 Engage 30

Concept Building — ENGAGING
Bravo!
"Today we are going to play a fast-paced comparing game with numerals."

Follow the instructions on the Activity Card to play **Bravo!** As children play, ask questions about what is happening in the activity.

Purpose Children compare two numerals and determine which is bigger.

Activity Card 8

Monitoring Student Progress

If . . . children are having trouble comparing correctly,

Then . . . include a third child in the group and have the children alternate between playing the game and serving as a group leader who makes sure the players are comparing correctly.

 Teacher's Note If you wish, have children play a variation of **Bravo!** using a combination of Dot Set Cards and Number Cards.

Building Blocks For additional practice comparing numerals, children should complete **Building Blocks** Number Compare 1.

3 Reflect 10

Extended Response REASONING

Ask questions such as the following:

- **How can you check your answer?** Answers will vary, but may include counting corresponding sets of objects or dot sets, or counting to see which number is higher in the number sequence.
- **What does this remind you of?** Accept all reasonable answers.

4 Assess

Informal Assessment

Use the Student Assessment Record, **Assessment,** page 100, to record informal observations.

COMPUTING	ENGAGING
Catch the Teacher	**Bravo!**
Did the child	Did the child
❑ respond accurately?	❑ pay attention to the contributions of others?
❑ respond quickly?	
❑ respond with confidence?	❑ contribute information and ideas?
❑ self-correct?	
	❑ improve on a strategy?
	❑ reflect on and check accuracy of work?

Matching Dot Sets to Numerals • Lesson 4   71

Week 4 — Lesson 5
Matching Dot Sets to Numerals
Review

Objective
Children review the material presented in Week 4 of *Number Worlds*.

Prepare Ahead
Warm Up
Two columns on the board, one labeled *Students* and one labeled *Teacher*

Creating Context
Discuss with children multiple meanings of the word *free*. Children may know that *free* means "no cost" but may not know that *free choice* means they can choose which activity they would like to do.

1 Warm Up

Skill Practice — COMPUTING
Catch the Teacher
Before beginning the **Free-Choice** activity use the **Catch the Teacher** activity with the whole group.

Purpose Children master the counting sequence and understand that each number occupies a fixed position in the sequence.

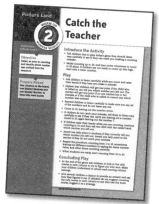

Warm-Up Card 2

2 Engage

Concept Building — APPLYING
Free-Choice Activity

For the last day of the week, allow children to choose an activity from the previous weeks. Some activities they may choose are the following:
- Object Land: **Count and Compare**
- Object Land: **Food Fun**
- Object Land: **Party**

Make a note of the activities children select. Do they prefer easy or challenging activities? Continue to provide Challenge opportunities for children who have mastered the basic activities.

Monitoring Student Progress

If...	Then...
children would benefit from extra practice on specific skills,	choose an activity for them.

Reflect 10

Extended Response REASONING

Ask questions such as the following:
- **What did you like about playing** Concentration to 20?
- **Was there anything about playing this game you didn't like?**
- **Did this game help you do something you couldn't do before? What did it help you do?**
- **What was easy when you were playing** Bravo?
- **What was hard when you were playing** Bravo?
- **What do you remember most about this game?**
- **What was your favorite part?**

Using Student Pages

Have each child complete **Workbook,** page 16. Is the child correctly applying the skills they learned in Week 4?

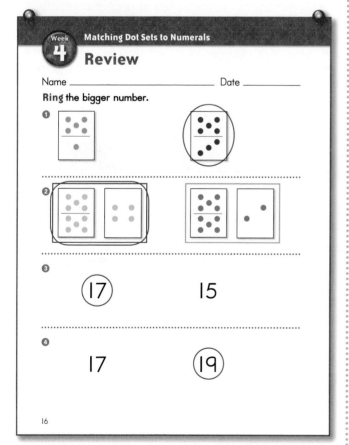

Workbook, p. 16

 Assess 10

A Gather Evidence

Formal Assessment
Have students complete the weekly test on **Assessment,** page 31. Record formal assessment scores on the Student Assessment Record, **Assessment,** page 100.

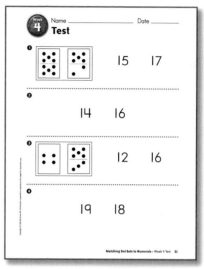

Assessment, p. 31

B Summarize Findings

Review the Student Assessment Records. Determine whether students have Minimal, Basic, or Secure understanding of the concepts presented in Week 4.

C Differentiate Instruction

Based on your observations, use these teaching strategies next week to follow up.

Minimal Understanding
- Repeat the Warm-Up and Engage activities in subsequent weeks to develop basic counting and numeral recognition concepts.
- Repeat the Picture Land activities from this week. Observe children's understanding of these concepts and reinforce as necessary.
- Use **Building Blocks** computer activities beginning with Count to develop and reinforce numeration concepts.

Basic Understanding
- Repeat group activities in subsequent weeks to reinforce children's understanding of the teen numbers.
- Use **Building Blocks** computer activities beginning with Memory Number 1 to reinforce numeration and addition concepts.

Secure Understanding
- Use Challenge variations of **Catch the Teacher.**
- Use **Building Blocks** computer activities beginning with Number Compare 2.

Matching Dot Sets to Numerals • Lesson 5 Review 73

Week 5: Number Sequence and Number Lines

Week at a Glance

This week children begin **Number Worlds,** Week 5, and are introduced to Line Land.

Background

In Line Land, numbers are represented as positions on a line, and change in quantity is represented as movement along a line. When a child is asked "Where are you now?" he or she might think of the position on the number line, the set size that corresponds to that position, or the numeral that shows the position on the number line and indicates the set size it marks.

How Children Learn

As children begin this week's lesson, they should be comfortable counting from 1 to 20.

At the end of the week, children should be able to compare positions on a number line and should begin to associate those positions with set size.

Skills Focus
- Identify and sequence numerals 1–20
- Compare and order numbers
- Identify or compute set size

Teaching for Understanding

As children engage in these activities, they will begin to use the spatial terms *before* and *after* with the number sequence and will gradually learn to associate position on a number line with set size and with the numeral that represents that set size.

Observe closely while evaluating the Engage activities assigned for this week.

- Can children sequence numbers to 20?
- Are students using *before* and *after* correctly when comparing positions on a number line?
- Do students associate position on a number line with set size?

Math at Home
Give one copy of the Letter to Home, page A5, to each child. Complete the activity in class, and then encourage children to share it with their caregivers.

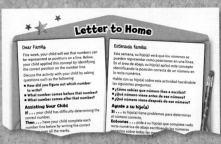

Letter to Home, Teacher Edition, p. A5

Math Vocabulary
before In front of; to come first
after Behind; to come later
near Close to
far Not close to; a long way away

English Learners
SPANISH COGNATES

English	Spanish
order	orden
number/numeral	número
positions	posiciones

ALTERNATE VOCABULARY
mail carrier Person who delivers letters

Week 5 Planner — Number Sequence and Number Lines

PACING	LESSON	LEARNING GOALS	MATERIALS	TECHNOLOGY
DAY 1	Warm Up 3 Line Land: The Neighborhood*	Children use the spatial terms *before* and *after* with the number sequence.	**Program Materials** Neighborhood Number Line (1–10 and 11–20); Sections 21–100 for a Challenge **Additional Materials** Self-adhesive notes	Building Blocks Before and After Math
	Activity 9 Line Land: Delivering Mail to the Neighborhood*	Children use spatial terms with the number sequence and discuss the locations of different numbers.	**Program Materials** • Neighborhood Number Line (1–10 and 11–20); Sections 21–100 for a Challenge • Pawn **Additional Materials** • Small stickers • Self-adhesive notes • Five envelopes, each with a different number 1–20 on the outside	
DAY 2	Warm Up 3 Line Land: The Neighborhood*	Children continue to sequence numbers using the terms *before* and *after*.	Same as Lesson 1	Building Blocks Before and After Math
	Activity 10 Line Land: Magnetic Number Line Game	Children identify set size and associate set size with position on a number line.	**Program Materials** • Magnetic Number Line, 1 per player • Magnetic Chips in different colors for each player • Dot Cube	
DAY 3	Warm Up 3 Line Land: The Neighborhood*	Children continue to use spatial terms with the number sequence.	Same as Lesson 1	Building Blocks Before and After Math
	Activity 9 Line Land: Delivering Mail to the Neighborhood*	Children further discuss the locations of different numbers.	Same as Lesson 1	
DAY 4	Warm Up 3 Line Land: The Neighborhood*	Children continue to sequence numbers using the terms *before* and *after*.	Same as Lesson 1	Building Blocks Before and After Math
	Activity 10 Line Land: Magnetic Number Line Game	Children associate set size with position on a number line.	**Program Materials** Same as Lesson 2	
DAY 5	Warm Up 3 Line Land: The Neighborhood*	Children refine their understanding of the number sequence.	Same as Lesson 1	Building Blocks Review previous activities
	Review and Assess	Children review favorite activities.	Materials will be selected from those used in previous weeks.	

*Includes Challenge Variations

Week 5 — Number Sequence and Number Lines

Lesson 1

Objective
Children use the spatial terms *before* and *after* with the number sequence and use clues from the number sequence to discuss the location of different numbers.

Program Materials

Warm Up
Neighborhood Number Line (1–10 and 11–20); Sections 21–100 for a Challenge

Engage
- Neighborhood Number Line (1–10 and 11–20); Sections 21–100 for a Challenge
- Pawn

Additional Materials

Warm Up
Self-adhesive notes placed over each number before activity begins

Engage
- Small stickers
- Self-adhesive notes placed over each number before activity begins
- Five envelopes, each with a different number 1–20 written on the outside; higher numbers written on the outside for a Challenge

Access Vocabulary
near Close to
far Not close to; a long way away

Creating Context
Some children receive mail at a post office box or at a central mail area rather than receiving mail at an individual house. Review a mail carrier's route with the children so they understand the premise of the game.

1 Warm Up

Skill Building — COMPUTING
The Neighborhood
Before beginning the whole-class activity **Delivering Mail to the Neighborhood,** use the **The Neighborhood** activity with the whole group.

Purpose Children identify and sequence numbers and use the spatial terms *before* and *after* with the number sequence.

Warm-Up Card 3

Teacher's Note Hang the first two sections of the Neighborhood Number Line on the wall so children can plainly see and have access to them. Leave plenty of room to add additional sections of the number line to 100.

2 Engage

Concept Building — UNDERSTANDING
Delivering Mail to the Neighborhood
"Today you are going to be a mail carrier and deliver letters."

Follow the instructions on the Activity Card to play **Delivering Mail to the Neighborhood.** As children play, ask questions about what is happening in the activity.

Purpose Children recognize the spatial terms *before* and *after* and use them with the number sequence. They also use clues from the number sequence to discuss the locations of other numbers.

Activity Card 9

Monitoring Student Progress

If . . . children need prompting to identify the location of a number,	Then . . . ask the mail carrier to say and write that number on the board. Ask the class how many tens and how many ones are in the number. Help them use this information to determine to which block of houses the letter should be delivered.

 MathTools Use the Number Line Tool to demonstrate and explore number sequence.

Teacher's Note Some children may begin counting before the pawn makes its first move. Demonstrate that when children move along the number line they should start counting on the first *new* space and not on their starting numbers.

 For additional practice with number sequence, children should complete **Building Blocks** Before and After Math.

3 Reflect — 10

Extended Response REASONING

Ask questions such as the following:

- **Which way did you need to go? How did you figure that out?** Answers will vary but should show an understanding of the relative position of numbers in the counting sequence.
- **Did it take a long time to get there? How did you know?** Answers will vary but should reflect whether the next number was close to or far from the current number in the number sequence.

Using Student Pages

Have each child complete **Workbook,** page 17. Did the child correctly order the numbers from smallest to biggest?

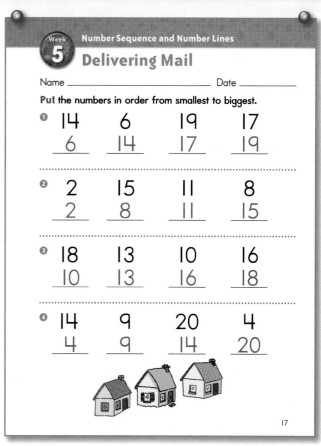

Workbook, p. 17

Assess

Informal Assessment

Use the Student Assessment Record, **Assessment,** page 100, to record informal observations.

COMPUTING	UNDERSTANDING
The Neighborhood Did the child ❏ respond accurately? ❏ respond quickly? ❏ respond with confidence? ❏ self-correct?	**Delivering Mail to the Neighborhood** Did the child ❏ make important observations? ❏ extend or generalize learning? ❏ provide insightful answers? ❏ pose insightful questions?

Number Sequence and Number Lines • Lesson 1 77

Week 5 — Lesson 2: Number Sequence and Number Lines

Objective
Children associate set size with position on a number line and recognize the spatial terms *before* and *after*.

Program Materials
Warm Up
Neighborhood Number Line (1–10 and 11–20); Sections 21–100 for a Challenge

Engage
- Magnetic Number Line, 1 per player
- Magnetic Chips in different colors for each player
- Dot Cube

Additional Materials
Warm Up
Self-adhesive notes placed over each number before activity begins

Access Vocabulary
before Coming first
after Coming next

Creating Context
English prepositions present a challenge to English Learners, but acting out their meaning can help make them understandable. Have several children line up and point out who is in line before and after several children. Then go to a number line and show a number that comes before or after another number.

---

1 Warm Up 5

Concept Building — COMPUTING
The Neighborhood
Before beginning the small-group activity **Magnetic Number Line Game,** use the **The Neighborhood** activity with the whole group.

Purpose Children identify and sequence numbers and use the spatial terms *before* and *after* with the number sequence.

Warm-Up Card 3

Monitoring Student Progress
| If... children have trouble sequencing the numbers 11–20, | Then... give them extra practice sequencing numbers 1–10 before gradually adding numbers 11–20. |

---

2 Engage 30

Concept Building — ENGAGING
Magnetic Number Line Game
"Today we are going to see who can get to 25 the fastest."

Follow the instructions on the activity card to play the **Magnetic Number Line Game.** As children play, ask questions about what is happening in the activity.

Purpose Children will associate set size with position on a number line, and will associate an increase in quantity with forward movement on a number line.

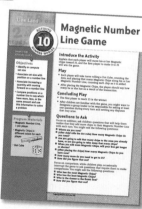

Activity Card 10

78 Number Worlds • Week 5

Monitoring Student Progress

If . . . a child counts all the Magnetic Chips on the line to determine how many he or she now has,

Then . . . tell the child that he or she can also look at the number on which the last chip rests to figure out the answer.

 Building Blocks For additional practice with number sequence, children should complete **Building Blocks** Before and After Math.

3 Reflect 10

Extended Response REASONING

Ask children questions that will help them realize that they added more Magnetic Chips to their Magnetic Number Lines with each turn. You might ask questions such as the following:

- **After you rolled the Dot Cube, how many Magnetic Chips did you need?** Answers will vary, but should match the number rolled on the Dot Cube.
- **Did you add that many Magnetic Chips to the ones you already have, or did you take away that many chips?** Add.
- **When you added more Magnetic Chips, did your line get longer or shorter?** Longer.

Using Student Pages

Have each child complete **Workbook,** page 18. Did the child correctly interpret the information on the number line?

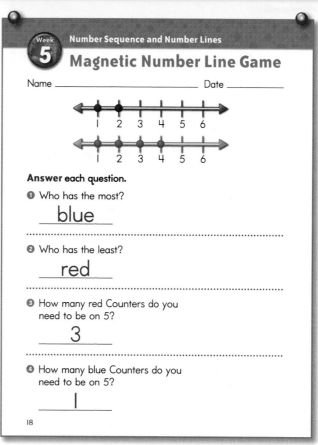

Workbook, p. 18

4 Assess

Informal Assessment

Use the Student Assessment Record, **Assessment,** page 100, to record informal observations.

COMPUTING	ENGAGING
The Neighborhood Did the child	**Magnetic Number Line Game** Did the child
❑ respond accurately? ❑ respond quickly? ❑ respond with confidence? ❑ self-correct?	❑ pay attention to the contributions of others? ❑ contribute information and ideas? ❑ improve on a strategy? ❑ reflect on and check accuracy of work?

Number Sequence and Number Lines • Lesson 2 79

Week 5 — Number Sequence and Number Lines
Lesson 3

Objective
Children use the spatial terms *before* and *after* with the number sequence and use clues from the number sequence to discuss the location of other numbers.

Program Materials
Warm Up
Neighborhood Number Line (1–10 and 11–20); Sections 21–100 for a Challenge

Engage
- Neighborhood Number Line (1–10 and 11–20); Sections 21–100 for a Challenge
- Pawn

Additional Materials
Warm Up
Self-adhesive notes placed over each number before activity begins

Engage
- Small stickers
- Self-adhesive notes placed over each number before activity begins
- Five envelopes, each with a different number 1–20 written on the outside; higher numbers written on the outside for a Challenge

Access Vocabulary
mail carrier Person who delivers letters

Creating Context
To help children understand why it is more efficient to deliver mail in sequential order, have six children stand in a circle in front of the class. Have a child deliver mail to the children in sequential order, then in order of the shortest child to the tallest. Decide which way was the quickest.

1 Warm Up 5

Skill Building — COMPUTING
The Neighborhood
Before beginning the whole-class activity **Delivering Mail to the Neighborhood,** use the **The Neighborhood** activity with the whole group.

Purpose Children will use the spatial terms *before* and *after* with the number sequence, and for a challenge will identify and sequence the numerals 21–100.

Warm-Up Card 3

Monitoring Student Progress
If . . . children	Then . . .
can fluently identify and sequence numbers from 1–20,	play the challenge variation of the activity with numbers 21–100.

2 Engage 30

Concept Building — REASONING
Delivering Mail to the Neighborhood
"Today you are going to be a mail carrier and deliver letters."

Follow the instructions on the Activity Card to play **Delivering Mail to the Neighborhood.** As children play, ask questions about what is happening in the activity.

Activity Card 9

Purpose Children recognize the spatial terms *before* and *after* and use them with the number sequence. They also use clues from the number sequence to discuss the locations of other numbers.

80 Number Worlds • Week 5

Monitoring Student Progress

If . . . children are ready for a challenge,

Then . . . add additional sections of the Neighborhood Number Line throughout the first half of the school year, and repeat the activity with each new section.

Teacher's Note Encourage children to use this week's vocabulary words as they engage in the activities, discuss math concepts, and make predictions.

 For additional practice with number sequence, children should complete **Building Blocks** Before and After Math.

4 Assess

Informal Assessment

Use the Student Assessment Record, **Assessment,** page 100, to record informal observations.

COMPUTING	REASONING
The Neighborhood	**Delivering Mail to the Neighborhood**
Did the child	Did the child
❏ respond accurately?	❏ provide a clear explanation?
❏ respond quickly?	❏ communicate reasons and strategies?
❏ respond with confidence?	
❏ self-correct?	❏ choose appropriate strategies?
	❏ argue logically?

3 Reflect 10

Extended Response REASONING

Ask questions such as the following:

- **Would this letter go to a house that is before (point to the lower numbers) or after (point to the higher numbers) this house?** Answers will vary.
- **How did you figure that out?** Accept all reasonable answers. Encourage children to discuss the relationship between the tens digit of their current location and the new location.

Week 5

Number Sequence and Number Lines

Objective
Children associate set size with position on a number line and recognize the spatial terms *before* and *after*.

Program Materials
Warm Up
Neighborhood Number Line (1–10 and 11–20); Sections 21–100 for a Challenge

Engage
- Magnetic Number Line, 1 per player
- Magnetic Chips in different colors for each player
- Dot Cube

Additional Materials
Warm Up
Self-adhesive notes placed over each number before activity begins

Access Vocabulary
close Near; next to
far Not near; a long way away

Creating Context
Ask if any children have ever participated in a race. How did they know how far to run? What did they use for the finish line to show where the race ended? Discuss how the **Magnetic Number Line Game** is like a race.

1 Warm Up

Skill Building COMPUTING
The Neighborhood
Before beginning the small-group activity **Magnetic Number Line Game,** use the **The Neighborhood** activity with the whole group.

Purpose Children will identify and sequence numbers and use the spatial terms *before* and *after* with the number sequence.

Warm-Up Card 3

Monitoring Student Progress
If . . . children can fluently identify and sequence numbers,
Then . . . have children skip count as they play the Challenge variation of the activity.

2 Engage

Concept Building ENGAGING
Magnetic Number Line Game

"Today we are going to see who can get to 25 the fastest."

Follow the instructions on the Activity Card to play the **Magnetic Number Line Game.** As children play, ask questions about what is happening in the activity.

Purpose Children will associate set size with position on a number line and will associate an increase in quantity with forward movement on a number line.

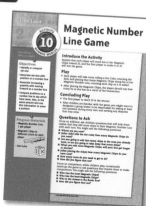
Activity Card 10

82   Number Worlds • Week 5

Monitoring Student Progress

If . . . children count all their Magnetic Chips to determine their position on the Number Line,

Then . . . encourage them to use the addition process to answer this question (e.g., "I had 15; I added 5; now I have 20 because 15 plus 5 is 20.").

 For additional practice with number sequence, children should complete **Building Blocks** Before and After Math.

3 Reflect

Extended Response REASONING

Point to a number on the children's number line.

- **When you were on this number, how many more did you need to get to 25?** Answers will vary.
- **How did you figure that out?** Accept all reasonable answers. Children may count on to 25 from the number that matches their last Magnetic Chip.

4 Assess

Informal Assessment

Use the Student Assessment Record, **Assessment,** page 100, to record informal observations.

COMPUTING	ENGAGING
The Neighborhood Did the child ❏ respond accurately? ❏ respond quickly? ❏ respond with confidence? ❏ self-correct?	**Magnetic Number Line Game** Did the child ❏ pay attention to the contributions of others? ❏ contribute information and ideas? ❏ improve on a strategy? ❏ reflect on and check accuracy of work?

Number Sequence and Number Lines • Lesson 4

Week 5: Number Sequence and Number Lines

Lesson 5
Review

Objective
Children review the material presented in Week 5 of *Number Worlds*.

Program Materials
Warm Up
Neighborhood Number Line (1–10 and 11–20); Sections 21–100 for a Challenge

Additional Materials
Warm Up
Self-adhesive notes placed over each number before activity begins

Access Vocabulary
before In front of; to come first
after Behind; to come later

Creating Context
Practice the different comparative adjectives that are used with the number line (*more, less, bigger, smaller,* and so on). Say, "Show a number greater than 5. Show a number smaller than 19." Have children use a number line to show their answers.

1 Warm Up 5

Skill Practice COMPUTING
The Neighborhood
Before beginning the **Free-Choice** activity, use the **The Neighborhood** activity with the whole group.

Purpose Children should associate increasing a quantity with moving forward on a number line; compare positions on a number line to say which have more, less, or the same amount; and use this information to solve a problem.

Warm-Up Card 3

Teacher's Note Remember to hang the first two sections of the Classroom Neighborhood Number Line on the wall so the children can plainly see them.

2 Engage 20

Concept Building APPLYING
Free-Choice Activity

For the last day of the week, allow children to choose an activity from the previous weeks. Some activities they may choose include the following:
- Picture Land: **Catch the Teacher**
- Picture Land: **Concentration to 20**
- Picture Land: **Bravo!**

Make a note of the activities children select. Do they prefer easy or challenging activities? Continue to provide challenge opportunities for students who have mastered the basic game.

Monitoring Student Progress
| If . . . children would benefit from extra practice on specific skills, | Then . . . choose an activity for them. |

Reflect   10

Extended Response REASONING
Ask questions such as the following:
- **What did you like about playing** Delivering Mail to the Neighborhood?
- **Was there anything about playing this game you didn't like?**
- **Did this game help you do something you couldn't do before? What did it help you do?**
- **What was easy when you were playing** Magnetic Number Line Game?
- **What was hard when you were playing** Magnetic Number Line Game?
- **What do you remember most about this game?**
- **What was your favorite part?**

Using Student Pages
Have each child complete **Workbook,** page 19. Is the child correctly applying the skills they learned in Week 5?

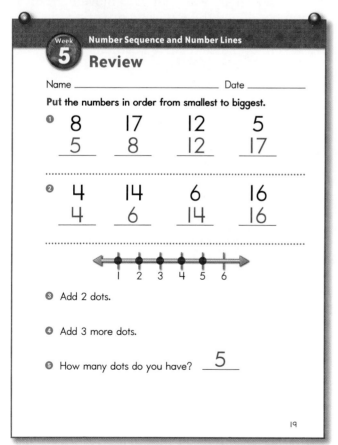

Workbook, p. 19

 ## Assess   10

A Gather Evidence
Formal Assessment
Have students complete the weekly test on *Assessment,* page 33. Record formal assessment scores on the Student Assessment Record, *Assessment,* page 100.

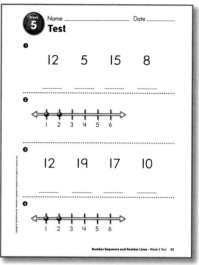

Assessment, p. 33

B Summarize Findings
Review the Student Assessment Records. Determine whether students have Minimal, Basic, or Secure understanding of the concepts presented in Week 5.

C Differentiate Instruction
Based on your observations, use these teaching strategies next week to follow up.

Minimal Understanding
- Repeat the Warm-up and Group activities in subsequent weeks to develop children's understanding of the number sequence 1–100.
- Repeat the Line Land activities from this week. Observe children's understanding of these concepts and reinforce as necessary.
- Use **Building Blocks** computer activities beginning with Count to develop and reinforce numeration and addition concepts.

Basic Understanding
- Repeat Engage activities in subsequent weeks to reinforce number sequence concepts.
- Use **Building Blocks** computer activities beginning with Before and After Math to reinforce numeration and addition concepts.

Secure Understanding
- Use Challenge variations of **The Neighborhood.**
- Use **Building Blocks** computer activities beginning with School Supply Shop.

Number Sequence and Number Lines • Lesson 5 Review   85

Week 6: More Number Sequence and Number Lines

Week at a Glance

This week children begin **Number Worlds,** Week 6, and continue to explore Line Land.

Background

In Line Land, numbers are represented as positions on a line, and change in quantity is represented as movement along a line. When a child is asked, "Where are you now?" he or she might think of the position on the number line, the size of the set that corresponds to that position, or the numeral that shows the position on the number line.

How Children Learn

As children begin this week's lessons, they should already have experience sequencing numbers on a number line and using spatial terms such as *before* and *after.*

By the end of the week, children should begin to associate forward movement along a number line with an increase in quantity and backward movement with a decrease in quantity.

Skills Focus
- Compare and order numbers
- Identify or compute set size
- Add whole numbers
- Subtract whole numbers

Teaching for Understanding

As children engage in these activities, they will continue to use the terms *before* and *after* with the number sequence, while associating position on a number line with set size and movement along the number line as an increase or decrease in set size.

Observe closely while evaluating the Engage activities assigned for this week.

- Do children associate position on a number line with set size?
- Do children associate forward movement on a number line with increasing quantity and backward movement with decreasing quantity?
- Can children use relative positions on a number line to solve problems?

Math at Home

Give one copy of the Letter to Home, page A6, to each child. Complete the activity in class, and then encourage children to share it with their caregivers.

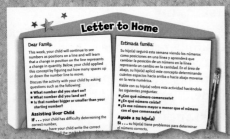

Letter to Home, Teacher Edition, p. A6

Math Vocabulary

before In front of; to come first

after Behind; to come later

plus sign Symbol that means "add"

minus sign Symbol that means "take away"

English Learners

SPANISH COGNATES

English	Spanish
minus	menos
different	diferente
cards	cartas

ALTERNATE VOCABULARY

near Close to

far Not close to; a long way away

Week 6 Planner: More Number Sequence and Number Lines

PACING	LESSON	LEARNING GOALS	MATERIALS	TECHNOLOGY
DAY 1	Warm Up 3 — Line Land: The Neighborhood*	Children use the spatial terms *before* and *after* with the number sequence.	**Program Materials**: Neighborhood Number Line (1–20) **Additional Materials**: Self-adhesive notes	Building Blocks — Before and After Math
	Activity 9 — Line Land: Delivering Mail to the Neighborhood*	Children use spatial terms with the number sequence and discuss the locations of different numbers.	**Program Materials**: • Neighborhood Number Line (1–20) • Pawn **Additional Materials**: • Small stickers • Self-adhesive notes • Five envelopes, each with a different number 1–20 on the outside	
DAY 2	Warm Up 3 — Line Land: The Neighborhood*	Children continue to sequence numbers using the terms *before* and *after*.	Same as Lesson 1	Building Blocks — Barkley's Bones 1
	Activity 11 — Line Land: Plus-Minus Game	Children predict how many they will have if they add or subtract Counters from a number line.	**Program Materials**: • Magnetic Number Line, 1 per player • Magnetic Chips in different colors • Line Land +1\−1 Cards • Dot Cube	
DAY 3	Warm Up 3 — Line Land: The Neighborhood*	Children continue to use spatial terms with the number sequence.	Same as Lesson 1	Building Blocks — Before and After Math
	Activity 9 — Line Land: Delivering Mail to the Neighborhood*	Children further explore the spatial terms *before* and *after* and discuss the locations of different numbers.	Same as Lesson 1	
DAY 4	Warm Up 3 — Line Land: The Neighborhood*	Children associate increasing a quantity with moving forward on a number line.	Same as Lesson 1	Building Blocks — Barkley's Bones 1
	Activity 11 — Line Land: Plus-Minus Game	Children predict how many they will have if they add or subtract Counters from a number line.	Same as Lesson 2	
DAY 5	Warm Up 3 — Line Land: The Neighborhood*	Children refine their understanding of the number sequence.	Same as Lesson 1	Building Blocks — Review previous activities
	Review and Assess	Children review their favorite activities.	Materials will be selected from those used in previous weeks.	

* Includes Challenge Variations

Week 6 — More Number Sequence and Number Lines

Lesson 1

Objective
Children use spatial terms with the number sequence and discuss the locations of different numbers.

Program Materials
Warm Up
Neighborhood Number Line (1–10 and 11–20); Sections 21–100 for a Challenge

Engage
- Neighborhood Number Line (1–10 and 11–20); Sections 21–100 for a Challenge
- Pawn

Additional Materials
Warm Up
Self-adhesive notes placed over each number before activity begins

Engage
- Small stickers
- Self-adhesive notes placed over each number before activity begins
- Five envelopes, each with a different number 1–20 written on the outside; higher numbers written on the outside for a Challenge

Access Vocabulary
before In front of; to come first
after Behind; to come later

Creating Context
Prepositions can pose some difficulty for first graders, and especially for English Learners. To help children understand the terms *before* and *after*, have children work with a partner to list three things they do before school and three things they do after school. Have children share their lists with the class.

1 Warm Up 5

Skill Building COMPUTING

The Neighborhood
Before beginning the whole-class activity **Delivering Mail to the Neighborhood,** use the **The Neighborhood** activity with the entire class.

Purpose Children will identify and sequence numbers and use the spatial terms *before* and *after* to describe the magnitude of numbers in the number sequence.

Warm-Up Card 3

Teacher's Note Hang the first two sections of the Neighborhood Number Line on the wall so the children can plainly see and have access to it. Leave plenty of room to add additional sections of the number line to 100.

2 Engage 30

Concept Building UNDERSTANDING

Delivering Mail to the Neighborhood
"Today you will pretend to be mail carriers and deliver letters."

Follow the instructions on the Activity Card to play **Delivering Mail to the Neighborhood.** As children play, ask questions about what is happening in the activity.

Activity Card 9

88 Number Worlds • Week 6

Purpose Children recognize the spatial terms *before* and *after* and use them with the number sequence. They also use cues from the number sequence to discuss the locations of other numbers.

Monitoring Student Progress

If . . . children need prompting to identify the location of a number,

Then . . . ask the mail carrier to say and to write that number on the board. Ask the class how many tens and how many ones are in the number. Help them use this information to determine to which block of houses the letter should be delivered.

 For additional practice with number sequence, children should complete **Building Blocks** Before and After Math.

3 Reflect

Extended Response REASONING

Ask questions such as the following:

- **Which way did you need to go? How did you figure that out?** Answers will vary, but should show an understanding of the relative position of numbers in the counting sequence.
- **Did it take a long time to get there? Why?** Answers will vary, but should reflect whether the next number was close to or far from the current number in the number sequence.

Using Student Pages

Have each child complete **Workbook,** page 20. Did the child ring the correct number?

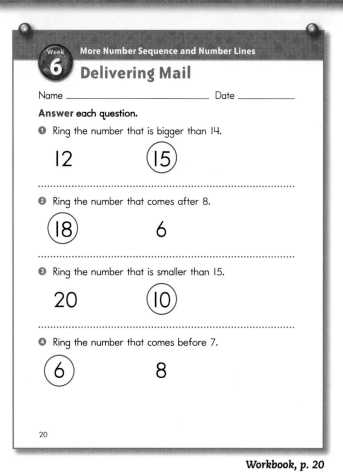

Workbook, p. 20

4 Assess

Informal Assessment

Use the Student Assessment Record, **Assessment,** page 100, to record informal observations.

COMPUTING	UNDERSTANDING
The Neighborhood	**Delivering Mail to the Neighborhood**
Did the child	Did the child
❏ respond accurately?	❏ make important observations?
❏ respond quickly?	❏ extend or generalize learning?
❏ respond with confidence?	❏ provide insightful answers?
❏ self-correct?	❏ pose insightful questions?

More Number Sequence and Number Lines • Lesson 1

Week 6

More Number Sequence and Number Lines

Lesson 2

Objective
Children use spatial terms with the number sequence and predict how many they will have if they add or subtract Counters from a number line.

Program Materials
Warm Up
First two sections of the Neighborhood Number Line Game Board (1–10 and 11–20); remaining sections (21–100), for a Challenge

Engage
- Magnetic Number Line, 1 per player
- Magnetic Chips in different colors for each player
- Line Land +1\−1 Cards, 1 set (Line Land +2\−2 Cards for a Challenge)
- Dot Cube

Additional Materials
Warm Up
Self-adhesive notes placed over each number before the activity begins

Access Vocabulary
plus sign Symbol that means "add"
minus sign Symbol that means "take away"

Creating Context
Be sensitive to the fact that some children may not have experience with board games. Discuss the routines used in board games such as determining who starts the game, taking turns, and knowing how the game is won.

1 Warm Up   5

Concept Building COMPUTING

The Neighborhood
Before beginning the small-group activity **Plus-Minus Game**, use the **The Neighborhood** activity with the whole class.

Purpose Children identify and sequence numerals and use the spatial terms *before* and *after* to describe the magnitude of numbers in the number sequence.

Warm-Up Card 3

Monitoring Student Progress

| **If . . .** children need prompting to identify the location of a number, | **Then . . .** ask the class how many tens and how many ones are in the number. Help children use this information to determine to which block of houses to deliver the letter. |

2 Engage   30

Concept Building ENGAGING

Plus-Minus Game
"Today you will try to be the first player to reach 25."

Follow the instructions on the Activity Card to play **Plus-Minus Game**. As children play, ask questions about what is happening in the game.

Purpose Children associate increases and decreases in set size with moving forward and backward on the number line.

Activity Card 11

90 Number Worlds • Week 6

Monitoring Student Progress

If . . . a child counts all the Magnetic Chips on the number line to determine how many he or she has,

Then . . . tell the child that he or she can also look at the number on which the last chip rests to find the answer.

 Teacher's Note Encourage children to use this week's vocabulary words as they engage in the activities, discuss math concepts, and make predictions.

Building Blocks For additional practice with addition and subtraction, children should complete **Building Blocks** Barkley's Bones 1.

3 Reflect

Extended Reponse REASONING

Ask questions such as the following:

- **Where were you when you started?** Answers will vary.
- **What did the card tell you to do? What number were you on when you did that?** Answers will vary.
- **Which direction did you need to move?** Answers will vary, but should indicate movement up or down the number line.

Using Student Pages

Have each child complete **Workbook,** page 21. Tell children, "Write what number you will be on if you start on the first number and pick the card shown." Did the child name the correct sum?

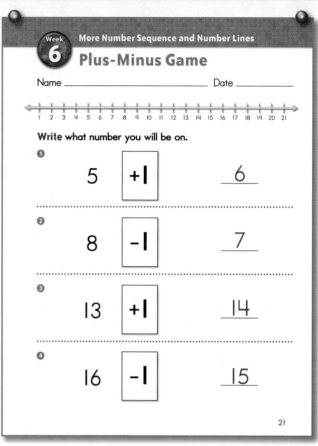

Workbook, p. 21

4 Assess

Informal Assessment

Use the Student Assessment Record, **Assessment,** page 100, to record informal observations.

COMPUTING	ENGAGING
The Neighborhood Did the child	**Plus-Minus Game** Did the child
❏ respond accurately?	❏ pay attention to the contributions of others?
❏ respond quickly?	
❏ respond with confidence?	❏ contribute information and ideas?
❏ self-correct?	
	❏ improve on a strategy?
	❏ reflect on and check accuracy of work?

More Number Sequence and Number Lines • Lesson 2   **91**

Week 6 — Lesson 3
More Number Sequence and Number Lines

Objective
Children use spatial terms with the number sequence and discuss the locations of different numbers.

Program Materials
Warm Up
Neighborhood Number Line (1–10 and 11–20); Sections 21–100 for a Challenge

Engage
- Neighborhood Number Line (1–10 and 11–20); Sections 21–100 for a Challenge
- Pawn

Additional Materials
Warm Up
Self-adhesive notes placed over each number before activity begins

Engage
- Small stickers
- Self-adhesive notes placed over each number before activity begins
- Five envelopes, each with a different number 1–20 written on the outside; higher numbers written on the outside for a Challenge

Access Vocabulary
near Close to
far Not close to; a long way away

Creating Context
Practice the concepts of *before* and *after* with a calendar. Discuss the facts that yesterday was before today, 10 comes before 11, and so on.

1 Warm Up — 5

Skill Building — COMPUTING
The Neighborhood
Before beginning the whole-class activity **Delivering Mail to the Neighborhood**, use the **The Neighborhood** activity with the whole class.

Purpose Children will identify and sequence numbers and use the spatial terms *before* and *after* to describe the magnitude of numbers in the number sequence.

Warm-Up Card 3

Monitoring Student Progress
If . . . children can count fluently to 20,
Then . . . play **The Neighborhood** Challenge 1 variation.

2 Engage — 30

Concept Building — REASONING
Delivering Mail to the Neighborhood
"Today you will pretend to be mail carriers and deliver letters."

Follow the instructions on the activity card to play **Delivering Mail to the Neighborhood.** As children play, ask questions about what is happening in the activity.

Activity Card 9

Purpose Children recognize the spatial terms *before* and *after* and use them with the number sequence. They also use cues from the number sequence to discuss the location of other numbers.

Monitoring Student Progress

If . . . children are ready for a challenge,

Then . . . add sections to the Neighborhood Number Line throughout the first half of the school year, and repeat the activity with each new section.

eMathTools Use the Number Line tool to demonstrate and explore number sequence.

 Teacher's Note Some children may begin counting before the pawn makes its first move. Demonstrate that when children move along the number line they should start counting on their first *new* space and not on their starting number.

 For additional practice with number sequence, children should complete **Building Blocks** Before and After Math.

3 Reflect 10

Extended Response REASONING
Ask questions such as the following:
- **Which way did you need to go?** Answers will vary.
- **Would this letter go to a house that is before (point to the lower numbers) or after (point to the higher numbers) this house?** Answers should indicate an understanding of relative position on the number line.

4 Assess

Informal Assessment
Use the Student Assessment Record, **Assessment**, page 100, to record informal observations.

COMPUTING	REASONING
The Neighborhood Did the child ❑ respond accurately? ❑ respond quickly? ❑ respond with confidence? ❑ self-correct?	**Delivering Mail to the Neighborhood** Did the child ❑ provide a clear explanation? ❑ communicate reasons and strategies? ❑ choose appropriate strategies? ❑ argue logically?

More Number Sequence and Number Lines • Lesson 3

Week 6: More Number Sequence and Number Lines

Lesson 4

Objective
Children use spatial terms with the number sequence and predict how many they will have if they add or subtract Counters from a number line.

Program Materials
Warm Up
First two sections of the Neighborhood Number Line (1–10 and 11–20); remaining sections (21–100), for a Challenge

Engage
- Magnetic Number Line, 1 per player
- Magnetic Chips in different colors for each player
- Line Land +1\−1 Cards, 1 set; (Line Land +2\−2 Cards for a Challenge)
- Dot Cube

Additional Materials
Warm Up
Self-adhesive notes placed over each number before the activity begins

Access Vocabulary
add Put together
subtract Take away

Creating Context
Playing games is an excellent way to give English Learners practice listening to and speaking English. The natural repetition of procedural and counting language replaces tedious drill with authentic, active experience. Wait time is built into the process, and the low-stakes environment makes the use of games an enjoyable and efficient learning tool in math.

1 Warm Up — 5

Skill Building COMPUTING
The Neighborhood
Before beginning the small-group activity **Plus-Minus Game**, use the **The Neighborhood** activity with the whole class.

Purpose Children identify and sequence numerals and use the spatial terms *before* and *after* to describe the magnitude of numbers in the number sequence.

Warm-Up Card 3

Monitoring Student Progress
If...	Then...
children are ready for a challenge,	play **The Neighborhood** Challenge 2 variation.

2 Engage — 30

Concept Building ENGAGING
Plus-Minus Game
"Today you will try to be the first player to reach 25."

Follow the instructions on the Activity Card to play **Plus-Minus Game**. As children play, ask questions about what is happening in the game.

Purpose Children associate increases and decreases in set size with moving forward and backward on the number line.

Activity Card 11

94 Number Worlds • Week 6

Monitoring Student Progress

If . . . children count all their Magnetic Chips to determine their position on the number line,

Then . . . encourage them to use the addition process to answer this question (e.g., "I had 15; I added 5; now I have 20 because 15 plus 5 is 20.").

 Building Blocks For additional practice with addition and subtraction, children should complete **Building Blocks** Barkley's Bones 1.

3 Reflect — 10

Extended Response REASONING

Toward the end of the game, ask questions such as the following:

- **Who is the closest to the end of the number line?** The student with the most Magnetic Chips.
- **What number must you roll to reach the end of the number line?** Answers will vary.
- **How do you know?** Encourage the children to explain their answers by referring to the number of spaces they have already progressed through, the numeral on which their last Magnetic Chip sits, and the number of spaces left to go.

4 Assess

Informal Assessment

Use the Student Assessment Record, **Assessment**, page 100, to record informal observations.

COMPUTING	ENGAGING
The Neighborhood Did the student ❑ respond accurately? ❑ respond quickly? ❑ respond with confidence? ❑ self-correct?	**Plus-Minus Game** Did the student ❑ pay attention to the contributions of others? ❑ contribute information and ideas? ❑ improve on a strategy? ❑ reflect on and check accuracy of work?

More Number Sequence and Number Lines • Lesson 4

Week 6: More Number Sequence and Number Lines

Lesson 5 Review

Objective
Children review the material presented in Week 6 of *Number Worlds*.

Program Materials
Warm Up
First two sections of the Neighborhood Number Line (1–10 and 11–20); remaining sections (21–100), for a Challenge

Additional Materials
Warm Up
Self-adhesive notes placed over each number before the activity begins

Creating Context
Discuss with children multiple meanings of the word *free*. Children may know that *free* means "no cost" but may not know that *free choice* means they can choose which activity they would like to do.

✓ Cumulative Assessment
After Lesson 5 is complete, students should complete the Cumulative Assessment, **Assessment**, pages 85–86. Using the key on **Assessment**, page 84, identify incorrect responses. Reteach and review the suggested activities to reinforce concept understanding.

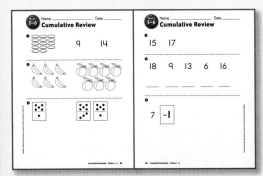
Assessment, pp. 85–86

1 Warm Up 5

Skill Building — COMPUTING
The Neighborhood
Before beginning the **Free-Choice** activity, use the **The Neighborhood** activity with the whole class.

Purpose Children identify and sequence numerals and use the spatial terms *before* and *after* to describe the magnitude of numbers in the number sequence.

Warm-Up Card 3

Teacher's Note Hang the first two sections of the Neighborhood Number Line on the wall so the children can plainly see them.

2 Engage 20

Concept Building — APPLYING
Free-Choice Activity
For the last day of the week, allow children to choose an activity from the previous weeks. Some activities they may choose are the following:
- Picture Land: **Catch the Teacher**
- Object Land: **Party!**
- Picture Land: **Bravo!**

Make a note of the activities children select. Do they prefer easy or challenging activities? Continue to provide challenge opportunities for children who have mastered the basic games.

Monitoring Student Progress
| If . . . children would benefit from extra practice on specific skills, | Then . . . choose an activity for them. |

Reflect 10

Extended Response REASONING
Ask questions such as the following:
- **What did you like about playing** Delivering Mail to the Neighborhood?
- **Was there anything about playing this game you didn't like?**
- **Did this game help you do something you couldn't do before? What did it help you do?**
- **What was easy when you were playing** Plus-Minus Game?
- **What was hard when you were playing** Plus-Minus Game?
- **What do you remember most about this game?**
- **What was your favorite part?**

Using Student Pages
Have each child complete **Workbook,** page 22. Is the child correctly applying the skills they learned in Week 6?

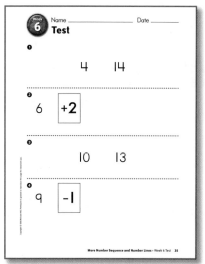

Workbook, p. 22

 Assess 10

A Gather Evidence
Formal Assessment
Have students complete the weekly test on *Assessment,* page 35. Record formal assessment scores on the Student Assessment Record, *Assessment,* page 100.

Assessment, p. 35

B Summarize Findings
Review the Student Assessment Records. Determine whether students have Minimal, Basic, or Secure understanding of the concepts presented in Week 6.

C Differentiate Instruction
Based on your observations, use these teaching strategies next week to follow up.

Minimal Understanding
- Repeat the Warm-Up and Engage activities in subsequent weeks to develop numeration concepts.
- Repeat the Line Land activities from this week. Observe children's understanding, and reinforce as necessary.
- Use **Building Blocks** computer activities beginning with Number Compare 1.

Basic Understanding
- Repeat Engage activities in subsesquent weeks to reinforce basic counting concepts.
- Use **Building Blocks** computer activities beginning with Before and After Math to reinforce numeration and addition concepts.

Secure Understanding
- Use the Challenge variations of **Delivering Mail to the Neighborhood** and the **Plus-Minus Game.**
- Use **Building Blocks** computer activities beginning with Barkley's Bones 1.

More Number Sequence and Number Lines • Lesson 5 Review

Week 7: Number Neighborhoods

Week at a Glance

This week children begin **Number Worlds,** Week 7, and are introduced to Sky Land.

Background

In Sky Land, children learn about movement along a vertical plane. This helps them learn to use the language of height in addition to the language they learned in Line Land, and it helps children begin to understand how the terminology can be used interchangeably.

How Children Learn

As children begin this week's lesson, they should be comfortable counting to 20 and using spatial terms to describe numbers on a number line.

At the end of this week, children should be able to count down from a given number and should begin to understand that +1 and −1 indicate movement along a number line.

Skills Focus
- Count to 20
- Count down from 20
- Compare and order numbers
- Add and subtract whole numbers

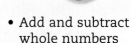

Teaching for Understanding

As children engage in these activities, their understanding of number relationships will become more sophisticated. They will see that movement along a number line can be vertical as well as horizontal and will use number relationships to target missing numbers and describe visible numbers.

Observe closely while evaluating the Engage activities assigned for this week.

- Can children use the terms *before, after,* and *between* correctly with the number sequence?
- Do children understand that +1 and −1 indicate movement up and down the number sequence?

Math at Home

Give one copy of the Letter to Home, page A7, to each child. Complete the activity in class, and then encourage children to share it with their caregivers.

Letter to Home, Teacher Edition, p. A7

Math Vocabulary

pawn Game piece
roll Drop a Dot Cube onto a game board
before In front of; to come first
after Behind; to come later
between In the middle of two things

English Learners

SPANISH COGNATES

English	Spanish
addition	adición
distance	distancia
group	grupo
number	número
thermometer	termómetro

ALTERNATE VOCABULARY

thermometer Tool used to measure how hot or cold something is

Week 7 Planner: Number Neighborhoods

PACING	LESSON	LEARNING GOALS	MATERIALS	TECHNOLOGY
DAY 1	Warm Up 4 — Sky Land: Thermometer Counting*	Children count from 1 up to 20 and from 20 down to 1.	**Program Materials** Thermometer Picture **Additional Materials** Erasable red marker	Building Blocks — Rocket Blast 1
DAY 1	Activity 12 — Line Land: Numbering the Neighborhood	Children identify and sequence numbers 1–20 and determine the location of a number.	**Program Materials** • Neighborhood Number Line (1–20) • Number Tiles (1–20) **Additional Materials** Self-stick notes	
DAY 2	Warm Up 4 — Sky Land: Thermometer Counting*	Children continue to count from 1 up to 20, and from 20 down to 1.	**Program Materials** Thermometer Picture **Additional Materials** Erasable red marker	Building Blocks — Before and After Math
DAY 2	Activity 13 — Line Land: The Biggest Neighborhood*	Children move forward and backward on a number line.	**Program Materials** • Neighborhood Number Line (1–20) • Number Cards (1–20) • Dot Cube, Pawns • +1\−1 Cards (Challenge) **Additional Materials** Colored stickers	
DAY 3	Warm Up 4 — Sky Land: Thermometer Counting*	Children count from 1 up to 20 and from 20 down to 1.	**Program Materials** Thermometer Picture **Additional Materials** Erasable red marker	Building Blocks — Rocket Blast 1
DAY 3	Activity 12 — Line Land: Numbering the Neighborhood	Children continue to identify and sequence numbers 1–20.	**Program Materials** Same as Lesson 1	
DAY 4	Warm Up 4 — Sky Land: Thermometer Counting*	Children count from 1 up to 20 and from 20 down to 1.	**Program Materials** Thermometer Picture **Additional Materials** Erasable red marker	Building Blocks — Before and After Math
DAY 4	Activity 13 — Line Land: The Biggest Neighborhood*	Children move forward and backward on a number line.	Same as Lesson 2	
DAY 5	Warm Up 4 — Sky Land: Thermometer Counting*	Children count from 1 up to 20 and from 20 down to 1.	**Program Materials** Thermometer Picture **Additional Materials** Erasable red marker	Building Blocks — Review previous activities
DAY 5	Review and Assess	Children review their favorite activities to improve their understanding.	Materials will be selected from those used in previous weeks.	

* Includes Challenge Variations

Week 7 — Lesson 1: Number Neighborhoods

Objective
Children identify and sequence the numbers 1–20 and determine the locations of numbers in the number sequence.

Program Materials
Warm Up
Thermometer Picture

Engage
- First two sections of the Neighborhood Number Line (1–10 and 11–20)
- Number Tiles (1–20)

Additional Materials
Warm Up
Erasable red marker

Engage
Self-stick notes placed over each number on the Neighborhood Number Line before the activity begins

Access Vocabulary
thermometer Tool used to measure how hot or cold something is
between In the middle of two things

Creating Context
Practice the different comparative adjectives that are used with the number line (*more, less, bigger, smaller,* and so on). Say, "Show a number greater than 5. Show a number smaller than 19." Have children use a number line to show their answers.

---

1 Warm Up — 5

Skill Building — COMPUTING
Thermometer Counting
Before beginning the whole-class activity **Numbering the Neighborhood,** use the **Thermometer Counting** activity with the whole class.

Purpose Children count from 1 up to 20 and from 20 down to 1.

Warm-Up Card 4

Monitoring Student Progress
If . . . children have trouble counting to 20,
Then . . . have them count up to and down from 10, and gradually add higher numbers.

Teacher's Note **Thermometer Counting** may also be used to record attendance at the beginning of each day or may be used as a cross-curricular extension to science and weather.

2 Engage — 30

Concept Building — UNDERSTANDING
Numbering the Neighborhood

"Today you are going to use Number Tiles to number the neighborhood."

Follow the instructions on the Activity Card to play **Numbering the Neighborhood.** As children play, ask questions about what is happening in the game.

Activity Card 12

100 Number Worlds • Week 7

Purpose Children identify and sequence the numbers 1–20 while determining the location of a number in the number sequence.

Monitoring Student Progress

| If . . . certain children still need to count to each house, beginning with 1, | Then . . . allow them to do so, but verify their decisions with reference to the number's position in the sequence. You might say, "You're right, that is house number 7. See, it comes one house after 6." |

Teacher's Note If a child places a number tile out of order, allow it. Improper placement is fine as long as the tile is in the correct neighborhood and is close to the correct position. This will provide the basis for a good discussion when it has to be moved later in the game.

 For additional practice with number sequence, children should complete **Building Blocks** Rocket Blast 1.

3 Reflect 10

Extended Response REASONING
Ask questions such as the following:

- **How did you figure out what numbers were missing?** Answers will vary, but should indicate that children looked at neighboring numbers to determine the missing numbers in the sequence.
- **What number did you put on the last house?** 20 **Why?** Possible answers: It is the highest number; there are two blocks of 10 houses, which makes 20.

Using Student Pages
Have each child complete **Workbook**, page 23. Did the child correctly identify the missing numbers?

 Number Neighborhoods

Numbering the Neighborhood

Name _____ Date _____

Write the missing number.

① __1__ 2 3 4 5

② 6 7 8 __9__ 10

③ 12 13 __14__ 15 16

④ 16 __17__ 18 19 20

23

Workbook, p. 23

4 Assess

Informal Assessment
Use the Student Assessment Record, **Assessment,** page 100, to record informal observations.

COMPUTING	UNDERSTANDING
Thermometer Counting Did the child ❑ respond accurately? ❑ respond quickly? ❑ respond with confidence? ❑ self-correct?	**Numbering the Neighborhood** Did the child ❑ make important observations? ❑ extend or generalize learning? ❑ provide insightful answers? ❑ pose insightful questions?

Number Neighborhoods • Lesson 1   101

Week 7: Number Neighborhoods
Lesson 2

Objective
Children continue to explore the counting sequence and move forward and backward along a number line.

Program Materials
Warm Up
Thermometer Picture

Engage
- Neighborhood Number Line (1–10 and 11–20)
- Number Cards (1–20)
- Dot Cube (or Spinner)
- Pawns, one for each student
- Line Land +1\−1 Cards, for a Challenge

Additional Materials
Warm Up
Erasable red marker

Engage
Colored stickers

Access Vocabulary
pawn Game piece
roll Drop a Dot Cube onto a game board

Creating Context
In the **Thermometer Counting** activity children are asked to raise their hands if they think another child has made a mistake. Some children may think this is unkind and may not want to participate. Provide alternative ways for children to show they have detected a mistake, such as writing down a missing or repeated number.

1 Warm Up

Concept Building — COMPUTING
Thermometer Counting
Before beginning the small-group activity **The Biggest Neighborhood**, use the **Thermometer Counting** activity with the whole class.

Purpose Children count from 1 up to 20 and from 20 down to 1.

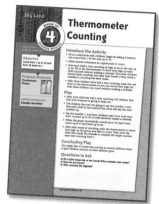
Warm-Up Card 4

Monitoring Student Progress
If . . . children are counting fluently,
Then . . . gradually increase the numbers.

2 Engage

Concept Building — ENGAGING
The Biggest Neighborhood
"Today you will try to build the biggest neighborhood by claiming the most houses in a row."

Follow the instructions on the Activity Card to play **The Biggest Neighborhood**. As children play, ask questions about what is happening in the game.

Purpose Children move forward and backward along a number line.

Activity Card 13

102 Number Worlds • Week 7

Monitoring Student Progress

If . . . children seem to move only forward,

Then . . . remind them that the game is not a race to the end but a competition to see who can get the most consecutively numbered houses, or "the longest line of houses in a row."

eMathTools Use the Number Line Tool to demonstrate and explore number sequence.

 Teacher's Note Encourage children to use this week's vocabulary words as they engage in the activities, discuss math concepts, and make predictions.

Building Blocks For additional practice with number sequence, children should complete **Building Blocks** Before and After Math.

3 Reflect 10

Extended Response REASONING

After a player rolls the cube, you might ask questions such as the following:

- **Which direction are you going to move? Why do you want to go that way?** Answers will vary but should indicate that children considered the position of houses they had already claimed.
- **Where will you be after you move?** Answers will vary.

Using Student Pages

Have each child complete **Workbook,** page 24. Did the child ring the correct card?

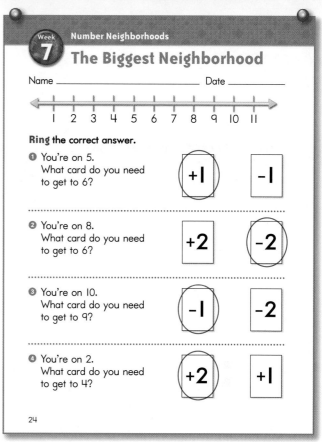

Workbook, p. 24

4 Assess

Informal Assessment

Use the Student Assessment Record, **Assessment,** page 100, to record informal observations.

COMPUTING	ENGAGING
Thermometer Counting Did the child ❏ respond accurately? ❏ respond quickly? ❏ respond with confidence? ❏ self-correct?	**The Biggest Neighborhood** Did the child ❏ pay attention to the contributions of others? ❏ contribute information and ideas? ❏ improve on a strategy? ❏ reflect on and check accuracy of work?

Number Neighborhoods • Lesson 2 103

Week 7 — Number Neighborhoods
Lesson 3

Objective
Children identify and sequence the numbers 1–20 and determine the locations of numbers in the number sequence.

Program Materials
Warm Up
Thermometer Picture

Engage
- First two sections of the Neighborhood Number Line (1–10 and 11–20)
- Number Tiles (1–20)

Additional Materials
Warm Up
Erasable red marker

Engage
Self-stick notes placed over each number on the Neighborhood Number Line before the activity begins

Access Vocabulary
before In front of; to come first
after Behind; to come later

Creating Context
Prepositions are not in the beginning level of language proficiency. Help English Learners practice prepositions such as *before*, *after*, and *between* when they line up for recess. Have children identify who is before them in line, who is after them, and who is between two other children.

1 Warm Up   5

Skill Building   COMPUTING
Thermometer Counting
Before beginning the whole-class activity **Numbering the Neighborhood,** use the **Thermometer Counting** activity with the whole class.

Purpose Children count from 1 up to 20 and from 20 down to 1.

Warm-Up Card 4

Monitoring Student Progress
If . . . children are ready for a challenge,
Then . . . play the **Thermometer Counting** Challenge variation.

2 Engage   30

Concept Building   REASONING
Numbering the Neighborhood
"Today you are going to use Number Tiles to number the neighborhood."

Follow the instructions on the Activity Card to play **Numbering the Neighborhood.** As children play, ask questions about what is happening in the game.

Activity Card 12

Purpose Children identify and sequence the numbers 1–20 while determining the location of a number in the sequence.

104 Number Worlds • Week 7

Monitoring Student Progress

| If... children can fluently identify and sequence numbers 1–20, | Then... add the next section of the Neighborhood Number Line and play with Number Tiles (1–30). |

 For additional practice with number sequence, children should complete **Building Blocks** Rocket Blast 1.

3 Reflect 10

Extended Response REASONING

Ask questions such as the following:

- **What numbers were missing? How did you know?** Answers will vary but should indicate that children used their knowledge of the number sequence to determine the missing number.

- **What number did you put on the first house?** 1 **Why?** Possible answers: It is the starting place; it is the lowest number.

4 Assess

Informal Assessment

Use the Student Assessment Record, **Assessment**, page 100, to record informal observations.

COMPUTING	REASONING
Thermometer Counting Did the child ❏ respond accurately? ❏ respond quickly? ❏ respond with confidence? ❏ self-correct?	**Numbering the Neighborhood** Did the child ❏ provide a clear explanation? ❏ communicate reasons and strategies? ❏ choose appropriate strategies? ❏ argue logically?

Number Neighborhoods • Lesson 3 105

Week 7 — Number Neighborhoods
Lesson 4

Objective
Children continue to explore the counting sequence and move forward and backward along a number line.

Program Materials
Warm Up
Thermometer Picture

Engage
- Neighborhood Number Line (1–10 and 11–20)
- Number Cards (1–20)
- Dot Cube (or Spinner)
- Pawns, one for each student
- Line Land +1\−1 Cards, for a Challenge

Additional Materials
Warm Up
Erasable red marker

Engage
Colored stickers

Access Vocabulary
compare To think about how things are alike and different
near Close to

Creating Context
To reinforce the concepts of *near* and *far*, have children work in pairs to generate a list of places that are near to or far from the school. Make a chart with columns labeled *Near* and *Far* to record the answers, and have children illustrate or add a symbol for each location.

1 Warm Up 5

Concept Building COMPUTING
Thermometer Counting
Before beginning the small-group activity **The Biggest Neighborhood**, use the **Thermometer Counting** activity with the whole class.

Purpose Children count from 1 up to 20 and from 20 down to 1.

Warm-Up Card 4

Monitoring Student Progress
If... children are counting fluently, **Then...** gradually increase the numbers.

2 Engage 30

Concept Building ENGAGING
The Biggest Neighborhood
"Today you will try to build the biggest neighborhood by claiming the most houses in a row."

Follow the instructions on the Activity Card to play **The Biggest Neighborhood**. As children play, ask questions about what is happening in the game.

Purpose Children move forward and backward along a number line and develop the understanding that +1 indicates moving one number up in a sequence and that −1 indicates moving one number down in a sequence.

Activity Card 13

106 Number Worlds • Week 7

Monitoring Student Progress

If . . . children are comfortable with the activity procedure,

Then . . . have them play the Challenge 1 variation of the activity.

 Teacher's Note Play this game with the children until they are familiar with the procedures and questions. Then designate one child to be the group leader, who distributes the Number Cards, asks at least one question during every turn, and settles any disputes that arise.

Building Blocks For additional practice with number sequence, children should complete **Building Blocks** Before and After Math.

3 Reflect

Extended Response REASONING

When a player completes all moves, ask questions such as the following:

- **Where are you now? Who are your neighbors?**
  Answers will vary.

- **Whose house is the closest to where you are now? Can you give me directions to that house?**
  Encourage the children to explain their answers by referring to the number of spaces that a particular player has already progressed, the numeral on which their pawn sits, and the number of spaces left to go.

4 Assess

Informal Assessment

Use the Student Assessment Record, **Assessment**, page 100, to record informal observations.

COMPUTING	ENGAGING
Thermometer Counting Did the child ❏ respond accurately? ❏ respond quickly? ❏ respond with confidence? ❏ self-correct?	**The Biggest Neighborhood** Did the child ❏ pay attention to the contributions of others? ❏ contribute information and ideas? ❏ improve on a strategy? ❏ reflect on and check accuracy of work?

Number Neighborhoods • Lesson 4

Week 7 — Number Neighborhoods

Lesson 5
Review

Objective
Children review the material presented in *Number Worlds,* week 7.

Program Materials
Warm Up
Thermometer Picture

Additional Materials
Warm Up
Erasable red marker

Creating Context
To help English Learners at early proficiency levels, remember to use questions during assessment that are not heavily dependent on language. Assessment activities that tell children to "Show me how many" or "Point to the set of 5 books" are good ways to include early English Learners.

1 Warm Up — 5

Skill Practice COMPUTING
Thermometer Counting
Before beginning the **Free-Choice** activity, use the **Thermometer Counting** activity with the whole class.

Purpose Children count from 1 up to 20 and from 20 down to 1.

Warm-Up Card 4

2 Engage — 20

Concept Building APPLYING
Free-Choice Activity
For the last day of the week, allow children to choose an activity from the previous weeks. Some activities they may choose are as follows:
- Line Land: **The Neighborhood**
- Line Land: **Delivering Mail to the Neighborhood**
- Line Land: **Plus-Minus Game**

Make a note of the activities children select. Do they prefer easy or challenging activities? Continue to provide Challenge opportunities for children who have mastered the basic games.

Monitoring Student Progress

| If . . . children would benefit from extra practice on specific skills, | Then . . . choose an activity for them. |

108 Number Worlds • Week 7

3 Reflect

Extended Response REASONING
Ask questions such as the following:
- **What did you like about playing** Numbering the Neighborhood?
- **Was there anything about playing this game you didn't like?**
- **Did this game help you do something you couldn't do before? What did it help you do?**
- **What was easy when you were playing** The Biggest Neighborhood?
- **What was hard when you were playing** The Biggest Neighborhood?
- **What do you remember most about this game?**
- **What was your favorite part?**

Using Student Pages
Have each child complete the skill practice on **Workbook**, page 25. Is the child correctly applying the skills he or she learned in Week 7?

Workbook, p. 25

4 Assess

A Gather Evidence
Formal Assessment
Have students complete the weekly test on **Assessment,** page 37. Record formal assessment scores on the Student Assessment Record, **Assessment,** page 100.

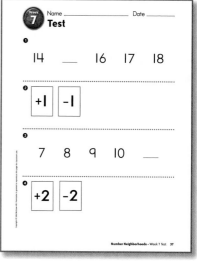

Assessment, p. 37

B Summarize Findings
Review the Student Assessment Records. Determine whether students have Minimal, Basic, or Secure understanding of the concepts presented in Week 7.

C Differentiate Instruction
Based on your observations, use these teaching strategies next week to follow up.

Minimal Understanding
- Repeat the Warm-Up and Engage activities in subsequent weeks to develop concepts of counting objects.
- Repeat the Line and Sky Land activities from this week. Observe children's understanding of these concepts, and reinforce as necessary.
- Use **Building Blocks** computer activities beginning with Count to develop and reinforce numeration and addition concepts.

Basic Understanding
- Repeat Engage activities in subsequent weeks to reinforce basic counting concepts.
- Use **Building Blocks** computer activities beginning with Before and After Math to reinforce numeration and addition concepts.

Secure Understanding
- Use the Challenge variations of **Thermometer Counting** and the **The Biggest Neighborhood.**
- Use **Building Blocks** computer activities beginning with Rocket Blast 2.

Number Neighborhoods • Lesson 5 Review 109

Week 8: More Number Neighborhoods

Week at a Glance

This week children begin **Number Worlds,** Week 8, and continue to explore Picture Land.

Background

Numerals help children forge links between the world of real quantities (objects) and the world of formal symbols (written numerals). In Picture Land, children have opportunities to identify numerals and use the formal number system.

How Children Learn

As children begin this week's lesson they should be comfortable counting from 1 to 20 but may not understand the relationships of individual numbers within the sequence.

At the end of this week, children should begin to understand number relationships within the number sequence, and should understand that +1\−1 and +2\−2 represent movement up and down the sequence.

Skills Focus

- Count up or down from a number
- Use the spatial terms *before* and *after* with the number sequence

Teaching for Understanding

As children engage in these activities, they will gradually come to see the relationships among small groups of numbers in the number sequence.

Observe closely while evaluating the Engage activities assigned for this week.

- Can children count up or down from a given number?
- Can children correctly use *before* and *after* to describe the positions of numbers in the number sequence?

Math at Home

Give one copy of the Letter to Home, page A8, to each child. Complete the activity in class, and then encourage children to share it with their caregivers.

Letter to Home, Teacher Edition, p. A8

Math Vocabulary

between In the middle of two things

before In front of; to come first

after Behind; to come later

discard Set cards aside where they won't be used anymore

stack Pile of flat things such as cards or books

English Learners

SPANISH COGNATES

English	Spanish
count	contar
distance	distancia
number	número
thermometer	termómetro

ALTERNATE VOCABULARY

thermometer Tool used to measure how hot or cold something is.

Week 8 Planner — More Number Neighborhoods

PACING	LESSON	LEARNING GOALS	MATERIALS	TECHNOLOGY
DAY 1	Warm Up 4 **Sky Land:** Thermometer Counting*	Children count from 1 up to 20 and from 20 down to 1.	**Program Materials** Thermometer Picture **Additional Materials** Erasable red marker	**Building Blocks** Before and After Math
	Activity 14 **Picture Land:** Meet Your Neighbors	Children identify and sequence the numbers 1–20 and explore number relationships.	**Program Materials** Number Cards (1–20)	
DAY 2	Warm Up 4 **Sky Land:** Thermometer Counting*	Children count from 1 up to 20 and from 20 down to 1.	**Program Materials** Thermometer Picture **Additional Materials** Erasable red marker	**Building Blocks** Before and After Math
	Activity 15 **Picture Land:** Building a Neighborhood*	Children determine which numbers are in the same neighborhood as the focus number.	**Program Materials** Number Cards (1–20), shuffled	
DAY 3	Warm Up 4 **Sky Land:** Thermometer Counting*	Children count from 1 up to 20 and from 20 down to 1.	**Program Materials** Thermometer Picture **Additional Materials** Erasable red marker	**Building Blocks** Before and After Math
	Activity 14 **Picture Land:** Meet Your Neighbors	Children identify and sequence the numbers 1–20 and explore number relationships.	**Program Materials** Number Cards (1–20)	
DAY 4	Warm Up 4 **Sky Land:** Thermometer Counting*	Children count from 1 up to 20 and from 20 down to 1.	**Program Materials** Thermometer Picture **Additional Materials** Erasable red marker	**Building Blocks** Before and After Math
	Activity 15 **Picture Land:** Building a Neighborhood*	Children continue to determine which numbers are in the same neighborhood as the focus number.	**Program Materials** Number Cards (1–20), shuffled	
DAY 5	Warm Up 4 **Sky Land:** Thermometer Counting*	Children count from 1 up to 20 and from 20 down to 1.	**Program Materials** Thermometer Picture **Additional Materials** Erasable red marker	**Building Blocks** Review previous activities
	Review and Assess	Children review their favorite activities to improve their understanding of more difficult concepts.	Materials will be selected from those used in previous weeks.	

* Includes Challenge Variations

Week 8 — Lesson 1: More Number Neighborhoods

Objective
Children identify and sequence the numbers 1–20 and count from 1 up to 20 and from 20 down to 1.

Program Materials
Warm Up
Thermometer Picture

Engage
Number Cards (1–20), one for each child

Additional Materials
Warm Up
Erasable red marker

Access Vocabulary
before In front of; to come first
after Behind; to come later

Creating Context
English prepositions present a challenge to English Learners, but acting out their meaning can help make them understandable. Have several children line up and point out who is in line before and after them. Then go to a number line and show a number that comes before or after another number.

1 Warm Up — 5

Skill Building COMPUTING
Thermometer Counting

Before beginning the whole-class activity **Meet Your Neighbors**, use the **Thermometer Counting** activity with the whole class.

Purpose Children count from 1 up to 20 and from 20 down to 1.

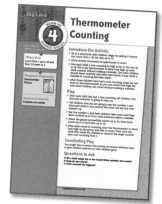

Warm-Up Card 4

Monitoring Student Progress
If . . . children have trouble counting up to or down from 20,

Then . . . have them count up to or down from 10 and gradually add higher numbers.

 Teacher's Note **Thermometer Counting** may be used to record attendance at the beginning of each day.

2 Engage — 30

Concept Building UNDERSTANDING
Meet Your Neighbors

"Today you are going to be part of a neighborhood and meet your neighbors."

Follow the instructions on the Activity Card to play **Meet Your Neighbors.** As children play, ask questions about what is happening in the game.

Activity Card 14

112 Number Worlds • Week 8

Purpose Children identify and sequence the numbers 1–20 while using the spatial terms *before* and *after* with the number sequence.

Monitoring Student Progress

If . . . a child is having trouble identifying his or her neighbors,

Then . . . have all children in the neighborhood count from 1 until they reach that child.

 For additional practice with counting and number sequence, children should complete **Building Blocks** Before and After Math.

3 Reflect — 10

Extended Response REASONING
Ask questions such as the following:
- **What was your number?** Answers will vary.
- **Who were your next-door neighbors? How did you figure that out?** Answers will vary; children may add and subtract or count up and count down to identify the neighboring numbers in the sequence.

Using Student Pages
Have each child complete **Workbook**, page 26. Did the child correctly identify the next-door neighbor?

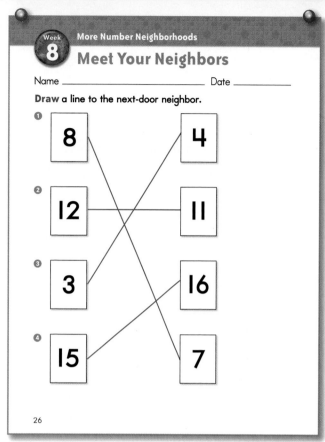

Workbook, p. 26

4 Assess

Informal Assessment
Use the Student Assessment Record, **Assessment**, page 100, to record informal observations.

COMPUTING	UNDERSTANDING
Thermometer Counting Did the child ❏ respond accurately? ❏ respond quickly? ❏ respond with confidence? ❏ self-correct?	**Meet Your Neighbors** Did the child ❏ make important observations? ❏ extend or generalize learning? ❏ provide insightful answers? ❏ pose insightful questions?

More Number Neighborhoods • Lesson 1   113

Week 8 — More Number Neighborhoods
Lesson 2

Objective
Children identify and sequence the numbers 1–20 and determine which numbers are in the same neighborhood as the focus number.

Program Materials
Warm Up
Thermometer Picture

Engage
Number Cards (1–20), shuffled

Additional Materials
Warm Up
Erasable red marker

Access Vocabulary
discard Set cards aside where they won't be used anymore
stack Pile of flat things such as cards or books

Creating Context
Playing games is an excellent way to give English Learners practice listening to and speaking English. The natural repetition of procedural and counting language replaces tedious drill with authentic, active experience. Wait time is built into the process, and the low-stakes environment makes the use of games an enjoyable and efficient learning tool in math.

1 Warm Up — 5

Concept Building COMPUTING
Thermometer Counting
Before beginning the small-group activity **Building a Neighborhood,** use the **Thermometer Counting** activity with the whole class.

Purpose Children count from 1 up to 20 and from 20 down to 1.

Warm-Up Card 4

Monitoring Student Progress
If . . . children have trouble remembering the sequence of numbers,
Then . . . model counting for them throughout the day.

2 Engage — 30

Concept Building ENGAGING
Building a Neighborhood
"Today you will try to get three cards in a row."

Follow the instructions on the Activity Card to play **Building a Neighborhood.** As children play, ask questions about what is happening in the game.

Purpose Children enter the number sequence at any point 1–20 and determine which numbers are in the same neighborhood as the focus number.

Activity Card 15

Monitoring Student Progress

| If . . . children are having trouble playing this game, | Then . . . have them play the first several rounds with their cards faceup on the table. |

 For additional practice with counting and number sequence, children should complete **Building Blocks** Before and After Math.

3 Reflect 10

Extended Response REASONING

Ask questions such as the following:

- **What card did you need?** Answers will vary, but children should indicate that they needed cards next to their numbers in the number sequence.
- **Which cards did you discard? Why?** Children should have discarded cards that were not near their cards in the number sequence.

Using Student Pages

Have each child complete **Workbook,** page 27. Did the child correctly identify the number that would give him or her three in a row?

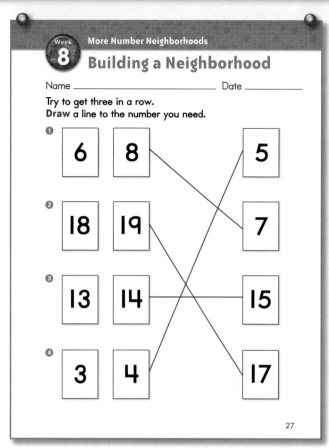

Workbook, p. 27

Assess

Informal Assessment

Use the Student Assessment Record, **Assessment,** page 100, to record informal observations.

COMPUTING	ENGAGING
Thermometer Counting Did the child ❏ respond accurately? ❏ respond quickly? ❏ respond with confidence? ❏ self-correct?	**Building a Neighborhood** Did the child ❏ pay attention to the contributions of others? ❏ contribute information and ideas? ❏ improve on a strategy? ❏ reflect on and check accuracy of work?

More Number Neighborhoods • Lesson 2   **115**

Week 8: More Number Neighborhoods

Objective
Children identify and sequence the numbers 1–20; they count from 1 up to 20 and 20 down to 1.

Program Materials
Warm Up
Thermometer Picture

Engage
Number Cards (1–20), one for each child

Additional Materials
Warm Up
Erasable red marker

Access Vocabulary
near Close to
far Not close to; a long way away

Creating Context
Prepositions can pose some difficulty for first graders, and especially for English Learners. To help children understand the terms *before* and *after*, have children work with a partner to list three things they do before school and three things they do after school. Have children share their lists with the class.

1 Warm Up

Skill Building — COMPUTING
Thermometer Counting
Before beginning the whole-class activity **Meet Your Neighbors**, use the **Thermometer Counting** activity with the whole class.

Purpose Children count from 1 up to 20 and from 20 down to 1.

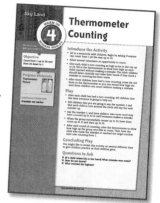
Warm-Up Card 4

Monitoring Student Progress
If . . . children are ready for a challenge,
Then . . . play the **Thermometer Counting** Challenge variation.

2 Engage

Concept Building — REASONING
Meet Your Neighbors
"Today you are going to be part of a neighborhood and meet your neighbors."

Follow the instructions on the Activity Card to play **Meet Your Neighbors**. As children play, ask questions about what is happening in the game.

Purpose Children identify and sequence the numbers 1–20 while using the spatial terms *before* and *after* with the number sequence.

Activity Card 14

116 Number Worlds • Week 8

Monitoring Student Progress

If . . . a child is having trouble identifying his or her neighbors,

Then . . . have all children in the neighborhood count from 1 until they reach that child.

 Teacher's Note Encourage children to use this week's vocabulary words as they engage in the activities, discuss math concepts, and make predictions.

Building Blocks For additional practice with counting and number sequence, children should complete **Building Blocks** Before and After Math.

3 Reflect — 10

Extended Response REASONING

Ask questions such as the following:
- **Who were your close neighbors? How did you figure that out?** Children should indicate numbers that are 2 above or 2 below their number in the counting sequence.
- **What number comes next?** Answers will vary.

4 Assess

Informal Assessment

Use the Student Assessment Record, **Assessment,** page 100, to record informal observations.

COMPUTING	REASONING
Thermometer Counting	**Meet Your Neighbors**
Did the child	Did the child
❑ respond accurately?	❑ provide a clear explanation?
❑ respond quickly?	❑ communicate reasons and strategies?
❑ respond with confidence?	❑ choose appropriate strategies?
❑ self-correct?	❑ argue logically?

More Number Neighborhoods • Lesson 3

Week 8 — More Number Neighborhoods

Lesson 4

Objective
Children identify and sequence the numbers 1–20 and determine which numbers are in the same neighborhood as the focus number.

Program Materials
Warm Up
Thermometer Picture

Engage
Number Cards (1–20), shuffled

Additional Materials
Warm Up
Erasable red marker

Access Vocabulary
faceup Cards on a table with the dots or numerals up (showing)
facedown Cards on a table with the dots or numerals down (not showing)

Creating Context
When using the game cards, explain the concept of *faceup* and *facedown*. These idioms may not be familiar to children. The main, or most meaningful, side of the card is the face. A card that is faceup communicates specific information, while a facedown card gives little information.

---

1 Warm Up — 5

Concept Building — COMPUTING
Thermometer Counting
Before beginning the small-group activity **Building a Neighborhood,** use the **Thermometer Counting** activity with the whole class.

Purpose Children count from 1 up to 20 and from 20 down to 1.

Warm-Up Card 4

Monitoring Student Progress
If . . . children are counting fluently,
Then . . . gradually increase the numbers to 24.

2 Engage — 30

Concept Building — ENGAGING
Building a Neighborhood
"Today you will try to get three cards in a row."

Follow the instructions on the Activity Card to play **Building a Neighborhood.** As children play, ask questions about what is happening in the game.

Purpose Children enter the number sequence at any point 1–20 and determine which numbers are in the same neighborhood as the focus number.

Activity Card 15

118 Number Worlds • Week 8

Monitoring Student Progress

| If . . . children are having trouble playing the game, | Then . . . have them play the first several rounds with their cards faceup on the table. |

 MathTools Use the Number Line Tool to demonstrate and explore number sequence.

 For additional practice with counting and number sequence, children should complete **Building Blocks** Before and After Math.

3 Reflect 10

Extended Response REASONING

Ask questions such as the following:

- **What patterns do you see?** Accept all reasonable answers. Encourage children to notice similarities based on place value; for example, 5, 6, and 7 make a number neighborhood as do 15, 16, and 17.

- **Did anything about this problem surprise you? Why?** Accept all reasonable answers.

4 Assess

Informal Assessment

Use the Student Assessment Record, **Assessment,** page 100, to record informal observations.

COMPUTING	ENGAGING
Thermometer Counting Did the child ❑ respond accurately? ❑ respond quickly? ❑ respond with confidence? ❑ self-correct?	**Building a Neighborhood** Did the child ❑ pay attention to the contributions of others? ❑ contribute information and ideas? ❑ improve on a strategy? ❑ reflect on and check accuracy of work?

More Number Neighborhoods • Lesson 4   119

Week 8 — More Number Neighborhoods

Lesson 5

Review

Objective
Children review the material presented in Week 8 of *Number Worlds*.

Program Materials
Warm Up
Thermometer Picture

Additional Materials
Warm Up
Erasable red marker

Creating Context
Discuss with children different meanings of the word *like*. Make a two-column chart on the board. Label one column *I Like* and the other column *Because*. Have children give reasons they like something as you complete the chart on the board. This will help them describe why they like certain **Number Worlds** activities.

1 Warm Up

Skill Practice — COMPUTING
Thermometer Counting

Before beginning the **Free-Choice** activity, use the **Thermometer Counting** activity with the whole class.

Purpose Children count from 1 up to 20 and from 20 down to 1.

Warm-Up Card 4

Monitoring Student Progress

| If . . . children are counting fluently, | Then . . . gradually increase the numbers to 24. |

2 Engage

Concept Building — APPLYING
Free-Choice Activity

For the last day of the week, allow children to choose an activity from the previous weeks. Some activities they may choose are as follows:
- Line Land: **The Neighborhood**
- Line Land: **Delivering Mail to the Neighborhood**
- Line Land: **Numbering the Neighborhood**

Make a note of the activities children select. Do they prefer easy or challenging activities? Continue to provide Challenge opportunities for children who have mastered the basic games.

Monitoring Student Progress

| If . . . children would benefit from extra practice on specific skills, | Then . . . choose an activity for them. |

Reflect 10

Extended Response REASONING
Ask questions such as the following:
- What did you like about playing Meet Your Neighbor?
- Was there anything about playing this game you didn't like?
- Did this game help you do something you couldn't do before? What did it help you do?
- What was easy when you were playing Building a Neighborhood?
- What was hard when you were playing Building a Neighborhood?
- What do you remember most about this game?
- What was your favorite part?

Using Student Pages
Have each child complete the skill practice on **Workbook,** page 28. Is the child correctly applying the skills he or she learned in Week 8?

Workbook, p. 28

 Assess 10

A Gather Evidence
Formal Assessment
Have students complete the weekly test on *Assessment,* page 39. Record formal assessment scores on the Student Assessment Record, *Assessment,* page 100.

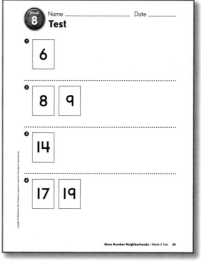

Assessment, p. 39

B Summarize Findings
Review the Student Assessment Records. Determine whether students have Minimal, Basic, or Secure understanding of the concepts presented in Week 8.

C Differentiate Instruction
Based on your observations, use these teaching strategies next week to follow up.

Minimal Understanding
- Repeat the Warm-Up and Engage activities in subsequent weeks to develop concepts of counting and number sequence.
- Repeat the Picture Land activities from this week. Observe children's understanding of these concepts, and reinforce as necessary.
- Use **Building Blocks** computer activities beginning with Count to develop and reinforce numeration concepts.

Basic Understanding
- Repeat Engage activities in subsequent weeks to reinforce basic number sequence concepts.
- Use **Building Blocks** computer activities beginning with Before and After Math to reinforce number sequence concepts.

Secure Understanding
- Use the Challenge variations of **Thermometer Counting** and **Building a Neighborhood.**
- Use **Building Blocks** computer activities beginning with Math-O-Scope.

More Number Neighborhoods • Lesson 5 Review   121

Week 9: Adding Numbers

Week at a Glance

This week children begin **Number Worlds,** Week 9, and continue to explore Picture Land.

Background

Numerals help children forge links between the world of real quantities (objects) and the world of formal symbols (written numerals). In Picture Land children have opportunities to identify numerals and use the formal number system.

How Children Learn

As children begin this week, they should understand number relationships within the number sequence and should understand that +1 and −1 represent movement up and down the sequence.

By the end of the week, children should deepen their understanding of number relationships and should begin to add small quantities to single-digit numbers by counting or using objects.

Skills Focus
- Compare and order numbers
- Add whole numbers
- Subtract whole numbers

Teaching for Understanding

As children engage in these activities, they will transition from adding in the context of the number line to adding small quantities to single-digit numbers in the context of a game.

Observe closely while evaluating the Engage activities assigned for this week.

- Are students comfortable using position words to describe numbers in the number sequence?
- Can students add single-digit numbers?

Math at Home

Give one copy of the Letter to Home, page A9, to each child. Complete the activity in class, and then encourage children to share it with their caregivers.

Letter to Home, Teacher Edition, p. A9

Math Vocabulary

add Put together
plus sign Symbol that means "add"
pawn Game piece
altogether In all; total

English Learners

SPANISH COGNATES

English	Spanish
addition	adición
compare	compare
number	número
signs	signos

ALTERNATE VOCABULARY

more A bigger number or amount
less A smaller number or amount
quest A search or adventure

122 Number Worlds • Week 9

Week 9 Planner: Adding Numbers

PACING	LESSON	LEARNING GOALS	MATERIALS	TECHNOLOGY
DAY 1	**Warm Up 5** Object Land: What Number Am I?*	Children add and subtract small quantities from single- and double-digit numbers.	**Program Materials** Neighborhood Number Line	**Building Blocks** Rocket Blast 1
	Activity 16 Picture Land: Can You Guess My Number?	Children give clues about the location of a number in the number sequence.	**Program Materials** • Number Tiles, one for each child • Neighborhood Number Line **Additional Materials** Self-stick notes	
DAY 2	**Warm Up 5** Object Land: What Number Am I?*	Children add and subtract small quantities.	**Program Materials** Neighborhood Number Line	**Building Blocks** Counting Activity 1
	Activity 17 Picture Land: Dragon Quest 1*	Children add small quantities to single-digit numbers.	**Program Materials** • Dragon Quest Game Board • Dragon Quest Cards (+1 through +4) • Pawns, Spinner, Learning Links	
DAY 3	**Warm Up 5** Object Land: What Number Am I?*	Children add and subtract small quantities.	**Program Materials** Neighborhood Number Line	**Building Blocks** Rocket Blast 1
	Activity 16 Picture Land: Can You Guess My Number?	Children give clues about the location of a number in the number sequence.	**Program Materials** • Number Tiles, one for each child • Neighborhood Number Line **Additional Materials** Self-stick notes	
DAY 4	**Warm Up 5** Object Land: What Number Am I?*	Children add and subtract small quantities.	**Program Materials** Neighborhood Number Line	**Building Blocks** Counting Activity 1
	Activity 17 Picture Land: Dragon Quest 1*	Children add small quantities to single-digit numbers.	**Program Materials** • Dragon Quest Game Board • Dragon Quest Cards (+1 through +4) • Pawns, Spinner, Learning Links	
DAY 5	**Warm Up 5** Object Land: What Number Am I?*	Children add and subtract small quantities.	**Program Materials** Neighborhood Number Line	**Building Blocks** Review previous activities
	Review and Assess	Children review their favorite activities to improve their understanding of more difficult concepts.	Materials will be selected from those used in previous weeks.	

* Includes Challenge Variations

Week 9 — Adding Numbers
Lesson 1

Objective
Children give and use clues about the location and value of a number in the number sequence and add and subtract small quantities from single- and double-digit numbers.

Program Materials
Warm Up
Neighborhood Number Line

Engage
- Number Tiles, one for each child, beginning at 1
- Neighborhood Number Line (first two or three sections)

Additional Materials
Engage
Self-stick notes covering the numbers on the Neighbhorhood Number Line

Access Vocabulary
before In front of; to come first
after Behind; to come later

Creating Context
Acquiring prepositions is an intermediate-to-advanced skill on the continuum of English proficiency, so some English Learners may find them difficult. Because not all languages share the same structure, the notion of prepositions may be unfamiliar to some children. Have children demonstrate and practice *before* and *after* by lining up and saying who comes before or after each child in line.

1 Warm Up (5)

Skill Building — COMPUTING
What Number Am I?
Before beginning the whole-class activity **Can You Guess My Number?** use the **What Number Am I?** activity with the whole class.

Purpose Children predict which number will come before a specified number and add and subtract small quantities from single- and double-digit numbers.

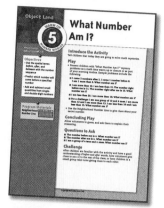

Warm-Up Card 5

Monitoring Student Progress
If . . . children have trouble remembering the clues,
Then . . . give them a paper number line and some Counters or a pencil to mark off eliminated numbers.

 Teacher's Note Encourage children to play this game at home with their caregivers.

2 Engage (30)

Concept Building — UNDERSTANDING
Can You Guess My Number?
"Today I am going to ask you some questions that will help me guess what number you have."

Follow the instructions on the Activity Card to play **Can You Guess My Number?** As children play, ask them questions about what is happening in the game.

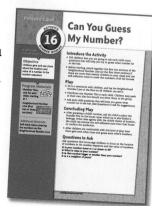

Activity Card 16

124 Number Worlds • Week 9

Purpose Children give and use clues about the location and value of a number in the number sequence.

Monitoring Student Progress

If . . . children have trouble describing the position of their number in the sequence,

Then . . . display a number line for reference.

Teacher's Note This game can be played with the whole class, or small groups can play on their own after children are familiar with the procedure. It also works well as a math time warm-up or closure or can be played quickly at any time during the day.

Building Blocks For additional practice with number sequence, children should complete **Building Blocks** Rocket Blast 1.

3 Reflect — 10

Extended Response REASONING

Ask questions such as the following:

- Was your number closer to 1 or closer to 20?
  Answers will vary.
- What number was close to your number?
  Answers will vary.

Using Student Pages

Have each child complete **Workbook,** page 29. Did the child ring the correct answers?

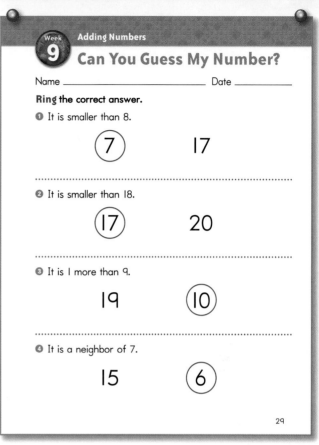

Workbook, p. 29

4 Assess

Informal Assessment

Use the Student Assessment Record, **Assessment,** page 100, to record informal observations.

COMPUTING	UNDERSTANDING
What Number Am I? Did the child	**Can You Guess My Number?** Did the child
❏ respond accurately?	❏ make important observations?
❏ respond quickly?	❏ extend or generalize learning?
❏ respond with confidence?	❏ provide insightful answers?
❏ self-correct?	❏ pose insightful questions?

Adding Numbers • Lesson 1 125

Week 9 — Adding Numbers
Lesson 2

Objective
Children will add small quantities to single- and double-digit numbers.

Program Materials
Warm Up
Neighborhood Number Line

Engage
- Dragon Quest Game Board
- Pawns, one for each child
- Spinner
- Dragon Quest Cards (+1 through +4 only)
- Learning Links
- Picture Land Activity Sheet 1, for a Challenge

Access Vocabulary
add Put together
plus sign Symbol that means "add"

Creating Context
Work with children to brainstorm the many ways we describe the process of adding together objects or numbers. For example, we might say "1 plus 2," "1 and 2," "1 and 1 more," and so on. Make a word bank and display it in the classroom so children can add other ways to talk about addition.

1 Warm Up — 5

Concept Building — COMPUTING
What Number Am I?
Before beginning the small-group activity **Dragon Quest 1**, use the **What Number Am I?** activity with the whole class.

Purpose Children predict which number will come before a specified number and add and subtract small quantities from single- and double-digit numbers.

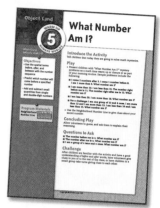

Warm-Up Card 5

Monitoring Student Progress
If . . . children have trouble remembering the sequence of numbers,
Then . . . model counting for them throughout the day.

Teacher's Note This activity may also be used to record attendance at the beginning of each day.

2 Engage — 30

Concept Building — ENGAGING
Dragon Quest 1
"Today you will be heroes and save a village from a dragon."

Follow the instructions on the Activity Card to play **Dragon Quest 1**. As children play, ask questions about what is happening in the game.

Activity Card 17

126 Number Worlds • Week 9

Purpose Children will add small quantities to single-digit numbers.

Monitoring Student Progress

| If... children have trouble keeping track of how many buckets of water they have, | Then... encourage them to write the corresponding numeral after each turn. |

 For additional practice with number sequence and addition skills, children should complete **Building Blocks** Road Race Counting Game.

3 Reflect 10

Extended Response REASONING

When children are near the end of the game, ask questions such as the following:

- **How many buckets did you have before you picked that card?** Answers will vary.
- **How many buckets do you think you will have altogether after you add those buckets?** Children may count on or add to find the correct answer.

Using Student Pages

Have each child complete **Workbook,** page 30. Tell children, "Pretend that you collected the buckets shown on two separate turns. Write the number story to tell how many buckets you have altogether." Did the child write the correct number story?

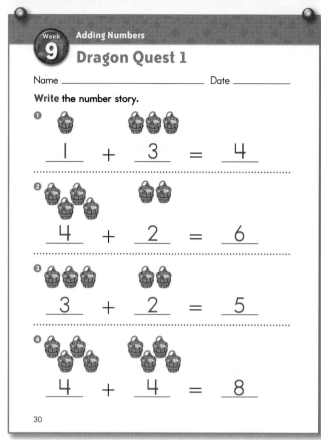

Workbook, p. 30

4 Assess

Informal Assessment

Use the Student Assessment Record, **Assessment,** page 100, to record informal observations.

COMPUTING	ENGAGING
What Number Am I? Did the child ❏ respond accurately? ❏ respond quickly? ❏ respond with confidence? ❏ self-correct?	**Dragon Quest 1** Did the child ❏ pay attention to the contributions of others? ❏ contribute information and ideas? ❏ improve on a strategy? ❏ reflect on and check accuracy of work?

Adding Numbers • Lesson 2 127

Week 9: Adding Numbers

Objective
Children give and use clues about the location and value of a number in the number sequence and add and subtract small quantities from single- and double-digit numbers.

Program Materials

Warm Up
Neighborhood Number Line

Engage
- Number Tiles, one for each child, beginning at 1
- Neighborhood Number Line (first two or three sections)

Additional Materials

Engage
Self-stick notes covering the numbers on a number line

Access Vocabulary
more A bigger number or amount
less A smaller number or amount

Creating Context
Some English Learners may not have enough English proficiency to describe their understanding of the word *more*. Use comprehension checks that do not rely heavily on language by having children use manipulatives or by using prompts such as "Show me thumbs up if you need more."

1. Warm Up

Skill Building — COMPUTING
What Number Am I?
Before beginning the whole-class activity **Can You Guess My Number?** use the **What Number Am I?** activity with the whole class.

Purpose Children predict which number will come before a specified number and add and subtract small quantities from single- and double-digit numbers.

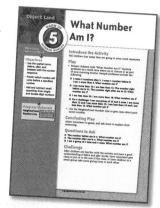

Warm-Up Card 5

Monitoring Student Progress
If . . . children are ready for a challenge,
Then . . . play the **What Number Am I?** Challenge variation.

2. Engage

Concept Building — REASONING
Can You Guess My Number?

"Today I am going to ask you some questions that will help me guess what number you have."

Follow the instructions on the Activity Card to play **Can You Guess My Number?** As children play, ask them questions about what is happening in the game.

Purpose Children give and use clues about the location and value of a number in the number sequence.

Activity Card 16

128 Number Worlds • Week 9

Monitoring Student Progress

If . . . children are comfortable with this activity,

Then . . . have them play in small groups and ask each other questions about their numbers.

eMathTools Use the Number Line Tool to demonstrate and explore number sequence.

 Teacher's Note Encourage children to use this week's vocabulary words as they engage in the activities, discuss math concepts, and make predictions.

Building Blocks For additional practice with number sequence, children should complete **Building Blocks** Rocket Blast 1.

3 Reflect 10

Extended Response REASONING

Ask questions such as the following:

- **What patterns do you see?** Accept all reasonable answers.

- **(if children play in small groups) How did you solve this problem? Why did you ask those questions?** Children's answers should reflect an understanding of the number sequence; for example, "I knew the number was closer to 20 than to 1, so I asked if it was higher than 15."

4 Assess

Informal Assessment

Use the Student Assessment Record, **Assessment,** page 100, to record informal observations.

COMPUTING	REASONING
What Number Am I? Did the child ❑ respond accurately? ❑ respond quickly? ❑ respond with confidence? ❑ self-correct?	**Can You Guess My Number?** Did the child ❑ provide a clear explanation? ❑ communicate reasons and strategies? ❑ choose appropriate strategies? ❑ argue logically?

Adding Numbers • Lesson 3 129

Week 9 — Adding Numbers
Lesson 4

Objective
Children will add small quantities to single- and double-digit numbers.

Program Materials
Warm Up
Neighborhood Number Line

Engage
- Dragon Quest Game Board
- Pawns, one for each child
- Spinner
- Dragon Quest Cards (+1 through +4 only)
- Learning Links
- Picture Land Activity Sheet 1, for a Challenge

Access Vocabulary
pawn Game piece
altogether In all; total

Creating Context
Some English Learners may know what the + and = signs mean, but may not know how to articulate this understanding. Be sure to include questions that let children show comprehension by pointing or by manipulating objects to demonstrate meaning. Model a sentence frame for children to use as they learn to talk about equations: "_____ + _____ = _____."

1 Warm Up — 5

Skill Building COMPUTING
What Number Am I?
Before beginning the small-group activity **Dragon Quest 1,** use the **What Number Am I?** activity with the whole class.

Purpose Children predict which number will come before a specified number and add and subtract small quantities from single- and double-digit numbers.

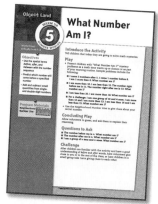

Warm-Up Card 5

Monitoring Student Progress
If . . . children can consistently identify the correct number,

Then . . . gradually use higher numbers.

2 Engage — 30

Concept Building ENGAGING
Dragon Quest 1
"Today you will be heroes and save a village from a dragon."

Follow the instructions on the Activity Card to play **Dragon Quest 1.** As children play, ask questions about what is happening in the game.

Purpose Children will add small quantities to single-digit numbers.

Activity Card 17

130 Number Worlds • Week 9

Monitoring Student Progress

| If . . . children are comfortable with **Dragon Quest 1,** | Then . . . have them play the Challenge 1 variation. |

 For additional practice with number sequence and addition skills, children should complete **Building Blocks** Road Race Counting Game.

3 Reflect

Extended Response REASONING

When children are near the end of the game, ask questions such as the following:

- **How many more buckets do you need to put out the dragon's fire?** Children may count on or subtract to find the correct answer.
- **What does this game remind you of?** Accept all reasonable answers.

4 Assess

Informal Assessment

Use the Student Assessment Record, **Assessment,** page 100, to record informal observations.

COMPUTING	ENGAGING
What Number Am I? Did the child ❑ respond accurately? ❑ respond quickly? ❑ respond with confidence? ❑ self-correct?	**Dragon Quest 1** Did the child ❑ pay attention to the contributions of others? ❑ contribute information and ideas? ❑ improve on a strategy? ❑ reflect on and check accuracy of work?

Adding Numbers • Lesson 4   131

Week 9 — Adding Numbers

Lesson 5
Review

Objective
Children review the material presented in Week 9 of *Number Worlds*.

Program Materials
Warm Up
Neighborhood Number Line

Creating Context
Although it is important to ask beginning English Learners questions that have a lower language demand, this does not mean that questions should have a lower cognitive load. Ask English Learners to demonstrate their understanding with manipulatives, models, and illustrations. Check understanding by asking children to show, demonstrate, or point to the answer.

1 Warm Up — 5

Skill Practice COMPUTING
What Number Am I?
Before beginning the **Free-Choice** activity, use the **What Number Am I?** activity with the whole class.

Purpose Children predict which number will come before a specified number and add and subtract small quantities from single- and double-digit numbers.

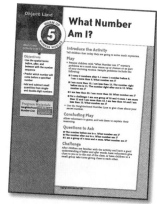

Warm-Up Card 5

2 Engage — 20

Concept Building APPLYING
Free-Choice Activity
For the last day of the week, allow children to choose an activity from the previous weeks. Some activities they may choose are as follows:
- Line Land: **Numbering the Neighborhood**
- Picture Land: **Meet Your Neighbors**
- Picture Land: **Building a Neighborhood**

Make a note of the activities children select. Do they prefer easy or challenging activities? Continue to provide Challenge opportunities for children who have mastered the basic games.

Monitoring Student Progress
If . . . children would benefit from extra practice on specific skills,
Then . . . choose an activity for them.

132 Number Worlds • Week 9

Reflect

Extended Response REASONING

Ask questions such as the following:

- **What did you like about playing** Can You Guess My Number?
- **Was there anything about playing this game you didn't like?**
- **Did this game help you do something you couldn't do before? What did it help you do?**
- **What was easy when you were playing** Dragon Quest 1?
- **What was hard when you were playing** Dragon Quest 1?
- **What do you remember most about this game?**
- **What was your favorite part?**

Using Student Pages

Have each child complete **Workbook,** page 31. For Questions 3 and 4 say, "Write the number story to tell how many buckets you collected in each turn and how many you have altogether." Is the child correctly applying the skills he or she learned in Week 9?

Workbook, p. 31

Assess

A Gather Evidence

Formal Assessment

Have students complete the weekly test on **Assessment,** page 41. Record formal assessment scores on the Student Assessment Record, **Assessment,** page 100.

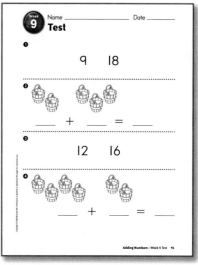

Assessment, p. 41

B Summarize Findings

Review the Student Assessment Records. Determine whether students have Minimal, Basic, or Secure understanding of the concepts presented in Week 9.

C Differentiate Instruction

Based on your observations, use these teaching strategies next week to follow up.

Minimal Understanding
- Repeat the Warm-Up and Engage activities in subsequent weeks to develop concepts of number sequence.
- Repeat the activities from this week. Observe children's understanding of these concepts, and reinforce as necessary.
- Use **Building Blocks** computer activities beginning with Number Compare 1 to develop and reinforce numeration concepts.

Basic Understanding
- Repeat Engage activities in subsequent weeks to reinforce basic number sequence concepts.
- Use **Building Blocks** computer activities beginning with Counting Activity 1 to reinforce numeration and addition concepts.

Secure Understanding
- Use the Challenge variations of **What Number Am I?** and **Dragon Quest 1.**
- Use **Building Blocks** computer activities beginning with Barkley's Bones 1.

Adding Numbers • Lesson 5 Review 133

Week 10: More Adding

Week at a Glance

This week children begin **Number Worlds,** Week 10, and continue to explore Object Land and Picture Land.

Skills Focus
- Add whole numbers
- Compare and order numbers

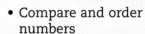

Background

Numerals help children forge links between the world of real quantities (objects) and the world of formal symbols (written numerals). This week's activities in Object Land and Picture Land will help children identify numerals and use the formal number system to write equations.

Teaching for Understanding

As children engage in these activities, they will gain practice with addition in the context of both number lines and real objects and will strengthen their understanding of the fact that addition results in an increase in quantity.

Observe closely while evaluating the Engage activities assigned for this week.

- Are children correctly comparing and ordering numbers?
- Are children correctly using position words and number relationships to describe the positions of numbers in the number sequence?
- Are children correctly adding numbers?

How Children Learn

As children begin this week, they should be familiar with the number sequence and should be able to add small quantities to single-digit numbers.

By the end of the week children should become more comfortable adding two quantities to determine the sum, and should begin to write equations or "number stories."

Math at Home

Give one copy of the Letter to Home, page A10, to each child. Complete the activity in class, and then encourage children to share it with their caregivers.

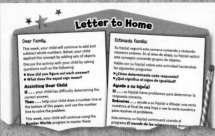

Letter to Home, Teacher Edition, p. A10

Math Vocabulary

add Put together

plus sign Symbol that means "add"

keep score To write the points each team earns in a game

between In the middle of two things

next The one after this one

English Learners

SPANISH COGNATES

English	Spanish
addition	adición
compare	compare
count	contar
group	grupo
number	número

ALTERNATE VOCABULARY

drop To let something fall

134 Number Worlds • Week 10

Week 10 Planner — More Adding

PACING	LESSON	LEARNING GOALS	MATERIALS	TECHNOLOGY
DAY 1	Warm Up 5 **Object Land:** What Number Am I?*	Children add and subtract small quantities from single- and double-digit numbers.	**Program Materials** Neighborhood Number Line	Building Blocks Rocket Blast 1
	Activity 16 **Picture Land:** Can You Guess My Number?	Children give clues about the location of a number in the number sequence.	**Program Materials** • Number Tiles, 1 per child • Neighborhood Number Line **Additional Materials** Self-stick notes	
DAY 2	Warm Up 5 **Object Land:** What Number Am I?*	Children add and subtract small quantities.	**Program Materials** Neighborhood Number Line	Building Blocks Double Compare 1
	Activity 18 **Object Land:** Drop and Add	Children add together two numbers between 0 and 10.	**Program Materials** Counters, 10 per child **Additional Materials** • Coffee cans, 2 per group • Paper and pencils	
DAY 3	Warm Up 5 **Object Land:** What Number Am I?*	Children add and subtract small quantities.	**Program Materials** Neighborhood Number Line	Building Blocks Rocket Blast 1
	Activity 16 **Picture Land:** Can You Guess My Number?	Children give clues about the location of a number in the number sequence.	**Program Materials** • Number Tiles, 1 per child • Neighborhood Number Line **Additional Materials** Self-stick notes	
DAY 4	Warm Up 5 **Object Land:** What Number Am I?*	Children add and subtract small quantities.	**Program Materials** Neighborhood Number Line	Building Blocks Double Compare 1
	Activity 18 **Object Land:** Drop and Add	Children add together two numbers between 0 and 10.	**Program Materials** Counters, 10 per child **Additional Materials** • Coffee cans, 2 per group • Paper and pencils	
DAY 5	Warm Up 5 **Object Land:** What Number Am I?*	Children add and subtract small quantities.	**Program Materials** Neighborhood Number Line	Building Blocks Review previous activities
	Review and Assess	Children review their favorite activities to improve understanding.	Materials will be selected from those used in previous weeks.	

* Includes Challenge Variations

Week 10 More Adding
Lesson 1

Objective
Children give and use clues about the location and value of a number in the number sequence and add and subtract small quantities from single- and double-digit numbers.

Program Materials
Warm Up
Neighborhood Number Line

Engage
- Number Tiles, 1 per child, beginning at 1
- Neighborhood Number Line (first two or three sections)

Additional Materials
Engage
Self-stick notes covering the numbers on the Neighborhood Number Line

Access Vocabulary
before In front of; to come first
after Behind; to come later

Creating Context
Prepositions can pose some difficulty for first graders, and especially for English Learners. To help children understand the terms *before* and *after,* have children work with a partner to list three things they do before school and three things they do after school. Have children share their lists with the class.

1 Warm Up 5

Skill Building COMPUTING
What Number Am I?
Before beginning the whole-class activity **Can You Guess My Number?** use the **What Number Am I?** activity with the whole class.

Purpose Children predict which number will come before a specified number and add and subtract small quantities from single- and double-digit numbers.

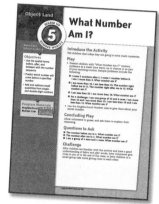

Warm-Up Card 5

Monitoring Student Progress
If . . . children have trouble remembering the clues,

Then . . . give them a paper number line and some Counters or a pencil to mark off eliminated numbers.

 Encourage children to play this game at home with their caregivers.

2 Engage 30

Concept Building UNDERSTANDING
Can You Guess My Number?
"Today I am going to ask you some questions that will help me guess what number you have."

Follow the instructions on the Activity Card to play **Can You Guess My Number?** As children play, ask them questions about what is happening in the game.

Activity Card 16

136 Number Worlds • Week 10

Purpose Children give and use clues about the location and value of a number in the number sequence.

Monitoring Student Progress

If . . . children have trouble describing the position of their number in the sequence,

Then . . . display a number line for reference.

Teacher's Note This game can be played with the whole class, or small groups can play on their own after the children are familiar with the procedure. It also works well as a math time warm-up or closure or can be played quickly at any time during the day.

Building Blocks For additional practice with number sequence, children should complete **Building Blocks** Rocket Blast 1.

3 Reflect 10

Extended Response REASONING
Ask questions such as the following:
- **Was your number closer to 1 or closer to 20?** Answers will vary.
- **What number is close to your number?** Answers will vary.

Using Student Pages
Have each child complete **Workbook**, page 32. Did the child ring the correct answers?

Week 10 — More Adding

Can You Guess My Number?

Name _____ Date _____

Ring the correct answer.

1. It is 1 more than 14.

   (15)   13

2. It is smaller than 15.

   (9)   19

3. It is a neighbor of 6.

   2   (7)

4. It is 1 less than 18.

   (17)   19

32

Workbook, p. 32

4 Assess

Informal Assessment
Use the Student Assessment Record, **Assessment**, page 100, to record informal observations.

COMPUTING	UNDERSTANDING
What Number Am I? Did the child	**Can You Guess My Number?** Did the child
☐ respond accurately?	☐ make important observations?
☐ respond quickly?	☐ extend or generalize learning?
☐ respond with confidence?	☐ provide insightful answers?
☐ self-correct?	☐ pose insightful questions?

More Adding • Lesson 1   137

Week 10 — More Adding
Lesson 2

Objective
Children add and subtract small quantities and add together two numbers between 1 and 10.

Program Materials
Warm Up
Neighborhood Number Line

Engage
Counters, 10 for each child

Additional Materials
Engage
- Coffee cans or other opaque containers, 1 for each team
- Paper and pencil for each team

Access Vocabulary
add Put together
plus sign Symbol that means "add"

Creating Context
Organize children into small groups and have them talk about things they do at home that rely on addition. Have children work together to create a group story, complete with illustrations to clearly show that addition was involved. Have children use manipulatives or props to act out their story.

1 Warm Up 5

Concept Building COMPUTING
What Number Am I?
Before beginning the small-group activity **Drop and Add**, use the **What Number Am I?** activity with the whole class.

Purpose Children predict which number will come before a specified number and add and subtract small quantities from single- and double-digit numbers.

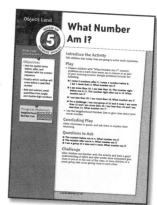

Warm-Up Card 5

Monitoring Student Progress
If . . . children have trouble remembering the sequence of numbers,
Then . . . model counting for them throughout the day.

 Teacher's Note What Number Am I? may also be used to record attendance at the beginning of each day.

2 Engage 30

Concept Building ENGAGING
Drop and Add
"Today you are going to try to drop as many Counters as possible into the can."

Follow the instructions on the Activity Card to play **Drop and Add.** As children play, ask questions about what is happening in the game.

Activity Card 18

138 Number Worlds • Week 10

Purpose Children add together two numbers between 0 and 10 and write the corresponding equations.

Monitoring Student Progress

If . . . team members have trouble remembering how many Counters each member dropped into the can,

Then . . . encourage them to use tally marks or Number Cards after each player's turn to tally the Counters that land in the can.

 For additional practice with addition and comparing sums, children should complete **Building Blocks** Double Compare 1.

3 Reflect 10

Extended Response REASONING

Ask questions such as the following:
- **Can you tell your number story?** Answers will vary but should include the terms *plus* and *equals*.
- **How do you know you won?** Possible answers: We had the most Counters; we had the highest number.

Using Student Pages

Have each child complete **Workbook,** page 33. Tell children, "Write the number story to show how many Counters are in each group and how many there are altogether." Did the child correctly write the number story?

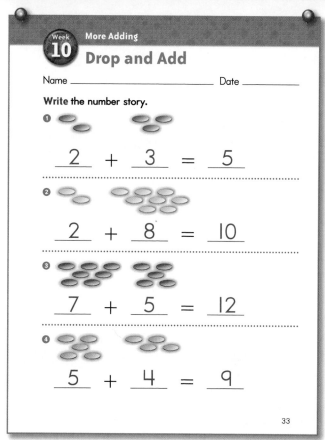

Workbook, p. 33

4 Assess

Informal Assessment

Use the Student Assessment Record, **Assessment,** page 100, to record informal observations.

COMPUTING	ENGAGING
What Number Am I? Did the child	**Drop and Add** Did the child
❏ respond accurately?	❏ pay attention to the contributions of others?
❏ respond quickly?	
❏ respond with confidence?	❏ contribute information and ideas?
❏ self-correct?	
	❏ improve on a strategy?
	❏ reflect on and check accuracy of work?

More Adding • Lesson 2 **139**

Week 10 — More Adding

Lesson 3

Objective

Children give and use clues about the location and value of a number in the number sequence and add and subtract small quantities from single- and double-digit numbers.

Program Materials

Warm Up
Neighborhood Number Line

Engage
- Number Tiles, 1 per child, beginning at 1
- Neighborhood Number Line (first two or three sections)

Additional Materials

Engage
Self-stick notes covering the numbers on the Neighborhood Number Line

Access Vocabulary

between In the middle of two things
next The one after this one

Creating Context

Teaching children to raise their hands to take turns or be called on by the teacher is not a universal practice in all cultures. Some children are taught to sit quietly as a sign of respect. Help all children learn this common signal used in U.S. schools

1 Warm Up

Skill Building — COMPUTING

What Number Am I?
Before beginning the whole-class activity **Can You Guess My Number?** use the **What Number Am I?** activity with the whole class.

Purpose Children predict which number will come before a specified number and add and subtract small quantities from single- and double-digit numbers.

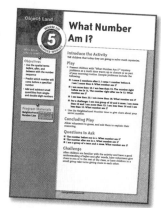

Warm-Up Card 5

Monitoring Student Progress

If . . . children are ready for a challenge,
Then . . . play the **What Number Am I?** Challenge variation.

2 Engage

Concept Building — REASONING

Can You Guess My Number?

"Today I am going to ask you some questions that will help me guess what number you have."

Follow the instructions on the Activity Card to play **Can You Guess My Number?** As children play, ask them questions about what is happening in the game.

Activity Card 16

Purpose Children give and use clues about the location and value of a number in the number sequence.

140 Number Worlds • Week 10

Monitoring Student Progress

| If . . . children are comfortable with this activity, | Then . . . have them play in small groups and ask each other questions about their numbers. |

eMathTools Use the Number Line Tool to demonstrate and explore number sequence.

 Teacher's Note Encourage children to use this week's vocabulary words as they engage in the activities, discuss math concepts, and make predictions.

Building Blocks For additional practice with number sequence, children should complete **Building Blocks** Rocket Blast 1.

3 Reflect — 10

Extended Response REASONING

Ask questions such as the following:

- **What patterns do you see?** Accept all reasonable answers.
- **(if children play in small groups) How did you solve this problem? Why did you ask those questions?** Children's answers should reflect an understanding of the number sequence; for example, "I knew the number was closer to 20 than to 1, so I asked if it was higher than 15."

4 Assess

Informal Assessment

Use the Student Assessment Record, **Assessment,** page 100, to record informal observations.

COMPUTING	REASONING
What Number Am I? Did the child ☐ respond accurately? ☐ respond quickly? ☐ respond with confidence? ☐ self-correct?	**Can You Guess My Number?** Did the child ☐ provide a clear explanation? ☐ communicate reasons and strategies? ☐ choose appropriate strategies? ☐ argue logically?

More Adding • Lesson 3   **141**

Week 10 More Adding

Objective
Children add and subtract small quantities and add together two numbers between 1 and 10.

Program Materials
Warm Up
Neighborhood Number Line

Engage
Counters, 10 for each child

Additional Materials
Engage
- Coffee cans or other opaque containers, 1 for each team
- Paper and pencil for each team

Access Vocabulary
keep score To write the points each team earns in a game

Creating Context
Help English Learners successfully write number stories by reading aloud a few models first. To focus on the math concept without worrying too much about the language needed to create a story, provide a story frame where children fill in key information. Adding illustrations will make it easier for English Learners to understand.

1 Warm Up

Skill Building COMPUTING
What Number Am I?
Before beginning the small-group activity **Drop and Add**, use the **What Number Am I?** activity with the whole class.

Purpose Children predict which number will come before a specified number, and add and subtract small quantities from single- and double-digit numbers.

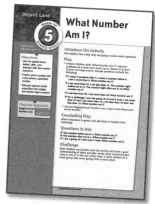

Warm-Up Card 5

Monitoring Student Progress
If . . . children can consistently identify the correct number,
Then . . . gradually use higher numbers.

2 Engage

Concept Building ENGAGING
Drop and Add
"Today you are going to try to drop as many Counters as possible into the can."

Follow the instructions on the Activity Card to play **Drop and Add**. As children play, ask questions about what is happening in the game.

Purpose Children add together two numbers between 0 and 10 and write the corresponding equations.

Activity Card 18

Monitoring Student Progress

If . . . children can consistently write the correct equations,

Then . . . gradually increase the number of Counters given to each player.

 For additional practice with addition and comparing sums, children should complete **Building Blocks** Double Compare 1.

3 Reflect 10

Extended Response REASONING
Ask questions such as the following:

- **Who had the biggest number? How did you figure that out?** Possible answers: They had the most Counters; their number is higher when you count up/higher on the number line.

- **How can you check the answer to your number story?** Possible answers: Count the total number of Counters in the can; use a number line to count up.

4 Assess

Informal Assessment
Use the Student Assessment Record, **Assessment,** page 100, to record informal observations.

COMPUTING	ENGAGING
What Number Am I? Did the child ❏ respond accurately? ❏ respond quickly? ❏ respond with confidence? ❏ self-correct?	**Drop and Add** Did the child ❏ pay attention to the contributions of others? ❏ contribute information and ideas? ❏ improve on a strategy? ❏ reflect on and check accuracy of work?

More Adding • Lesson 4

Week 10 — More Adding

Lesson 5
Review

Objective
Children review the material presented in Week 10 of **Number Worlds**.

Program Materials
Warm Up
Neighborhood Number Line

Creating Context
In addition to assessing and observing children using math concepts, complete an observation of children's language proficiency with academic English for mathematics. If your designated English Language Proficiency test has an observation checklist, use it to determine which English forms, functions, and vocabulary are still needed for academic language proficiency.

1 Warm Up — 5

Skill Practice COMPUTING
What Number Am I?
Before beginning the **Free-Choice** activity, use the **What Number Am I?** activity with the whole class.

Purpose Children predict which number will come before a specified number and add and subtract small quantities from single- and double-digit numbers.

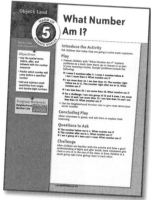

Warm-Up Card 5

2 Engage — 20

Concept Building APPLYING
Free-Choice Activity
For the last day of the week, allow children to choose an activity from the previous weeks. Some activities they may choose are as follows:
- Picture Land: **Building a Neighborhood**
- Picture Land: **Meet Your Neighbors**
- Picture Land: **Dragon Quest 1**

Make a note of the activities children select. Do they prefer easy or challenging activities? Continue to provide Challenge opportunities for children who have mastered the basic games.

Monitoring Student Progress
If . . . children would benefit from extra practice on specific skills,
Then . . . choose an activity for them.

Reflect

Extended Response **REASONING**
Invite volunteers to tell what they did and what they have learned. Ask questions such as the following:
- **What did you like about playing** Can You Guess My Number?
- **Did this game help you do something you couldn't do before? What did it help you do?**
- **What was hard when you were playing** Drop and Add?
- **What do you remember most about this game?**

Using Student Pages
Have each child complete **Workbook,** page 34. For Questions 3 and 4 say, "Write the number story to show how many Counters are in each group and how many there are altogether." Is the child correctly applying the skills they learned in Week 10?

Workbook, p. 34

 ## Assess

A Gather Evidence
Formal Assessment
Have students complete the weekly test on *Assessment,* page 43. Record formal assessment scores on the Student Assessment Record, *Assessment,* page 100.

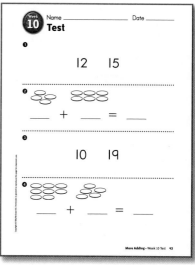
Assessment, p. 43

B Summarize Findings
Review the Student Assessment Records. Determine whether students have Minimal, Basic, or Secure understanding of the concepts presented in Week 10.

C Differentiate Instruction
Based on your observations, use these teaching strategies next week to follow up.

Minimal Understanding
- Repeat the Warm-Up and Engage activities in subsequent weeks to develop concepts of number sequence and addition.
- Repeat the activities from this week. Observe children's understanding of these concepts, and reinforce as necessary.
- Use **Building Blocks** computer activities beginning with Number Compare 1 to develop and reinforce numeration concepts.

Basic Understanding
- Repeat Engage activities in subsequent weeks to reinforce basic addition concepts.
- Use **Building Blocks** computer activities beginning with Counting Activity 1 to reinforce numeration and addition concepts.

Secure Understanding
- Use the Challenge variations of **What Number Am I?**
- Use **Building Blocks** computer activities beginning with Barkley's Bones 1.

More Adding • Lesson 5 Review 145

Week 11: Sequencing Numbers

Week at a Glance

This week children begin **Number Worlds,** Week 11, and continue to explore Picture Land and Object Land.

Background

Numerals help children forge links between the world of real quantities (objects) and the world of formal symbols (written numerals). This week's activities in Object Land and Picture Land will help children relate the number sequence to changes in set size.

How Children Learn

As children begin this week's lesson, they should be comfortable adding two quantities to determine the sum and should have begun to write equations, or "number stories."

At the end of this week, children should understand that movement up the number sequence represents an increase in quantity and should understand the role of zero in the number sequence.

Skills Focus

- Compare and order numbers
- Add whole numbers
- Subtract whole numbers
- Understand the value of zero

Teaching for Understanding

As children engage in these activities, they will gradually come to understand that each number up or down in the number sequence means that a set (quantity) has been increased or decreased by 1.

Observe closely while evaluating the Engage activities assigned for this week.

- Do students associate movement up the number sequence with an increase in quantity and set size?
- Do students associate movement down the number sequence with a decrease in quantity and set size?
- Do students understand the value of zero and its position in the number sequence?

Math at Home

Give one copy of the Letter to Home, page A11, to each child. Complete the activity in class, and then encourage children to share it with their caregivers.

Letter to Home, Teacher Edition, p. A11

Math Vocabulary

count on To continue counting

figure out To solve a problem

symbol A letter, number, or picture that has special meaning or stands for something else

English Learners

SPANISH COGNATES

English	Spanish
activity	actividad
symbol	símbolo
volunteer	voluntario
zero	cero

ALTERNATE VOCABULARY

keep score To write the points each team earns in a game

146 Number Worlds • Week 11

Week 11 Planner: Sequencing Numbers

PACING	LESSON	LEARNING GOALS	MATERIALS	TECHNOLOGY
DAY 1	Warm Up 2 **Picture Land:** Catch the Teacher*	Children detect an error in counting and identify the omitted number.	**Prepare Ahead** Two columns on the board, one labeled *Students* and one labeled *Teacher*	Building Blocks — Build Stairs 1
DAY 1	Activity 19 **Picture Land:** And One More Makes...*	Children associate movement in the counting sequence with an increase or decrease in set size.	**Program Materials** • Counters, 1 per child • Picture Land Activity Sheet 2, 1 per child, for a Challenge **Additional Materials** • Coffee can or other container • Pencils	
DAY 2	Warm Up 2 **Picture Land:** Catch the Teacher*	Children detect an error in counting and identify the omitted number.	**Prepare Ahead** Two columns on the board, one labeled *Students* and one labeled *Teacher*	Building Blocks — Rocket Blast 1
DAY 2	Activity 20 **Object Land:** More Food Fun	Children understand the value of zero when adding or subtracting small sets.	**Program Materials** Counters **Additional Materials** • Index cards, 1 per child • Pencils	
DAY 3	Warm Up 2 **Picture Land:** Catch the Teacher*	Children detect an error in counting and identify the omitted number.	**Prepare Ahead** Two columns on the board, one labeled *Students* and one labeled *Teacher*	Building Blocks — Build Stairs 1
DAY 3	Activity 19 **Picture Land:** And One More Makes...*	Children recognize that larger numbers are composed of groups of smaller numbers.	Same as Lesson 1	
DAY 4	Warm Up 2 **Picture Land:** Catch the Teacher*	Children detect an error in counting and identify the omitted number.	**Prepare Ahead** Two columns on the board, one labeled *Students* and one labeled *Teacher*	Building Blocks — Rocket Blast 1
DAY 4	Activity 20 **Object Land:** More Food Fun	Children know the symbol for zero and its position in the number sequence.	Same as Lesson 2	
DAY 5	Warm Up 2 **Picture Land:** Catch the Teacher*	Children detect an error in counting and identify the omitted number.	**Prepare Ahead** Two columns on the board, one labeled *Students* and one labeled *Teacher*	Building Blocks — Review previous activities
DAY 5	Review and Assess	Children review their favorite activities to improve their understanding.	Materials will be selected from those used in previous weeks.	

* Includes Challenge Variations

Week 11 — Sequencing Numbers
Lesson 1

Objective
Children strengthen their understanding of the number sequence and associate movement up or down the sequence with a change in quantity, or set size.

Program Materials
Engage
- Counters, 1 per child
- Picture Land Activity Sheet 2, 1 per child, for a Challenge

Additional Materials
Warm Up
2 columns on the board, one labeled *Students* and one labeled *Teacher*

Engage
- Coffee can or other opaque container
- Pencils, 1 per child

Access Vocabulary
predict To say what you think will happen
left out Not included

Creating Context
Teaching children to raise their hands to take turns or to be called on by the teacher is not a universal practice in all cultures. Some children are taught to sit quietly as a sign of respect. Help all children learn this common signal used in North American schools.

1. Warm Up

Skill Building — COMPUTING
Catch the Teacher
Before beginning the activity **And One More Makes . . .**, use the **Catch the Teacher** activity with the whole group.

Purpose Children learn to detect an error in counting and to identify which number was omitted from the sequence.

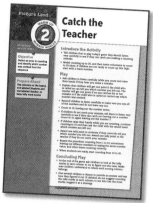

Warm-Up Card 2

Monitoring Student Progress
If . . . a child cannot remember which number was omitted,

Then . . . call on someone else whose hand is raised. If the child says a wrong number, ask the other children if they agree with the child's choice. If they do not agree, ask a volunteer which number was omitted. Then confirm this child's answer.

2. Engage

Concept Building — UNDERSTANDING
And One More Makes . . .
"Today you will predict the number of Counters in a can."
Follow the instructions on the Activity Card to play **And One More Makes. . . .** As children play, ask questions about what is happening in the game.

Purpose Children match set size to numerals.

Activity Card 19

148 Number Worlds • Week 11

Monitoring Student Progress

If . . . children have trouble recognizing the +1 pattern,

Then . . . have them use a number line to reinforce this concept.

 Building Blocks For additional practice with numeration and addition, children should complete **Building Blocks** Build Stairs 1.

3 Reflect   10

Extended Response REASONING

Ask questions such as the following:

- **How did you know how many Counters were in the can?** Accept all reasonable answers. Children may repeat the last number said or may count the number of children who dropped Counters into the can.

- **If one more person dropped his or her Counter, how many Counters would be in the can?** Answers will vary but should be one number higher than the current total.

Using Student Pages

Have each child complete **Workbook,** page 35. Did the child write the correct sum?

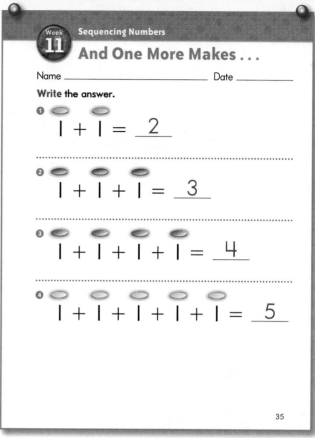

Workbook, p. 35

4 Assess

Informal Assessment

Use the Student Assessment Record, **Assessment,** page 100, to record informal observations.

COMPUTING	UNDERSTANDING
Catch the Teacher Did the child	**And One More Makes. . .** Did the child
❏ respond accurately?	❏ make important observations?
❏ respond quickly?	❏ extend or generalize learning?
❏ respond with confidence?	❏ provide insightful answers?
❏ self-correct?	❏ pose insightful questions?

Sequencing Numbers • Lesson 1   149

Week 11 — Sequencing Numbers
Lesson 2

Objective
Children strengthen their understanding of the number sequence and understand the value of zero.

Program Materials
Engage
Counters (used as pretend food)

Additional Materials
Warm Up
2 columns on the board, one labeled *Students* and one labeled *Teacher*

Engage
- Index cards, 1 per child
- Pencils, 1 per child

Access Vocabulary
zero None
tally marks Straight lines grouped in sets of five

Creating Context
To help English Learners at early proficiency levels, teach a sentence frame they can use when they catch the teacher in a counting mistake; for example, "The number _____ is missing." Some children may find it odd for the teacher to make such rudimentary mistakes. Make sure they understand that this is an exercise to make sure they are listening.

---

1. Warm Up — 5

Concept Building — COMPUTING
Catch the Teacher
Before beginning the small-group activity **More Food Fun,** use the **Catch the Teacher** activity with the whole class.

Purpose Children learn to detect an error in counting and to identify which number was omitted from the sequence.

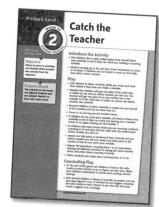
Warm-Up Card 2

Monitoring Student Progress
If . . . children have trouble identifying the missing number,

Then . . . practice counting with them throughout the day.

2. Engage — 30

Concept Building — ENGAGING
More Food Fun
"Today we will learn about the number zero."

Follow the instructions on the Activity Card to play **More Food Fun.** As children play, ask questions about what is happening in the game.

Purpose Children learn the symbol for zero and its position in the number sequence.

Activity Card 20

Monitoring Student Progress

| If... children have trouble determining how many more Counters they need to make 5, | Then... allow them to use a desktop number line. |

 For additional practice with number sequence, children should complete **Building Blocks** Rocket Blast 1.

3 Reflect 10

Extended Response REASONING

Toward the end of the game, ask questions such as the following:

- **How many do you have now? How many will you have if you add 0?** Answers will vary, but children should indicate that the number they have now will not change if they add 0.

- **How many will you have if you take away 0?** Answers will vary, but children should indicate that the number they have now will not change if they take away 0.

Using Student Pages

Have each child complete **Workbook,** page 36. Did the child write the correct sum?

Workbook, p. 36

4 Assess

Informal Assessment

Use the Student Assessment Record, **Assessment,** page 100, to record informal observations.

COMPUTING	ENGAGING
Catch the Teacher	**More Food Fun**
Did the child	Did the child
❏ respond accurately?	❏ pay attention to the contribution of others?
❏ respond quickly?	
❏ respond with confidence?	❏ contribute information and ideas?
❏ self-correct?	
	❏ improve on a strategy?
	❏ reflect on and check accuracy of work?

Sequencing Numbers • Lesson 2   **151**

Week 11: Sequencing Numbers

Objective

Children strengthen their understanding of the number sequence and associate movement up or down the sequence with a change in quantity, or set size.

Program Materials

Engage
- Counters, 1 per child
- Picture Land Activity Sheet 2, 1 per child, for a Challenge

Additional Materials

Warm Up
Two columns on the board, one labeled *Students* and one labeled *Teacher*

Engage
- Coffee can or other opaque container
- Pencils, 1 per child

Access Vocabulary

plus sign Symbol that means "add"
equal sign Symbol that means "having the same amount"

Creating Context

The word *make* is used in many English expressions. In this lesson's activity we say "and one more makes," which in this phrase means "equals." Help children understand that this is one of many ways to describe addition.

1. Warm Up (5)

Skill Building — COMPUTING

Catch the Teacher

Before beginning the activity **And One More Makes . . .** , use the **Catch the Teacher** activity with the whole group.

Purpose Children learn to detect an error in counting and to identify which number was omitted from the sequence.

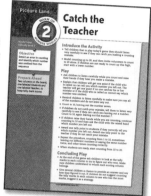

Warm-Up Card 2

Monitoring Student Progress

If . . . children are ready for a challenge,
Then . . . play **Catch the Teacher** backward.

2. Engage (30)

Concept Building — REASONING

And One More Makes . . .

"Today you will predict the number of Counters in a can."

Follow the instructions on the Activity Card to play **And One More Makes. . . .** As children play, ask questions about what is happening in the game.

Purpose Children match set size to numerals.

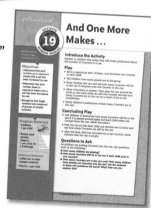

Activity Card 19

152 Number Worlds • Week 11

Monitoring Student Progress

If . . . children need extra practice determining increased set size,

Then . . . practice with them throughout the day. You can have them predict how many books there will be when you hand out one more, predict how many children there will be when one more gets in line for recess, and so on.

MathTools Use the Set Tool to demonstrate and explore counting.

 Teacher's Note Encourage children to use this week's vocabulary words as they engage in the activities, discuss math concepts, and make predictions.

Building Blocks For additional practice with numeration and addition, children should complete **Building Blocks** Build Stairs 1.

3 Reflect 10

Extended Response REASONING

Ask questions such as the following:

- **What patterns do you see?** Accept all reasonable answers.
- **What would happen if each person added 2 Counters to the can?** Possible answers: There would be more Counters in the can at the end; the can would still fill up; the can would fill up faster.

4 Assess

Informal Assessment

Use the Student Assessment Record, **Assessment**, page 100, to record informal observations.

COMPUTING	REASONING
Catch the Teacher Did the child ❏ respond accurately? ❏ respond quickly? ❏ respond with confidence? ❏ self-correct?	**And One More Makes. . .** Did the child ❏ provide a clear explanation? ❏ communicate reasons and strategies? ❏ choose appropriate strategies? ❏ argue logically?

Sequencing Numbers • Lesson 3 153

Week 11: Sequencing Numbers
Lesson 4

Objective
Children strengthen their understanding of the number sequence and understand the value of zero.

Program Materials
Engage
Counters (used as pretend food)

Additional Materials
Warm Up
2 columns on the board, one labeled *Students* and one labeled *Teacher*

Engage
- Index cards, 1 per child
- Pencils, 1 per child

Access Vocabulary
snack A little bit of food eaten between meals

Creating Context
Help English Learners expand their vocabulary by brainstorming different ways to say *zero*. Start a word wall with a large 0 at the top, and list the words used to describe this concept. Help early English Learners by asking questions that include physical cues, such as "Show me 0 fingers."

1 Warm Up — 5

Concept Building — COMPUTING
Catch the Teacher

Before beginning the small-group activity **More Food Fun,** use the **Catch the Teacher** activity with the whole class.

Purpose Children learn to detect an error in counting and to identify which number was omitted from the sequence.

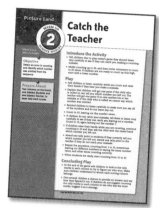

Warm-Up Card 2

Monitoring Student Progress
If . . . children are ready for a challenge, **Then** . . . have them play **Catch the Teacher** backward.

2 Engage — 30

Concept Building — APPLYING
More Food Fun

"Today we will learn about the number zero."

Follow the instructions on the Activity Card to play **More Food Fun.** As children play, ask questions about what is happening in the game.

Purpose Children learn the symbol for zero and its position in the number sequence.

Activity Card 20

154 Number Worlds • Week 11

Monitoring Student Progress

If . . . children need help understanding the value of zero,

Then . . . include it in your counting activities throughout the day. For example, before children line up for recess, note that there are 0 children in line, then have 1 child stand in line, and so on.

 For additional practice with number sequence, children should complete **Building Blocks** Rocket Blast 1.

3 Reflect 10

Extended Response REASONING

Ask questions such as the following:

- **What patterns do you see?** Encourage children to realize that any number plus or minus zero is still the original number.
- **What does this game remind you of?** Accept all reasonable answers.

4 Assess

Informal Assessment

Use the Student Assessment Record, **Assessment**, page 100, to record informal observations.

COMPUTING	APPLYING
Catch the Teacher	**More Food Fun**
Did the child	Did the child
❏ respond accurately?	❏ apply learning to new situations?
❏ respond quickly?	❏ contribute concepts?
❏ respond with confidence?	❏ contribute answers?
❏ self-correct?	❏ connect mathematics to the real world?

Week 11

Sequencing Numbers

Lesson 5

Review

Objective
Children review the material presented in Week 11 of *Number Worlds*.

Additional Materials
Warm Up
2 columns on the board, one labeled *Students* and one labeled *Teacher*

Creating Context
To help English Learners at early proficiency levels, remember to use questions during assessment that are not heavily dependent on language. Assessment activities that tell children to "Show me how many" or "Point to the set of 5 books" are good ways to include early English Learners.

1 Warm Up   5

Concept Building COMPUTING
Catch the Teacher
Before beginning the **Free-Choice** activity, use the **Catch the Teacher** activity with the whole class.

Purpose Children learn to detect an error in counting and to identify which number was omitted from the sequence.

Warm-Up Card 2

Teacher's Note Playing this game again and again will strengthen children's understanding of the number sequence.

2 Engage   20

Skill Building APPLYING
Free-Choice Activity
For the last day of the week, allow children to choose an activity from the previous weeks. Activities they may choose include the following:
- Object Land: **Food Fun**
- Object Land: **What Number Am I?**
- Picture Land: **Can You Guess My Number?**

Make a note of the activities children select. Do they prefer easy or challenging activities? Continue to provide Challenge opportunities for children who have mastered the basic game.

Monitoring Student Progress
| If... children would benefit from extra practice on specific skills, | Then... choose an activity for them. |

156   Number Worlds • Week 11

Reflect

Extended Response REASONING

Ask questions such as the following:
- **What did you like about playing** And One More Makes . . . ?
- Was there anything about playing this game you didn't like?
- Did this game help you do something you couldn't do before? What did it help you do?
- What was easy when you were playing More Food Fun?
- What was hard when you were playing More Food Fun?
- What do you remember most about this game?
- What was your favorite part?

Using Student Pages

Have each child complete the Skills Review, **Workbook,** page 37. Is the child correctly applying the skills learned in Week 11?

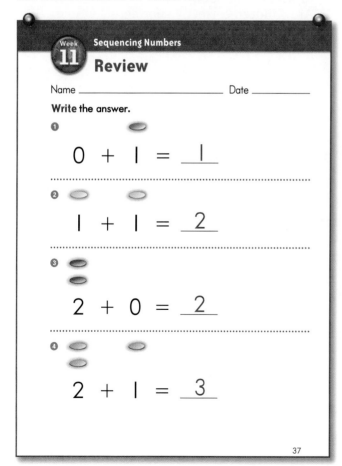

Workbook, p. 37

Assess

A Gather Evidence

Formal Assessment

Have students complete the weekly test on **Assessment,** page 45. Record formal assessment scores on the Student Assessment Record, **Assessment,** page 100.

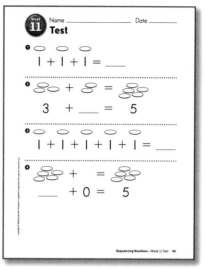

Assessment, p. 45

B Summarize Findings

Review the Student Assessment Records. Determine whether students have Minimal, Basic, or Secure understanding of the concepts presented in Week 11.

C Differentiate Instruction

Based on your observations, use these teaching strategies next week to follow up.

Minimal Understanding
- Repeat the Warm-Up and Engage activities in subsequent weeks to develop number sequence and counting concepts.
- Repeat the activities from this week. Observe children's understanding of these concepts and reinforce as necessary.
- Use **Building Blocks** computer activities beginning with Count to develop and reinforce numeration concepts.

Basic Understanding
- Repeat Engage activities in subsequent weeks to reinforce numeration and addition concepts.
- Use **Building Blocks** computer activities beginning with Build Stairs 1 to reinforce numeration and addition concepts.

Secure Understanding
- Use Challenge variations of **Catch the Teacher** and **And One More Makes. . . .**
- Use **Building Blocks** computer activities beginning with Rocket Blast 2.

Sequencing Numbers • Lesson 5 Review   157

Week 12 Writing Equations

Week at a Glance

This week children begin **Number Worlds**, Week 12, and continue to explore Object Land.

Background

In Object Land, a change in quantity is represented as an increase or decrease in the size of a group. The more objects added or taken away, the greater the difference will be between group sizes.

How Children Learn

As children begin this week's lesson, they should understand that movement up the number sequence represents an increase in quantity and should be fairly comfortable adding two quantities to determine the sum.

At the end of this week, children should understand that multiple-addend problems can be broken into smaller, more easily solved problems and should begin to write subtraction equations.

Skills Focus

- Add whole numbers
- Subtract whole numbers
- Write equations
- Solve equations with more than two addends

Teaching for Understanding

As children engage in these activities, they will come to see multiple-addend problems as an extension of the two-addend problems with which they are already familiar. Children will also expand their knowledge of addition equations and begin to gain experience with subtraction equations.

Observe closely while evaluating the Engage activities assigned for this week.

- Are children correctly adding combinations of whole numbers?
- Are children correctly subtracting whole numbers?
- Are children writing equations?

Math at Home

Give one copy of the Letter to Home, page A12, to each child. Complete the activity in class, and then encourage children to share it with their caregivers.

Letter to Home, Teacher Edition, p. A12

Math Vocabulary

few Small amount; not many
subtract Take away
minus sign Symbol that means "take away"

English Learners

SPANISH COGNATES

English	Spanish
concepts	conceptos
equation	ecuación
subtraction	substracción
sum	suma

ALTERNATE VOCABULARY

solve Figure out; find the answer

158 Number Worlds • Week 12

Week 12 Planner — Writing Equations

PACING	LESSON	LEARNING GOALS	MATERIALS	TECHNOLOGY
DAY 1	Warm Up 2 **Picture Land:** Catch the Teacher*	Children detect an error in counting and identify which number was omitted from the sequence.	**Prepare Ahead** Two columns on the board, one labeled *Students* and one labeled *Teacher*	**Building Blocks** Function Machine 1
	Activity 21 **Object Land:** They're Gone*	Children subtract small quantities and write subtraction equations.	**Program Materials** • Object Land Activity Sheet 3, 1 per child • Counters **Additional Materials** • Paper cup, 1 per child • Pencil, 1 per child	
DAY 2	Warm Up 2 **Picture Land:** Catch the Teacher*	Children detect errors in counting.	**Prepare Ahead** Two columns on the board, one labeled *Students* and one labeled *Teacher*	**Building Blocks** Number Snapshots 1
	Activity 22 **Object Land:** Add Them Up	Children flexibly solve equations with more than two addends.	**Program Materials** Counters, 3–5 for each child **Additional Materials** Paper and pencil	
DAY 3	Warm Up 2 **Picture Land:** Catch the Teacher*	Children continue to count higher as they master the counting sequence.	No additional materials required.	**Building Blocks** Function Machine 1
	Activity 21 **Object Land:** They're Gone*	Children continue to subtract small quantities and write subtraction equations.	Same as Lesson 1	
DAY 4	Warm Up 2 **Picture Land:** Catch the Teacher*	Children detect errors in counting.	**Prepare Ahead** Two columns on the board, one labeled *Students* and one labeled *Teacher*	**Building Blocks** Number Snapshots 1
	Activity 22 **Object Land:** Add Them Up	Children continue to flexibly solve equations with more than two addends.	**Program Materials** Counters, 3–5 for each child **Additional Materials** Paper and pencil	
DAY 5	Warm Up 2 **Picture Land:** Catch the Teacher*	Children continue to detect counting errors and identify the missing number.	**Prepare Ahead** Two columns on the board, one labeled *Students* and one labeled *Teacher*	**Building Blocks** Review previous activities
	Review and Assess	Children review their favorite activities to improve their understanding of more difficult concepts.	Materials will be selected from those used in previous weeks.	

* Includes Challenge Variations

Week 12 — Lesson 1: Writing Equations

Objective
Children strengthen their knowledge of the counting sequence, subtract small quantities, and write subtraction equations.

Program Materials
- Object Land Activity Sheet 3, 1 per child
- Counters, 5 per child

Additional Materials
- Paper cups, 1 per child
- Pencils, 1 per child

Access Vocabulary
subtract Take away
minus sign Symbol that means "take away"

Creating Context
English Learners often understand math concepts better when they can interact with other children who speak the same primary language. Allow children to work with partners or cross-age helpers who speak the same primary language so they can check understanding with one another in team situations.

1 Warm Up — 5

Skill Building — COMPUTING
Catch the Teacher

Before beginning the small-group activity **They're Gone,** use the **Catch the Teacher** activity with the whole class.

Purpose Children learn to detect an error in counting and identify which number was omitted from the sequence.

Warm-Up Card 2

Monitoring Student Progress
If . . . a child cannot remember which number was omitted,

Then . . . call on someone else whose hand is raised. If the child says a wrong number, ask the other children if they agree with the child's choice. If they do not agree, ask a volunteer which number was omitted. Then confirm this child's answer. Remember to allow wait time before calling on other children.

2 Engage — 30

Concept Building — UNDERSTANDING
They're Gone

"Today we are going to use Counters to talk about subtraction."

Follow the instructions on the Activity Card to play **They're Gone.** As children play, ask questions about what is happening in the game.

Purpose Children subtract small quantities from 5 and write subtraction equations.

Activity Card 21

Monitoring Student Progress

If . . . children have difficulty creating a subtraction number story,

Then . . . tell them the stories always have three parts. First, children should say how many Counters they had at the beginning. Next, they should say what happened to the Counters (children took some away and hid them under a cup). Finally, children should say how many Counters were left at the end.

 Teacher's Note This is a fun activity that works best with a small group and plenty of instructor involvement the first time you use it.

Building Blocks For additional practice distinguishing addition and subtraction, children should complete **Building Blocks** Function Machine 1.

3 Reflect (10)

Extended Response — REASONING
Ask questions such as the following:
- **How many did you start with?** Answers will vary.
- **How many did you have after you took away the red Counters?** Answers will vary.
- **Can you tell me a subtraction number story about what happened to the Counters?** Answers will vary, but should include correct vocabulary such as *minus* or *take away*.

Using Student Pages
Have each child complete **Workbook**, page 38. Did the child write the correct number story?

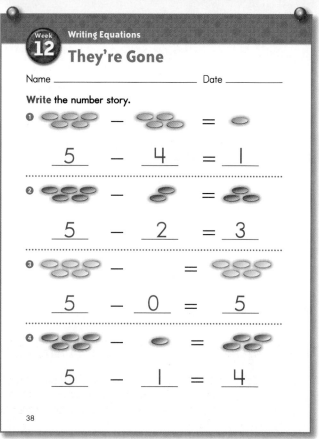

Workbook, p. 38

4 Assess

Informal Assessment
Use the Student Assessment Record, **Assessment**, page 100, to record informal observations.

COMPUTING	UNDERSTANDING
Catch the Teacher	**They're Gone**
Did the child	Did the child
❏ respond accurately?	❏ make important observations?
❏ respond quickly?	❏ extend or generalize learning?
❏ respond with confidence?	❏ provide insightful answers?
❏ self-correct?	❏ pose insightful questions?

Writing Equations • Lesson 1

Week 12 — Writing Equations
Lesson 2

Objective
Children reinforce their understanding of the number sequence and flexibly solve equations with more than two addends.

Program Materials
Counters, 3–5 for each student

Additional Materials
Paper and pencils

Access Vocabulary
plus sign Symbol that means "add"

Creating Context
Work with children to brainstorm the many ways we describe the process of adding objects or numbers. For example, we might say "1 plus 2," "1 and 2," "1 and 1 more," and so on. Make a word bank, and display it in the classroom so children can include other ways to talk about addition.

1 Warm Up — 5

Concept Building — COMPUTING
Catch the Teacher
Before beginning the small-group activity **Add Them Up**, use the **Catch the Teacher** activity with the whole class.

Purpose Children learn to detect an error in counting and identify which number was omitted from the sequence.

Warm-Up Card 2

Monitoring Student Progress
If . . . children have trouble identifying the missing number,

Then . . . practice counting with them throughout the day.

2 Engage — 30

Concept Building — ENGAGING
Add Them Up
"Today we will write number stories about addition."

Follow the instructions on the Activity Card to play **Add Them Up**. As children play, ask questions about what is happening in the game.

Purpose Children flexibly solve equations with more than two addends.

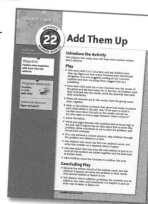
Activity Card 22

162 Number Worlds • Week 12

Monitoring Student Progress

If . . . children have trouble adding three or four groups of Counters (addends) at one time,

Then . . . begin by distributing Counters to only two children for each round of game play. The other children in the group can watch the game and answer the teacher's questions. When children are comfortable adding two addends, distribute Counters to three children in the group, then to four children.

 Teacher's Note This activity can be used several times with different quantities of Counters.

 Building Blocks For additional practice with combinations of numbers, children should complete **Building Blocks** Number Snapshots 1.

3 Reflect 10

Extended Response REASONING

Ask questions such as the following:

- **How can we figure out how many Counters were passed out altogether?** Possible answers: Put them together and count them; add them together.
- **If we wanted to write a number story about the pile in the middle, would we put plus signs or minus signs between these numbers?** Put plus signs.

Using Student Pages

Have each child complete **Workbook,** page 39. Did the child write the correct number story?

Workbook, p. 39

4 Assess

Informal Assessment

Use the Student Assessment Record, **Assessment,** page 100, to record informal observations.

COMPUTING	ENGAGING
Catch the Teacher	**Add Them Up**
Did the child	Did the child
❏ respond accurately?	❏ pay attention to the contributions of others?
❏ respond quickly?	
❏ respond with confidence?	❏ contribute information and ideas?
❏ self-correct?	
	❏ improve on a strategy?
	❏ reflect on and check accuracy of work?

Writing Equations • Lesson 2 **163**

Week 12 — Writing Equations

Objective
Children continue to subtract small quantities from 5 and 10 and write subtraction equations.

Program Materials
- Object Land Activity Sheet 3, 1 per child
- Counters, 5 per child

Additional Materials
- Paper cups, 1 per child
- Pencils, 1 per child

Access Vocabulary
equation Number story

Creating Context
The variety of words used to describe subtraction such as *take away, minus, less, subtracted from,* and so on may be puzzling for English Learners. Help children brainstorm a list of words and phrases used to describe subtraction. As you teach the lessons, use consistent vocabulary whenever possible to increase understanding.

1 Warm Up — 5

Skill Building — COMPUTING
Catch the Teacher
Before beginning the small-group activity **They're Gone**, use the **Catch the Teacher** activity with the whole class.

Purpose Children learn to detect an error in counting and identify which number was omitted from the sequence.

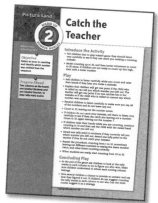
Warm-Up Card 2

Monitoring Student Progress
If . . . children are ready for a challenge,
Then . . . play **Catch the Teacher** backward.

2 Engage — 30

Concept Building — REASONING
They're Gone
"Today we are going to use counters to talk about subtraction."

Follow the instructions on the Activity Card to play **They're Gone**. As children play, ask questions about what is happening in the game.

Purpose Children subtract small quantities from 5 and write subtraction equations.

Activity Card 21

164 Number Worlds • Week 12

Monitoring Student Progress

If . . . children have trouble completing the equations on the Activity Sheet,	Then . . . create a worksheet that shows only one problem at a time. Label each element of the equation (e.g., *how many at start, what happened, how many left*), and help children write the numbers and symbols under each label (e.g., $5 - 3 = 2$). For now, write the equal sign for children and tell them we use this symbol before the answer.

 Teacher's Note If you think your students need a more gradual transition into writing subtraction problems, have them take away Counters the first few times they play without keeping a written record. Add the challenge of the Activity Sheet when children are ready.

Building Blocks For additional practice distinguishing addition and subtraction, children should complete **Building Blocks** Function Machine 1.

3 Reflect 10

Extended Response REASONING
Ask questions such as the following:
- **How can you check your answer?** Possible answer: Count the counters.
- **What does this remind you of?** Possible answers: counting down on the number line; the **More Food Fun** game when we took away Counters

4 Assess

Informal Assessment
Use the Student Assessment Record, **Assessment**, page 100, to record informal observations.

COMPUTING	REASONING
Catch the Teacher Did the child ❑ respond accurately? ❑ respond quickly? ❑ respond with confidence? ❑ self-correct?	**They're Gone** Did the child ❑ provide a clear explanation? ❑ communicate reasons and strategies? ❑ choose appropriate strategies? ❑ argue logically?

Writing Equations • Lesson 3   **165**

Week 12 — Writing Equations

Objective
Children continue to flexibly solve equations with more than two addends.

Program Materials
Counters, 3–5 for each child

Additional Materials
Paper and pencils

Access Vocabulary
equal sign Symbol that means "having the same amount"
solve Figure out; find the answer

Creating Context
Help English Learners successfully write number stories by first reading aloud a few models. To focus on the math concept without worrying too much about the language needed to create a story, provide a story frame in which children fill in key information. Adding illustrations will make it easier for English Learners to understand.

1 Warm Up — 5

Skill Building COMPUTING
Catch the Teacher
Before beginning the small-group activity **Add Them Up,** use the **Catch the Teacher** activity with the whole class.

Purpose Children learn to detect an error in counting and identify which number was omitted from the sequence.

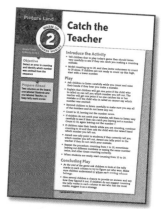

Warm-Up Card 2

Monitoring Student Progress
If . . . children are ready for a challenge,
Then . . . play **Catch the Teacher** backward.

2 Engage — 30

Concept Building APPLYING
Add Them Up
"Today we will write number stories about addition."

Follow the instructions on the Activity Card to play **Add Them Up.** As children play, ask questions about what is happening in the game.

Purpose Children flexibly solve equations with more than two addends.

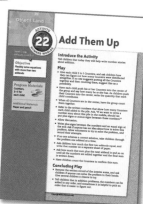

Activity Card 22

Monitoring Student Progress
If . . . children have difficulty adding the numbers,
Then . . . model the process of finding numbers that are easy to add (such as doubles) before tackling the harder numbers.

e MathTools Use the Set Tool to demonstrate and explore counting.

 Teacher's Note Encourage children to use this week's vocabulary words as they engage in the activities, discuss math concepts, and make predictions.

Building Blocks For additional practice with combinations of numbers, children should complete **Building Blocks** Number Snapshots 1.

3 Reflect

Extended Response REASONING

Ask questions such as the following:

- **How many different combinations of numbers can you add?** Answers will vary. Encourage children to realize that they can break multiple-addend problems into smaller parts.
- **Can you think of other times at school or at home when you have to add more than two numbers?** Accept all reasonable answers, such as setting the table, shopping at the grocery store, and so on.

4 Assess

Informal Assessment

Use the Student Assessment Record, **Assessment,** page 100, to record informal observations.

COMPUTING	APPLYING
Catch the Teacher	**Add Them Up**
Did the child	Did the child
❏ respond accurately?	❏ apply learning to new situations?
❏ respond quickly?	❏ contribute concepts?
❏ respond with confidence?	❏ contribute answers?
❏ self-correct?	❏ connect mathematics to the real world?

Writing Equations • Lesson 4   **167**

Week 12 Writing Equations
Lesson 5
Review

Objective
Children review the material presented in Week 12 of *Number Worlds*.

Program Materials
Warm Up
No materials required

Creating Context
Although it is important to ask beginning English Learners questions that have a lower language demand, this does not mean that questions should have a lower cognitive load. Ask English Learners to demonstrate their understanding with manipulatives, models, and illustrations. Check understanding by asking children to show, demonstrate, or point to the answer.

✓ Cumulative Assessment
After Lesson 5 is complete, students should complete the Cumulative Assessment, **Assessment,** pages 88–89. Using the key on **Assessment,** page 87, identify incorrect responses. Reteach and review the suggested activities to reinforce concept understanding.

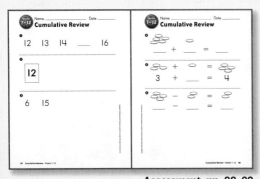

Assessment, pp. 88–89

1 Warm Up 5

Skill Practice COMPUTING
Catch the Teacher
Before beginning the **Free-Choice** activity, use the **Catch the Teacher** activity with the whole group.

Purpose Children learn to detect an error in counting and identify which number was omitted from the sequence.

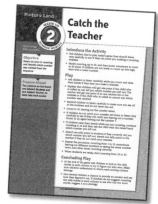

Warm-Up Card 2

2 Engage 20

Concept Building APPLYING
Free-Choice Activity

For the last day of the week, allow children to choose an activity from the previous weeks. Some activities they may choose include the following:
- Picture Land: **Can You Guess My Number?**
- Picture Land: **And One More Makes . . .**
- Object Land: **More Food Fun**

Make a note of the activities children select. Do they prefer easy or challenging activities? Continue to provide Challenge opportunities for children who have mastered the basic games.

Monitoring Student Progress

| If . . . children would benefit from extra practice on specific skills, | Then . . . choose an activity for them. |

168 Number Worlds • Week 12

Reflect 10

Extended Response REASONING
Ask questions such as the following:
- **What did you like about playing** They're Gone?
- **Was there anything about playing this game you didn't like?**
- **Did this game help you do something you couldn't do before? What did it help you do?**
- **What was easy when you were playing** Add Them Up?
- **What was hard when you were playing** Add Them Up?
- **What do you remember most about this game?**
- **What was your favorite part?**

Using Student Pages
Have each child complete **Workbook,** page 40. Is the child correctly applying the skills learned in Week 12?

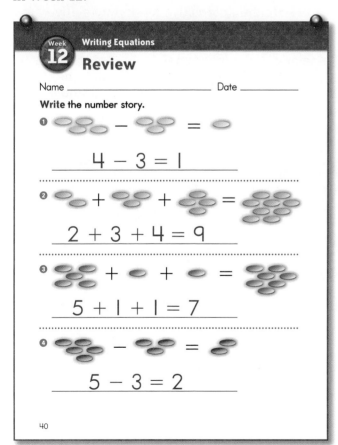

Workbook, p. 40

 Assess 10

A Gather Evidence
Formal Assessment
Have students complete the weekly test on *Assessment,* page 47. Record formal assessment scores on the Student Assessment Record, *Assessment,* page 100.

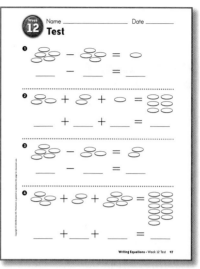

Assessment, p. 47

B Summarize Findings
Review the Student Assessment Records. Determine whether students have Minimal, Basic, or Secure understanding of the concepts presented in Week 12.

C Differentiate Instruction
Based on your observations, use these teaching strategies next week to follow up.

Minimal Understanding
- Repeat the Warm-Up and Engage activities in subsequent weeks to develop concepts of numeration.
- Repeat the Object Land activities from this week. Observe children's understanding of these concepts and reinforce as necessary.
- Use **Building Blocks** computer activities beginning with Count to develop and reinforce numeration concepts.

Basic Understanding
- Repeat Engage activities in subsequent weeks to reinforce basic concepts.
- Use **Building Blocks** computer activities beginning with Function Machine 1 to reinforce addition and subtraction concepts.

Secure Understanding
- Use Challenge variations of **Catch the Teacher.**
- Use **Building Blocks** computer activities beginning with Number Snapshots 2.

Writing Equations • Lesson 5 Review **169**

Week 13: Counting and Adding

Week at a Glance

This week, children begin **Number Worlds,** Week 13, and continue to explore Line Land.

Skills Focus
- Count on
- Compare and order numbers
- Solve problems

Background

In Line Land, a change in quantity is represented by movement along a line. For example, an increased set size and the larger number associated with it mean moving forward on the line.

Teaching for Understanding

As children engage in these activities, they will begin to realize that the physical addition and subtraction of objects they practiced in Object Land is equivalent to forward and backward movement along the number line. This understanding is essential for children to move from the world of countable objects to the world of abstract numbers.

Observe closely while evaluating the Engage activities assigned for this week.

- Are children counting to 20 and higher?
- Are children correctly identifying relative positions of numbers?
- Are children counting correctly as they move along the number line?
- Are children writing addition equations to indicate forward progress on the number line?

How Children Learn

As children begin this week's activities, they should already have experience moving forward and backward along the number line.

At the end of the week, children should understand that the greater the number, the greater the distance traveled along the number line. Children should also begin to write equations to explain forward movement along the number line.

Math at Home
Give one copy of the Letter to Home, page A13, to each child. Complete the activity in class, and then encourage children to share it with their caregivers.

Math Vocabulary
before In front of; to come first
after Behind; to come later
near Close to
far Not close to; a long way away

English Learners
SPANISH COGNATES

English	Spanish
activity	actividad
compare	comparar
movement	movimiento
position	posiciones
progress	progreso

ALTERNATE VOCABULARY
secret Not known

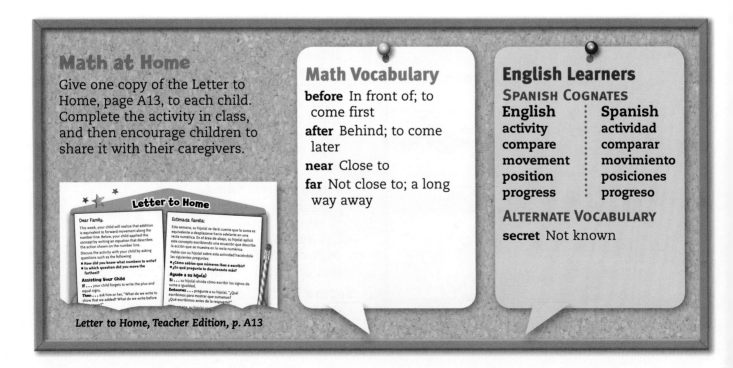

Letter to Home, Teacher Edition, p. A13

170 Number Worlds • Week 13

Week 13 Planner: Counting and Adding

PACING	LESSON	LEARNING GOALS	MATERIALS	TECHNOLOGY
DAY 1	Warm Up 3 Line Land: The Neighborhood*	Children use the spatial terms *before* and *after* with the number sequence.	**Program Materials** Neighborhood Number Line (1–20); remaining sections for a Challenge **Additional Materials** Self-stick notes over each number	Building Blocks Before and After Math eMathTools Number Line
	Activity 23 Line Land: Steve's New Bike*	Children identify and sequence numerals to 100 and relate number magnitude to distance traveled.	**Program Materials** • Neighborhood Number Line • Steve's Bike Pawn • Number Tiles (1–100) • Delivery Building Picture **Additional Materials** Small stickers	
DAY 2	Warm Up 3 Line Land: The Neighborhood*	Children use the spatial terms *before* and *after* with the number sequence.	Same as Lesson 1	Building Blocks Road Race: Shape Counting
	Activity 24 Line Land: Secret Number Game	Children associate addition with counting on and write equations to indicate progression on a number line.	**Program Materials** • Magnetic Number Line, 1 per child • Magnetic Chips • Number 1–6 Cube **Additional Materials** Sealed envelope with a number between 1 and 20 inside	
DAY 3	Warm Up 3 Line Land: The Neighborhood*	Children use the spatial terms *before* and *after* with the number sequence.	Same as Lesson 1	Building Blocks Before and After Math
	Activity 23 Line Land: Steve's New Bike*	Children relate number magnitude to distance traveled.	Same as Lesson 1	
DAY 4	Warm Up 3 Line Land: The Neighborhood*	Children use the spatial terms *before* and *after* with the number sequence.	Same as Lesson 1	Building Blocks Road Race: Shape Counting
	Activity 24 Line Land: Secret Number Game	Children write equations to indicate progression on a number line.	Same as Lesson 2	
DAY 5	Warm Up 3 Line Land: The Neighborhood*	Children use the spatial terms *before* and *after* with the number sequence.	Same as Lesson 1	Building Blocks Review previous activities
	Review and Assess	Children review their favorite activities to improve understanding.	Materials will be selected from those used in previous weeks.	

* Includes Challenge Variations

Week 13 — Counting and Adding

Lesson 1

Objective
Children work with numerals to 100 and relate number magnitude to distance traveled.

Program Materials
Warm Up
First two sections of the Neighborhood Number Line (1–10 and 11–20); remaining sections (21–100) for a Challenge

Engage
- Neighborhood Number Line (all sections)
- Steve's Bike Pawn
- Number Tiles (1–100)
- Delivery Building Picture

Additional Materials
Warm Up
Self-sticking notes placed over each number before activity begins

Engage
Small stickers

Access Vocabulary
near Close to
far Not close to; a long way away

Creating Context
Throughout this lesson children will make comparisons such as describing Steve's longest and shortest deliveries. Review comparative and superlative forms of the following words, and create a classroom chart with illustrations to aid understanding.

long—longer—longest
short—shorter—shortest
high—higher—highest
near—nearer—nearest
far—farther—farthest

1 Warm Up — 5

Concept Building COMPUTING
The Neighborhood
Before beginning the whole group activity **Steve's New Bike**, use **The Neighborhood** activity with the whole group.

Purpose Children use the spatial terms *before* and *after* with the number sequence.

Warm-Up Card 3

Monitoring Student Progress
| If... children are having trouble identifying numerals 1–20, | Then... have them work with only the first section of the Neighborhood Number Line (numerals 1–10). |

2 Engage — 30

Concept Building UNDERSTANDING
Steve's New Bike
"Today you will help Steve deliver messages in the neighborhood."

Follow the instructions on the Activity Card to play **Steve's New Bike**. As children play, ask questions about what is happening in the activity.

Purpose Children count, identify, and sequence numerals up to 100; determine the relative magnitude of double-digit numbers; and relate this understanding to distance traveled.

Activity Card 23

172 Number Worlds • Week 13

Monitoring Student Progress

If . . . a child delivers a message to the wrong address,

Then . . . ask the child to name the number he or she drew and write it on the board. Ask the child what block of houses the number is in and how the child knows. Then ask which house in that block the child will visit. If the child has trouble answering, call on other children to help.

 MathTools Use the Number Line Tool to demonstrate and explore number sequence.

Teacher's Note This activity can be played again and again. You might decide to focus on a specific section of the number sequence on a given day, such as 20–40 or 50–100, to fit the needs of your group.

Building Blocks For additional practice with the number sequence, children should complete **Building Blocks** Before and After Math.

3 Reflect 10

Extended Response REASONING
Ask questions such as the following:

- **What is one of the numbers that you picked?** Answers will vary based on the Number Tile selected.
- **Is that a long way or a short way from 0?** Accept all reasonable answers. Help children understand that higher numbers are farther from 0 than lower numbers.
- **What are some of the numbers that Steve passed along the way?** Answers will vary based on the Number Tile selected.

Using Student Pages
Have each child complete **Workbook,** page 41. Did the child circle the correct answer?

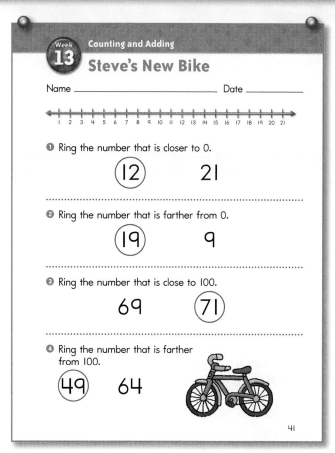

Workbook, p. 41

4 Assess

Informal Assessment
Use the Student Assessment Record, **Assessment,** page 100, to record informal observations.

COMPUTING	UNDERSTANDING
The Neighborhood	**Steve's New Bike**
Did the child	Did the child
❏ respond accurately?	❏ make important observations?
❏ respond quickly?	❏ extend or generalize learning?
❏ respond with confidence?	❏ provide insightful answers?
❏ self-correct?	❏ pose insightful questions?

Counting and Adding • Lesson 1 173

Week 13 Counting and Adding

Lesson 2

Objective
Children continue to work with the number sequence, associate addition with counting on, and write equations to indicate progression on a number line.

Program Materials
Warm Up
First two sections of the Neighborhood Number Line (1–10 and 11–20); remaining sections (21–100) for a Challenge

Engage
- Magnetic Number Line, 1 per child
- Magnetic Chips
- Number 1–6 Cube

Additional Materials
Warm Up
Self-sticking notes placed over each number before activity begins

Engage
Sealed envelope with a number between 1 and 20 inside

Access Vocabulary
secret Not known
add Combine; put things together

Creating Context
Acquiring prepositions is an intermediate-to-advanced skill on the continuum of English proficiency, so some English Learners may find them difficult. Because not all languages share the same structure, the notion of prepositions may be unfamiliar to some children. Have children demonstrate and practice *before* and *after* by lining up and saying who comes before or after each child in line.

---

1 Warm Up   5

Concept Building COMPUTING
The Neighborhood
Before beginning the small-group activity **Secret Number Game,** use **The Neighborhood** activity with the whole group.

Purpose Children use the spatial terms *before* and *after* with the number sequence.

Warm-Up Card 3

Monitoring Student Progress
If . . . children are fluently identifying and sequencing numerals 1–20,

Then . . . have them work with larger numerals by playing the Challenge variation.

2 Engage   30

Concept Building ENGAGING
Secret Number Game
"Today we will write number stories about addition."

Follow the instructions on the Activity Card to play the **Secret Number Game.** As children play, ask questions about what is happening in the activity.

Purpose Children associate addition with counting on and write equations to indicate forward progression on a number line.

Activity Card 24

174 Number Worlds • Week 13

Monitoring Student Progress

If . . . children have trouble remembering to include the plus and equal symbols in their equations,

Then . . . post these symbols on the classroom wall and practice writing number stories with children throughout the day. You can help the class write equations describing the number of children in line for recess, the number of books lined up on a bookshelf, and so on.

 Teacher's Note Encourage children to use this week's vocabulary words as they engage in the activities, discuss math concepts, and make predictions.

Building Blocks For additional practice with addition, children should complete **Building Blocks** Road Race: Shape Counting.

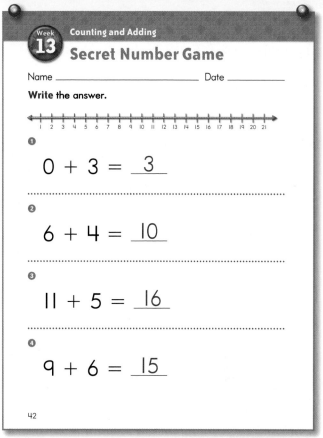

Workbook, p. 42

3 Reflect (10)

Extended Response REASONING

- **Look at the last turn on your Magnetic Number Line. What number did you start on?** Answers will vary.
- **How many did you add?** Answers will vary.
- **Where were you at the end of the turn?** Answers will vary.
- **Does the number story you wrote tell what happened during the turn?** Encourage children to explain how they decided the position of each number in the equation.

Using Student Pages

Have each child complete **Workbook,** page 42. Did the child write the correct answer?

4 Assess

Informal Assessment

Use the Student Assessment Record, **Assessment,** page 100, to record informal observations.

COMPUTING	ENGAGING
The Neighborhood	**Secret Number Game**
Did the child	Did the child
❏ respond accurately?	❏ pay attention to the contributions of others?
❏ respond quickly?	
❏ respond with confidence?	❏ contribute information and ideas?
❏ self-correct?	
	❏ improve on a strategy?
	❏ reflect on and check accuracy of work?

Counting and Adding • Lesson 2 175

Week 13 — Counting and Adding
Lesson 3

Objective
Children use numerals to 100 and relate number magnitude to distance traveled.

Program Materials
Warm Up
First two sections of the Neighborhood Number Line (1–10 and 11–20); remaining sections (21–100) for a Challenge

Engage
- Neighborhood Number Line (all sections)
- Steve's Bike Pawn
- Number Tiles (1–100)
- Delivery Building Picture

Additional Materials
Warm Up
Self-sticking notes placed over each number before activity begins

Engage
Small stickers

Access Vocabulary
pawn Game piece
deliver To take a message or package to someone

Creating Context
Using gestures is an excellent way to enhance comprehension for English learners. In **Steve's New Bike** children are asked questions about distance and comparison. Use hand gestures to help English Learners understand terms such as *nearest* and *farthest*.

1 Warm Up — 5

Skill Building COMPUTING
The Neighborhood
Before beginning the whole-group activity **Steve's New Bike,** use **The Neighborhood** activity with the whole group.

Purpose Children use the spatial terms *before* and *after* with the number sequence.

Warm-Up Card 3

Monitoring Student Progress
If . . . children are familiar with numbers to 100,
Then . . . have them skip count to 100 by tens in the Challenge 2 variation.

2 Engage — 30

Concept Building REASONING
Steve's New Bike
"Today you will help Steve deliver messages in the neighborhood."

Follow the instructions on the Activity Card to play, **Steve's New Bike.** As children play, ask questions about what is happening in the activity.

Purpose Children count, identify, and sequence numerals to 100; determine the relative magnitude of double-digit numbers; and relate this understanding to distance traveled.

Activity Card 23

176   Number Worlds • Week 13

Monitoring Student Progress

If . . . children have mastered the game,

Then . . . have them play the Challenge variation of **Steve's New Bike.**

Teacher's Note Children may enjoy playing this independently in small groups when they are familiar with the procedure. Have them figure out Steve's longest and shortest deliveries, how many deliveries he made altogether, and how much money Steve earned.

 Building Blocks For additional practice with the number sequence, children should complete **Building Blocks** Before and After Math.

4 Assess

Informal Assessment

Use the Student Assessment Record, **Assessment,** page 100, to record informal observations.

COMPUTING	REASONING
The Neighborhood	**Steve's New Bike**
Did the child	Did the child
❏ respond accurately?	❏ provide a clear explanation?
❏ respond quickly?	❏ communicate reasons and strategies?
❏ respond with confidence?	❏ choose appropriate strategies?
❏ self-correct?	❏ argue logically?

3 Reflect 10

Extended Response REASONING

Ask questions such as the following:

- **What was Steve's longest ride? How do you know?** Answers will vary. Help children understand that larger numbers indicate a longer distance traveled along the number line.

- **What was Steve's shortest ride? How did you figure that out?** Answers will vary. Help children understand that smaller numbers indicate a shorter distance traveled along the number line.

Counting and Adding • Lesson 3

Week 13 — Counting and Adding

Lesson 4

Objective

Children continue to work with the number sequence, associate addition with counting on, and write equations to indicate progression on a number line.

Program Materials

Warm Up
First two sections of the Neighborhood Number Line (1–10 and 11–20); remaining sections (21–100) for a Challenge

Engage
- Magnetic Number Line, 1 per child
- Magnetic Chips
- Number 1–6 Cube

Additional Materials

Warm Up
Self-sticking notes placed over each number before activity begins

Engage
Sealed envelope with a number between 1 and 20 inside

Access Vocabulary

plus sign Symbol that means "add"
equal sign Symbol that means "having the same amount"

Creating Context

English Learners often understand math concepts better when they can interact with other children who speak the same primary language. Allow children to work with partners or cross-age helpers who speak the same primary language so they can check understanding with one another in team situations.

1. Warm Up

Concept Building — COMPUTING

The Neighborhood
Before beginning the small-group activity **Secret Number Game**, use **The Neighborhood** activity with the whole group.

Purpose Children use the spatial terms *before* and *after* with the number sequence.

Warm-Up Card 3

Monitoring Student Progress

If . . . children are having trouble identifying numerals 1–20,

Then . . . have them work with only the first section of the Neighborhood Number Line (numerals 1–10).

2. Engage

Concept Building — APPLYING

Secret Number Game
"Today we will write number stories about addition."

Follow the instructions on the Activity Card to play **Secret Number Game**. As children play, ask questions about what is happening in the activity.

Purpose Children associate addition with counting on and write equations to indicate forward progression on the Number Line.

Activity Card 24

Monitoring Student Progress

If . . . children can fluently count on and write correct addition equations,

Then . . . challenge them to empty their Magnetic Number Lines and use only their written records to re-create the moves they made. Children can then pass their Magnetic Number Lines and written records to a partner who will check for accuracy.

 For further practice with addition, children should complete **Building Blocks** Road Race: Shape Counting.

3 Reflect   10

Extended Response   REASONING

Ask questions such as the following:

- **Did you land on the secret number? How can you check your answer?** Possible answers: by looking at the number after the equal sign in the number story; by counting the Magnetic Chips on the number line
- **When have we used these strategies before? Why does it make sense to use them now?** We counted on in Object Land and Picture Land and wrote number stories in Picture Land. Counting up on the number line is the same as adding objects to a set.

4 Assess

Informal Assessment

Use the Student Assessment Record, **Assessment,** page 100, to record informal observations.

COMPUTING	APPLYING
The Neighborhood	**Secret Number Game**
Did the child	Did the child
❏ respond accurately?	❏ apply learning to new situations?
❏ respond quickly?	❏ contribute concepts?
❏ respond with confidence?	❏ contribute answers?
❏ self-correct?	❏ connect mathematics to the real world?

Counting and Adding • Lesson 4

Week 13: Counting and Adding

Lesson 5
Review

Objective
Children review the material presented in Week 13 of **Number Worlds**.

Program Materials
Warm Up
First two sections of the Neighborhood Number Line (1–10 and 11–20); remaining sections (21–100) for a Challenge

Additional Materials
Warm Up
Self-sticking notes placed over each number before activity begins

Creating Context
To reinforce the concepts of near and far, have children work in pairs to generate a list of places that are near to or far from the school. Make a chart with columns labeled *Near* and *Far* to record the answers, and have children illustrate or add a symbol for each location.

1 Warm Up — 5

Skill Practice COMPUTING
The Neighborhood
Before beginning the **Free-Choice** activity, use **The Neighborhood** activity with the whole group.

Purpose Children use the spatial terms *before* and *after* with the number sequence.

Warm-Up Card 3

Monitoring Student Progress
If . . . children are fluently identifying and sequencing numerals 1–20,

Then . . . have them work with larger numerals by playing the Challenge variation.

2 Engage — 20

Skill Building APPLYING
Free-Choice Activity
For the last day of the week, allow children to choose an activity from the previous weeks. Some activities they may choose include the following:
- Picture Land: **Dragon Quest 1**
- Picture Land: **And One More Makes . . .**
- Object Land: **Add Them Up**

Make a note of the activities children select. Do they prefer easy or challenging activities? Continue to provide Challenge opportunities for children who have mastered the basic games.

Monitoring Student Progress
If . . . children would benefit from extra practice on specific skills,

Then . . . choose an activity for them.

Reflect

Extended Response REASONING
Ask questions such as the following:
- What did you like about playing Steve's New Bike?
- Was there anything about playing this game you didn't like?
- Did this game help you do something you couldn't do before? What did it help you do?
- What was easy when you were playing Secret Number Game?
- What was hard when you were playing Secret Number Game?
- What do you remember most about this game?
- What was your favorite part?

Using Student Pages
Have each child complete **Workbook**, page 43. Is the child correctly applying the concepts learned in Week 13?

Workbook, p. 43

 ## Assess

A Gather Evidence
Formal Assessment
Have students complete the weekly test on *Assessment,* page 49. Record formal assessment scores on the Student Assessment Record, *Assessment,* page 100.

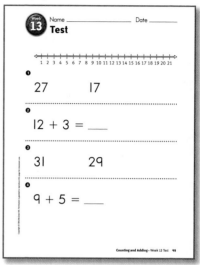

Assessment, p. 49

B Summarize Findings
Review the Student Assessment Records. Determine whether students have Minimal, Basic, or Secure understanding of the concepts presented in Week 13.

C Differentiate Instruction
Based on your observations, use these teaching strategies next week to follow up.

Minimal Understanding
- Repeat the Warm-Up and Engage activities in subsequent weeks to develop numeration concepts.
- Repeat the Line Land activities from this week. Observe children's understanding of these concepts, and reinforce as necessary.
- Use **Building Blocks** computer activities beginning with Count to develop and reinforce numeration concepts.

Basic Understanding
- Repeat Engage activities in subsequent weeks to reinforce basic concepts.
- Use **Building Blocks** computer activities beginning with Counting Activity 3 to reinforce numeration and addition concepts.

Secure Understanding
- Use the Challenge variation of **Steve's New Bike.**
- Use **Building Blocks** computer activities beginning with Counting Activity 4.

Counting and Adding • Lesson 5 Review **181**

Week 14: Making Equations

Week at a Glance

This week, children begin **Number Worlds**, Week 14, and continue to work in Line Land and Picture Land.

Background

In Picture Land, numbers are represented by sets of dots and later by numerals. In Line Land, children learn that numbers can also represent a sequence of steps along an ordered path.

How Children Learn

At the beginning of this week, children should have experience writing addition equations and should understand that the greater the number, the greater the distance traveled along the number line.

At the end of the week, children should be more comfortable writing addition equations and should understand the concept of equality and the use of the equal sign.

Skills Focus

- Compare and order numbers
- Add whole numbers
- Use math symbols
- Understand equality
- Write addition equations

Teaching for Understanding

As children engage in these activities, they will gradually realize the correspondence between movement along the number line, an increase or decrease in set size, and formal addition notation.

Observe closely while evaluating the Engage activities assigned for this week.

- Are children correctly ordering numbers?
- Can children add small whole numbers?
- Can children correctly use the plus and equal signs?
- Do children understand the concept of equality?

Math at Home

Give one copy of the Letter to Home, page A14, to each child. Complete the activity in class, and then encourage children to share it with their caregivers.

Letter to Home, Teacher Edition, p. A14

Math Vocabulary

symbol Letter, number, or picture that has special meaning or that stands for something else

plus sign Symbol that means "add"

equal sign Symbol that means "having the same amount"

English Learners

SPANISH COGNATES

English	Spanish
activity	actividad
equation	ecuación
order	orden
sequence	secuencia

ALTERNATE VOCABULARY

solve Figure out; find the answer

182 Number Worlds • Week 14

Week 14 Planner: Making Equations

PACING	LESSON	LEARNING GOALS	MATERIALS	TECHNOLOGY
DAY 1	Warm Up 3 — Line Land: The Neighborhood*	Children use the spatial terms *before* and *after* with the number sequence.	**Program Materials** Neighborhood Number Line (1–20); remaining sections for a Challenge **Additional Materials** Self-sticking notes	Building Blocks — Double Compare 1 • eMathTools
	Activity 25 — Picture Land: Making Equations for Beginners*	Children use manipulatives and Dot Set Cards to create simple addition equations.	**Program Materials** • Symbol Cards, Plus Sign and Equal sign • Dot Set Cards (1–5) • Counters • Number Cards, for the Challenge	100 Chart
DAY 2	Warm Up 3 — Line Land: The Neighborhood*	Children use the spatial terms *before* and *after* with the number sequence.	Same as Lesson 1	Building Blocks — Barkley's Bones 1
	Activity 26 — Picture Land: Making and Writing Equations	Children create addition equations using formal notation.	**Program Materials** Symbol Cards, Plus Sign and Equal Sign **Additional Materials** • Small items, several sets • Pencils and paper	
DAY 3	Warm Up 3 — Line Land: The Neighborhood*	Children use the spatial terms *before* and *after* with the number sequence.	Same as Lesson 1	Building Blocks — Double Compare 1
	Activity 25 — Picture Land: Making Equations for Beginners*	Children create simple addition equations.	Same as Lesson 1	
DAY 4	Warm Up 3 — Line Land: The Neighborhood*	Children use the spatial terms *before* and *after* with the number sequence.	Same as Lesson 1	Building Blocks — Barkley's Bones 1
	Activity 26 — Picture Land: Making and Writing Equations	Children create addition equations using formal notation.	Same as Lesson 2	
DAY 5	Warm Up 3 — Line Land: The Neighborhood*	Children use the spatial terms *before* and *after* with the number sequence.	Same as Lesson 1	Building Blocks — Review previous activities.
	Review and Assess	Children review their favorite activities to improve understanding.	Materials will be selected from those used in previous weeks.	

* Includes Challenge Variations

Week 14 — Making Equations
Lesson 1

Objective
Children continue to review the number sequence and use manipulatives and Dot Set Cards to create simple addition equations.

Program Materials
Warm Up
First two sections of the Neighborhood Number Line (1–10 and 11–20); remaining sections (21–100) for a Challenge

Engage
- Symbol Cards, Plus Sign and Equal Sign
- Dot Set Cards (1–5)
- Counters
- Number Cards, for the Challenge

Additional Materials
Warm Up
Self-sticking notes placed over each number before activity begins

Engage
Pan balance (optional)

Access Vocabulary
plus sign Symbol that means "add"
equal sign Symbol that means "having the same amount"

Building Background
Playing games is an excellent way to give English Learners practice listening to and speaking English. The natural repetition of procedural and counting language replaces tedious drills with authentic, active experience. Wait time is built into the process, and the low-stakes environment makes the use of games an enjoyable and efficient learning tool in math.

1 Warm Up (5)

Skill Building — COMPUTING
The Neighborhood
Before beginning the activity **Making Equations for Beginners,** use **The Neighborhood** activity with the whole group.

Purpose Children use the spatial terms *before* and *after* with the number sequence.

Warm-Up Card 3

Monitoring Student Progress
If . . . children are having trouble identifying numerals 1–20,
Then . . . have them work with only the first section of the Neighborhood Number Line (numerals 1–10).

eMathTools Use the 100 Table to demonstrate and explore place value.

2 Engage (30)

Concept Building — UNDERSTANDING
Making Equations for Beginners
"Today we are going to do the type of math the big kids do."

Follow the instructions on the Activity Card to play **Making Equations for Beginners.** As children play, ask questions about what is happening in the game.

Purpose Children use formal notation to add small quantities.

Activity Card 25

184 Number Worlds • Week 14

Monitoring Student Progress

If . . . children have trouble remembering which symbol to use,

Then . . . post them in the classroom and refer to them throughout the day. For example, you could discuss which symbol shows what happens as you call children to line up for recess, which symbol shows what happens when you have two books for two children, and so on.

 Building Blocks For additional practice with addition, children should complete **Building Blocks** Double Compare 1.

3 Reflect 10

Extended Response REASONING
Ask questions such as the following:

- **What symbol do we use to show that two things are the same?** the equal sign
- **Is the amount on this side of the equal sign the same as the amount on the other side? How did you figure that out?** Accept all reasonable answers. Children may count the dots or recognize the dot patterns on the cards to determine whether the total number on one side is equal to the number on the other side.

Using Student Pages
Have each child complete **Workbook,** page 44. Did the child match the Dot Set Cards to the correct number?

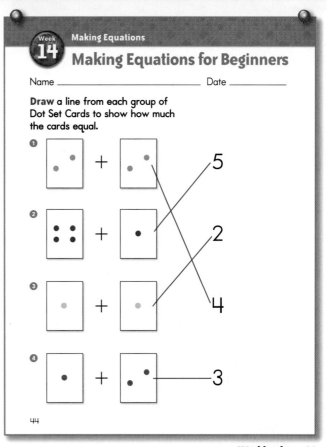

Workbook, p. 44

4 Assess

Informal Assessment
Use the Student Assessment Record, **Assessment,** page 100, to record informal observations.

COMPUTING	UNDERSTANDING
The Neighborhood Did the child	**Making Equations for Beginners** Did the child
❏ respond accurately?	❏ make important observations?
❏ respond quickly?	❏ extend or generalize learning?
❏ respond with confidence?	❏ provide insightful answers?
❏ self-correct?	❏ pose insightful questions?

Making Equations • Lesson 1   **185**

Week 14 Making Equations
Lesson 2

Objective
Children continue to identify and sequence numerals and create simple addition equations using formal notation.

Program Materials
Warm Up
First two sections of the Neighborhood Number Line (1–10 and 11–20); remaining sections (21–100) for a Challenge

Engage
Symbol Cards, Plus Sign and Equal Sign

Additional Materials
Warm Up
Self-sticking notes placed over each number before activity begins

Engage
- Small items, several sets that can be used for demonstrations
- Pencils and paper

Access Vocabulary
equation Number story
solve Figure out; find the answer

Building Background
Organize children into small groups to discuss things they do at home that require addition. Have children work together to create a group story, complete with illustrations, to show that addition was involved. Have children use manipulatives or props to act out their stories.

---

1 Warm Up 5

Concept Building COMPUTING
The Neighborhood
Before beginning the small-group activity **Making and Writing Equations,** use **The Neighborhood** activity with the whole group.

Purpose Children use the spatial terms *before* and *after* with the number sequence.

Warm-Up Card 3

Monitoring Student Progress
If . . . children are fluently identifying and sequencing numerals 1–20,

Then . . . have them work with larger numerals by playing the Challenge variation.

2 Engage 30

Concept Building ENGAGING
Making and Writing Equations

"Today you are going to write addition number stories."

Follow the instructions on the Activity Card to play **Making and Writing Equations.** As children play, ask questions about what is happening in the activity.

Purpose Children create addition equations using formal notation.

Activity Card 26

186 Number Worlds • Week 14

Monitoring Student Progress

If . . . children have trouble remembering the plus sign and equal sign,

Then . . . post the signs in the classroom and model making equations with them. Invite volunteers to stand in a line in front of the class, and have the rest of the class organize the volunteers into groups to make simple equations.

 Building Blocks For additional practice with addition, children should complete **Building Blocks** Barkley's Bones 1.

3 Reflect — 10

Extended Response REASONING

Ask questions such as the following:

- **How did you know what number goes to the right of the equal sign?** Accept all reasonable answers. Children should indicate that the number to the right of the equal sign should equal the sum of the numbers left of the equal sign, or that the amount on each side of the equal sign should be the same, or that the amounts on each side of the equal sign should balance.

- **How can you check your answer?** Possible answer: Count the items on each side of the equal sign.

Using Student Pages

Have each child complete **Workbook,** page 45. Did the child write the correct number story?

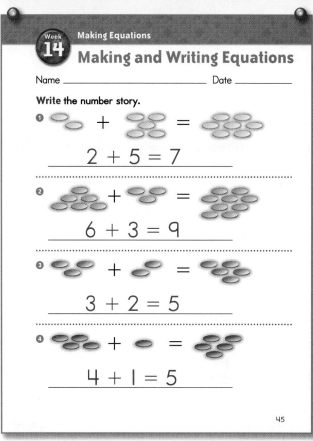

Workbook, p. 45

4 Assess

Informal Assessment

Use the Student Assessment Record, **Assessment,** page 100, to record informal observations.

COMPUTING **The Neighborhood** Did the child	ENGAGING **Making and Writing Equations** Did the child
❏ respond accurately? ❏ respond quickly? ❏ respond with confidence? ❏ self-correct?	❏ pay attention to the contributions of others? ❏ contribute information and ideas? ❏ improve on a strategy? ❏ reflect on and check accuracy of work?

Making Equations • Lesson 2   **187**

Week 14: Making Equations

Lesson 3

Objective

Children continue to review the number sequence and use manipulatives and Dot Set Cards to create simple addition equations.

Program Materials

Warm Up
First two sections of the Neighborhood Number Line (1–10 and 11–20); remaining sections (21–100) for a Challenge

Engage
- Symbol Cards, Plus Sign and Equal Sign
- Dot Set Cards (1–5)
- Counters
- Number Cards, for the Challenge

Additional Materials

Warm Up
Self-sticking notes placed over each number before activity begins

Engage
Pan balance (optional)

Access Vocabulary

figure out To think about a problem and solve it; discover the answer

Building Background

Using models and demonstrations is an excellent strategy to make instruction more comprehensible to beginning English Learners. This lesson suggests using a pan balance to demonstrate that two sides are equal. Allow children to demonstrate several equations using the balance. Give them a model sentence to use when describing their work: _____ plus _____ equals _____.

1 Warm Up — 5

Skill Building — COMPUTING

The Neighborhood

Before beginning the activity **Making Equations for Beginners,** use **The Neighborhood** activity with the whole group.

Purpose Children use the spatial terms *before* and *after* with the number sequence.

Warm-Up Card 3

Monitoring Student Progress

If . . . children are familiar with numbers to 100,

Then . . . have them skip count to 100 by tens in the Challenge 2 variation.

2 Engage — 30

Concept Building — REASONING

Making Equations for Beginners

"Today we are going to do the type of math the big kids do."

Follow the instructions on the Activity Card to play **Making Equations for Beginners.** As children play, ask questions about what is happening in the game.

Purpose Children use formal notation to add small quantities.

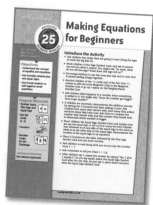

Activity Card 25

188   Number Worlds • Week 14

Monitoring Student Progress

| If . . . children are comfortable building equations, | Then . . . challenge them to use Number Cards instead of Dot Set Cards. |

Teacher's Note Encourage children to use this week's vocabulary words as they engage in the activities, discuss math concepts, and make predictions.

 For additional practice with addition, children should complete **Building Blocks** Double Compare 1.

3 Reflect 10

Extended Response REASONING
Ask questions such as the following:
- **Did anything about this problem surprise you? Why?** Accept all reasonable answers.
- **How would you explain this to someone who had never done it before?** Accept all reasonable answers. Children should indicate understanding of *plus* and *equals*.

4 Assess

Informal Assessment
Use the Student Assessment Record, **Assessment,** page 100, to record informal observations.

COMPUTING	REASONING
The Neighborhood Did the child ❑ respond accurately? ❑ respond quickly? ❑ respond with confidence? ❑ self-correct?	**Making Equations for Beginners** Did the child ❑ provide a clear explanation? ❑ communicate reasons and strategies? ❑ choose appropriate strategies? ❑ argue logically?

Making Equations • Lesson 3

Week 14 — Making Equations

Objective
Children continue to identify and sequence numerals and to create simple addition equations using formal notation.

Program Materials

Warm Up
First two sections of the Neighborhood Number Line (1–10 and 11–20); remaining sections (21–100) for the Challenge

Engage
Symbol Cards, Plus Sign and Equal Sign

Additional Materials

Warm Up
Self-sticking notes placed over each number before activity begins

Engage
- Small items, several sets that can be used for demonstrations
- Pencils and paper

Access Vocabulary
symbol Letter, number, or picture that has special meaning or that stands for something else

Creating Context
English Learners often understand math concepts better when they can interact with other children who speak the same primary language. Allow children to work with partners or cross-age helpers who speak the same primary language so they can check understanding with one another in game situations.

1 Warm Up — 5

Skill Building — COMPUTING

The Neighborhood
Before beginning the small-group activity **Making and Writing Equations,** use **The Neighborhood** activity with the whole group.

Purpose Children use the spatial terms *before* and *after* with the number sequence.

Warm-Up Card 3

Monitoring Student Progress

If . . . children are having trouble identifying numerals 1–20,

Then . . . model counting on the Neighborhood Number Line with them throughout the day.

2 Engage — 30

Concept Building — APPLYING

Making and Writing Equations

"Today you are going to write addition number stories."

Follow the instructions on the Activity Card to play **Making and Writing Equations.** As children play, ask questions about what is happening in the activity.

Purpose Children create addition equations using formal notation.

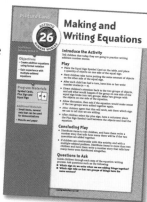
Activity Card 26

190 Number Worlds • Week 14

Monitoring Student Progress

If . . . children are comfortable creating and writing two-addend equations,

Then . . . challenge them to create equations with three or more addends.

 For additional practice with addition, children should complete **Building Blocks** Barkley's Bones 1.

3 Reflect 10

Extended Response REASONING

Ask questions such as the following:

- **How would you explain this to someone who had never done it before?** Accept all reasonable answers. Children should indicate that the amounts on the left and the right of the equal sign must be the same.

- **Can you think of other times at school or at home when you add?** Accept all reasonable answers, such as setting the table, having a snack with a sibling, and so on.

4 Assess

Informal Assessment

Use the Student Assessment Record, **Assessment,** page 100, to record informal observations.

COMPUTING	APPLYING
The Neighborhood Did the child ❏ respond accurately? ❏ respond quickly? ❏ respond with confidence? ❏ self-correct?	**Making and Writing Equations** Did the child ❏ apply learning to new situations? ❏ contribute concepts? ❏ contribute answers? ❏ connect mathematics to the real world?

Making Equations • Lesson 4   **191**

Week 14 — Making Equations

Lesson 5
Review

Objective
Children review the material presented in Week 14 of *Number Worlds*.

Program Materials
Warm Up
First two sections of the Neighborhood Number Line (1–10 and 11–20); remaining sections (21–100) for the Challenge

Additional Materials
Warm Up
Self-sticking notes placed over each number before activity begins

Building Background
In addition to assessing and observing children using math concepts, complete an observation of children's language proficiency with academic English for mathematics. If your designated English Language Proficiency Tests have an observation checklist, use it to determine which English forms, functions, and vocabulary are still needed for academic language proficiency.

1 Warm Up

Skill Practice — COMPUTING
The Neighborhood
Before beginning the **Free-Choice** activity, use **The Neighborhood** activity with the whole group.

Purpose Children use the spatial terms *before* and *after* with the number sequence.

Warm-Up Card 3

Monitoring Student Progress
If . . . children are familiar with numbers to 100, **Then . . .** have them skip count to 100 by tens, fives, or twos in the Challenge 2 activity.

2 Engage

Concept Building — APPLYING
Free-Choice Activity
For the last day of the week, allow children to choose an activity from the previous weeks. Some activities they may choose include the following:
- Object Land: **Add Them Up**
- Line Land: **Steve's New Bike**
- Line Land: **Secret Number Game**

Make a note of the activities children select. Do they prefer easy or challenging activities? Continue to provide Challenge opportunities for children who have mastered the basic activities.

Monitoring Student Progress
If . . . children would benefit from extra practice on specific skills, **Then . . .** choose an activity for them.

192 Number Worlds • Week 14

3 Reflect — 10

Extended Response REASONING
Ask questions such as the following:
- What did you like about playing **Making Equations for Beginners?**
- Was there anything about playing this game you didn't like?
- Did this game help you do something you couldn't do before? What did it help you do?
- What was easy when you were playing **Making and Writing Equations?**
- What was hard when you were playing **Making and Writing Equations?**
- What do you remember most about this game?
- What was your favorite part?

Using Student Pages
Have each child complete **Workbook,** page 46. Is the child correctly applying the concepts learned in Week 14?

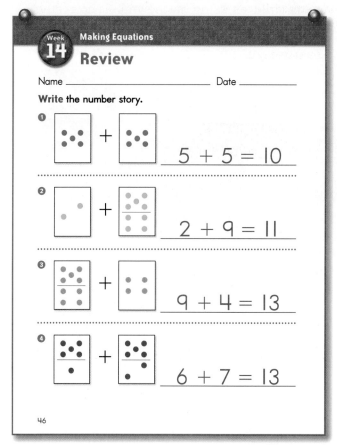

Workbook, p. 46

4 Assess — 10

A Gather Evidence
Formal Assessment
Have students complete the weekly test on **Assessment,** page 51. Record formal assessment scores on the Student Assessment Record, **Assessment,** page 100.

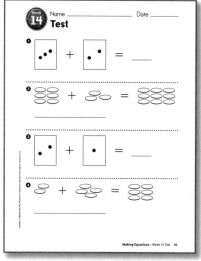

Assessment, p. 51

B Summarize Findings
Review the Student Assessment Records. Determine whether students have Minimal, Basic, or Secure understanding of the concepts presented in Week 14.

C Differentiate Instruction
Based on your observations, use these teaching strategies next week to follow up.

Minimal Understanding
- Repeat the Warm-Up and Engage activities in subsequent weeks to develop concepts of counting objects.
- Repeat the activities from this week. Observe children's understanding of these concepts and reinforce as necessary.
- Use **Building Blocks** computer activities beginning with Count to develop and reinforce numeration concepts.

Basic Understanding
- Repeat Engage activities in subsequent weeks to reinforce basic concepts.
- Use **Building Blocks** computer activities beginning with Double Compare to reinforce numeration and addition concepts.

Secure Understanding
- Use Challenge variations of this week's activities.
- Use **Building Blocks** computer activities beginning with Barkley's Bones 1.

Making Equations • Lesson 5 Review   193

Week 15: Graphing and Comparing Numbers

Week at a Glance

This week, children begin **Number Worlds,** Week 15, and continue to explore Sky Land and Picture Land.

Skills Focus
- Compare and order numbers
- Add whole numbers
- Use math symbols
- Understand equality

Background

In Sky Land, children learn about movement along a vertical plane. This helps them learn to use the language of height in addition to the language they learned in Line Land. This also helps children begin to understand how the terminology can be used interchangeably.

Teaching for Understanding

As children engage in this week's activities, they will come to understand that the number concepts they learned in the horizontal world of Line Land also apply to the vertical scales in Sky Land. Children will understand that the higher a number is on the scale, the bigger the set it represents.

Observe closely while evaluating the Engage activities assigned for this week.

- Are children comparing numbers correctly?
- Do children understand the concept of equality?
- Are children using math symbols correctly?

How Children Learn

As children begin this week's lessons, they should be comfortable writing addition equations and should understand the concept of equality.

At the end of this week, children should be able to read and understand a simple bar graph. They should also be able to compare set size and say whether sets have more, less, or the same amount.

Math at Home
Give one copy of the Letter to Home, page A15, to each child. Complete the activity in class, and then encourage children to share it with their caregivers.

Letter to Home, Teacher Edition, p. A15

Math Vocabulary

graph A special picture that shows how many you have

compare To think about how things are the same and different

equal Having the same amount

solve To figure out; to find the answer

English Learners

SPANISH COGNATES

English	Spanish
addition	adición
graph	gráfica
compare	comparar
equal	igual
group	grupo

ALTERNATE VOCABULARY

thermometer Tool used to measure how hot or cold something is

194 Number Worlds • Week 15

Week 15 Planner: Graphing and Comparing Numbers

PACING	LESSON	LEARNING GOALS	MATERIALS	TECHNOLOGY
DAY 1	**Warm Up 4** **Sky Land:** Thermometer Counting	Children count from 1 up to 20 and from 20 down to 1.	**Program Materials** Thermometer Picture **Additional Materials** Erasable red marker	**Building Blocks** Build Stairs 1
	Activity 27 **Sky Land:** Linking and Thinking	Children graph quantities and use both numerals and set size to compare them.	**Program Materials** • Sky Land Activity Sheet 1, 1 per child • Number Cards (8–15) • Learning Links, 50 of each color **Additional Materials** Crayons, pencils	
DAY 2	**Warm Up 4** **Sky Land:** Thermometer Counting	Children count from 1 up to 20 and from 20 down to 1.	**Program Materials** Thermometer Picture **Additional Materials** Erasable red marker	**Building Blocks** Barkley's Bones 1
	Activity 26 **Picture Land:** Making and Writing Equations	Children create addition equations using formal notation.	**Program Materials** Symbol Cards: Plus Sign and Equal Sign **Additional Materials** • Several sets of small items • Pencils and paper	
DAY 3	**Warm Up 4** **Sky Land:** Thermometer Counting	Children count from 1 up to 20 and from 20 down to 1.	**Program Materials** Thermometer Picture **Additional Materials** Erasable red marker	**Building Blocks** Build Stairs 1
	Activity 27 **Sky Land:** Linking and Thinking	Children graph and compare quantities.	Same as Lesson 1	
DAY 4	**Warm Up 4** **Sky Land:** Thermometer Counting	Children count from 1 up to 20 and from 20 down to 1.	**Program Materials** Thermometer Picture **Additional Materials** Erasable red marker	**Building Blocks** Barkley's Bones 1
	Activity 26 **Picture Land:** Making and Writing Equations	Children create addition equations using formal notation.	Same as Lesson 2	
DAY 5	**Warm Up 4** **Sky Land:** Thermometer Counting	Children count from 1 up to 20 and from 20 down to 1.	**Program Materials** Thermometer Picture **Additional Materials** Erasable red marker	**Building Blocks** Review previous activities
	Review and Assess	Children review their favorite activities to improve understanding.	Materials will be selected from those used in previous weeks.	

Week 15 — Graphing and Comparing Numbers

Lesson 1

Objective
Children identify and sequence the numbers 1–20 and compare quantities on a graph.

Program Materials
Warm Up
Thermometer Picture

Engage
- Sky Land Activity Sheet 1, 1 per child
- Number Cards (8–15)
- Learning Links, 50 of each color (red, blue, yellow, and green)

Additional Materials
Warm Up
Erasable red marker

Engage
- Crayons, 1 of each color (red, blue, yellow, and green)
- Pencils, 1 per child

Access Vocabulary
graph A special picture that shows how many you have
compare To think about how things are the same and different

Creating Context
Discuss the concepts of *most* and *least* with children. Review the chart of comparatives and superlatives from Week 1 to be sure children understand the words *most* and *least*. Ask them to point to the part of the classroom with the most books, the most jackets, the least number of children, and so on. Write an example sentence on the board for children to use as a model.

1 Warm Up 5

Skill Building COMPUTING
Thermometer Counting
Before beginning the small-group activity **Linking and Thinking**, use the **Thermometer Counting** activity with the whole class.

Purpose Children count from 1 up to 20 and from 20 down to 1.

Warm-Up Card 4

Monitoring Student Progress
If . . . children have trouble counting to 20,
Then . . . have them count up to and down from 10, and gradually add higher numbers.

Teacher's Note **Thermometer Counting** may also be used to record attendance at the beginning of each day.

2 Engage 30

Concept Building UNDERSTANDING
Linking and Thinking
"Today we are going to learn about graphs."

Follow the instructions on the Activity Card to play **Linking and Thinking**. As children play, ask questions about what is happening in the activity.

Purpose Children map quantities onto a graph and use both numerals and set size to say which have more, less, or the same amount.

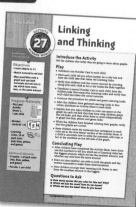
Activity Card 27

Monitoring Student Progress

If . . . children have trouble transferring the quantities to the graph,

Then . . . have them arrange the Learning Links on the desktop to match the graph format before counting the links and coloring the graph.

 For additional practice with numeration in a vertical format, children should complete **Building Blocks** Build Stairs 1.

3 Reflect 10

Extended Response REASONING
Ask questions such as the following:
- **How many red spaces did you color on the graph? How did you know how many to color?** Answers will vary, but children should have colored one space for each red Learning Link.
- **Which color has the most? How do you know?** Answers will vary. Children may compare the size of the colored areas or may compare the numerals on the graph to determine the answer.

Using Student Pages
Have each child complete **Workbook,** page 47. For Question 4 say, "Graph how many of each item you see in Questions 1, 2, and 3." Did the child correctly graph the quantities?

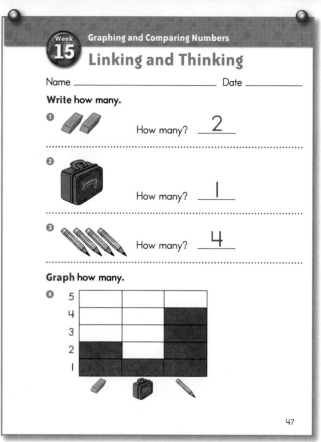

Workbook, p. 47

4 Assess

Informal Assessment
Use the Student Assessment Record, **Assessment,** page 100, to record informal observations.

COMPUTING	UNDERSTANDING
Thermometer Counting	**Linking and Thinking**
Did the child	Did the child
❏ respond accurately?	❏ make important observations?
❏ respond quickly?	❏ extend or generalize learning?
❏ respond with confidence?	❏ provide insightful answers?
❏ self-correct?	❏ pose insightful questions?

Graphing and Comparing Numbers • Lesson 1   197

Week 15 — Graphing and Comparing Numbers
Lesson 2

Objective
Children identify and sequence the numbers 1–20 and create simple addition equations.

Program Materials
Warm Up
Thermometer Picture

Engage
Symbol Cards: Plus Sign and Equal Sign

Additional Materials
Warm Up
Erasable red marker

Engage
- Several sets of small items that can be used for demonstrations
- Pencils and paper

Access Vocabulary
equal Having the same amount
equation Number story

Creating Context
Discuss the concept of equality with children. Organize groups of classroom objects into sets. Make some sets equal and some unequal. Discuss with children what makes two sets equal. Can children think of times when it is important for things to be equal?

1 Warm Up (5)

Concept Building — COMPUTING
Thermometer Counting
Before beginning the small-group activity **Making and Writing Equations,** use the **Thermometer Counting** activity with the whole class.

Purpose Children count from 1 up to 20 and from 20 down to 1.

Warm-Up Card 4

Monitoring Student Progress
If . . . children have trouble remembering the sequence of numbers,
Then . . . model counting for them throughout the day.

2 Engage (30)

Concept Building — ENGAGING
Making and Writing Equations
"Today you are going to write addition number stories."

Follow the instructions on the Activity Card to play **Making and Writing Equations.** As children play, ask questions about what is happening in the activity.

Purpose Children create addition equations using formal notation.

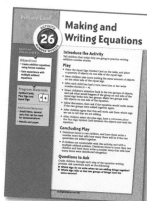
Activity Card 26

198 Number Worlds • Week 15

Monitoring Student Progress

If . . . children have trouble remembering the plus sign and equal sign,

Then . . . post the signs in the classroom and model making equations with them. Invite volunteers to stand in a line in front of the class, and have the rest of the class divide the volunteers into groups to make simple equations.

 Building Blocks For further practice with addition, children should complete **Building Blocks** Barkley's Bones 1.

3 Reflect 10

Extended Response REASONING

Ask questions such as the following:

- **How did you know what number to put on the right of the equal sign?** Accept all reasonable answers. Children should indicate that the number to the right of the equal sign should equal the sum of the numbers left of the equal sign, or that the amount on each side of the equal sign should be the same.

- **How can you check your answer?** Possible answer: count the items on each side of the equal sign.

Using Student Pages

Have each child complete **Workbook,** page 48. Say, "Write the number story to show how many Counters you have in each group and how many you have altogether when you add the Counters together." Did the child write the correct number story?

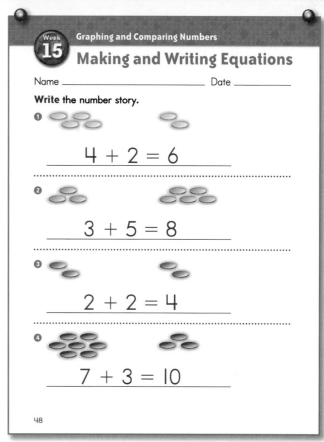

Workbook, p. 48

4 Assess

Informal Assessment

Use the Student Assessment Record, **Assessment,** page 100, to record informal observations.

COMPUTING	ENGAGING
Thermometer Counting Did the child ❏ respond accurately? ❏ respond quickly? ❏ respond with confidence? ❏ self-correct?	**Making and Writing Equations** Did the child ❏ pay attention to the contributions of others? ❏ contribute information and ideas? ❏ improve on a strategy? ❏ reflect on and check accuracy of work?

Graphing and Comparing Numbers • Lesson 2 199

Week 15: Graphing and Comparing Numbers
Lesson 3

Objective

Children identify and sequence the numbers 1–20 and compare quantities on a graph.

Program Materials

Warm Up
Thermometer Picture

Engage
- Sky Land Activity Sheet 1, 1 per child
- Number Cards (8–15)
- Learning Links, 50 of each color (red, blue, yellow, and green)

Additional Materials

Warm Up
Erasable red marker

Engage
- Crayons, 1 of each color (red, blue, yellow, and green)
- Pencils, 1 per child

Access Vocabulary

higher number Number with a greater value when counting up
more A bigger number or amount

Creating Context

In the **Thermometer Counting** activity children are asked to raise their hands if they think another child has made a mistake. Some children may think this is unkind and may not want to participate. Provide alternative ways for children to show they have detected a mistake, such as writing down a skipped or repeated number. Remind children that we learn from mistakes.

1 Warm Up — 5

Skill Building — COMPUTING

Thermometer Counting
Before beginning the small-group activity **Linking and Thinking**, use the **Thermometer Counting** activity with the whole class.

Purpose Children count from 1 up to 20 and from 20 down to 1.

Warm-Up Card 4

Monitoring Student Progress

If . . . children are ready for a challenge, | **Then** . . . play **Thermometer Counting** up to 24.

2 Engage — 30

Concept Building — REASONING

Linking and Thinking
"Today we are going to learn about graphs."

Follow the instructions on the Activity Card to play **Linking and Thining**. As children play, ask questions about what is happening in the activity.

Purpose Children map quantities on a graph and use both numerals and set size to say which have more, less, or the same amount.

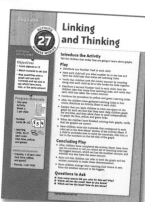

Activity Card 27

200 Number Worlds • Week 15

Monitoring Student Progress

If . . . children can compare quantities on the graph fluently,

Then . . . have them work with larger numbers.

eMathTools Use the Graphing Tool to demonstrate and explore graphs.

 Teacher's Note You might want to repeat this activity, allowing children to use different quantities of each color. As the children become more confident counting and comparing concrete quantities that vary in length (the sets of Learning Links), shift the focus of discussion to a comparison of numerals to judge which sets have the biggest and smallest amounts.

Building Blocks For additional practice with numeration in a vertical format, children should complete **Building Blocks** Build Stairs 1.

3 Reflect 10

Extended Response REASONING
Ask questions such as the following:
- **How would you explain this to someone who has never done it before?** Accept all reasonable answers.
- **What patterns do you see?** Accept all reasonable answers.

Assess

Informal Assessment
Use the Student Assessment Record, **Assessment**, page 100, to record informal observations.

COMPUTING	REASONING
Thermometer Counting	**Linking and Thinking**
Did the child	Did the child
☐ respond accurately?	☐ provide a clear explanation?
☐ respond quickly?	☐ communicate reasons and strategies?
☐ respond with confidence?	☐ choose appropriate strategies?
☐ self-correct?	☐ argue logically?

Graphing and Comparing Numbers • Lesson 3   **201**

Week 15: Graphing and Comparing Numbers

Lesson 4

Objective
Children identify and sequence the numbers 1–20 and create simple addition equations.

Program Materials

Warm Up
Thermometer Picture

Engage
Symbol Cards: Plus Sign and Equal Sign

Additional Materials

Warm Up
Erasable red marker

Engage
- Several sets of small items that can be used for demonstrations
- Pencils and paper

Access Vocabulary
symbol Letter, number, or picture that has a special meaning or that stands for something else
solve To figure out; to find the answer

Creating Context
English Learners often understand math concepts better when they can interact with other children who speak the same primary language. Allow children to work with partners or cross-age helpers who speak the same primary language so that English Learners can check understanding with one another in team situations.

1 Warm Up — 5

Concept Building COMPUTING
Thermometer Counting
Before beginning the small-group activity **Making and Writing Equations**, use the **Thermometer Counting** activity with the whole class.

Purpose Children count from 1 up to 20 and from 20 down to 1.

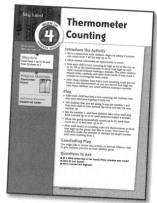

Warm-Up Card 4

Monitoring Student Progress
If . . . children are counting fluently,
Then . . . gradually increase the numbers to 24.

2 Engage — 30

Concept Building APPLYING
Making and Writing Equations
"Today you are going to write addition number stories."

Follow the instructions on the Activity Card to play **Making and Writing Equations**. As children play, ask questions about what is happening in the activity.

Purpose Children create addition equations using formal notation.

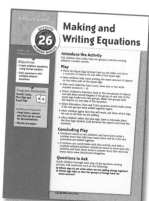

Activity Card 26

Monitoring Student Progress

If . . . children are comfortable creating and writing two-addend equations,

Then . . . challenge them to create equations with three or more addends.

 Teacher's Note Encourage children to use this week's vocabulary words as they engage in the activities, discuss math concepts, and make predictions.

Building Blocks For further practice with addition, children should complete **Building Blocks** Barkley's Bones 1.

3 Reflect   10

Extended Response   REASONING

Ask questions such as the following:

- **How would you explain this to someone who has never done it before?** Accept all reasonable answers. Children should indicate that the amounts on the left and the right of the equal sign must be the same.
- **Can you think of other times at school or at home when you add?** Accept all reasonable answers.

4 Assess

Informal Assessment

Use the Student Assessment Record, **Assessment,** page 100, to record informal observations.

COMPUTING	APPLYING
Thermometer Counting Did the child ❑ respond accurately? ❑ respond quickly? ❑ respond with confidence? ❑ self-correct?	**Making and Writing Equations** Did the child ❑ apply learning to new situations? ❑ contribute concepts? ❑ contribute answers? ❑ connect mathematics to the real world?

Graphing and Comparing Numbers • Lesson 4   203

Week 15: Graphing and Comparing Numbers

Lesson 5
Review

Objective
Children review the material presented in Week 15 of *Number Worlds*.

Program Materials
Warm Up
Thermometer Picture

Additional Materials
Warm Up
Erasable red marker

Creating Context
Although it is important to ask beginning English Learners questions that have a lower language demand, this does not mean that questions should have a lower cognitive load. Ask English Learners to demonstrate their understanding with manipulatives, models, and illustrations. Check understanding by asking children to show, demonstrate, or point to the answer.

1 Warm Up 5

Skill Practice — COMPUTING
Thermometer Counting
Before beginning the **Free-Choice** activity, use the **Thermometer Counting** activity with the whole class.

Purpose Children count from 1 up to 20 and from 20 down to 1.

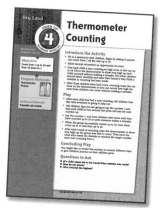

Warm-Up Card 4

Monitoring Student Progress
| If . . . children are counting fluently, | Then . . . gradually increase the numbers to 24. |

2 Engage 20

Concept Building — APPLYING
Free-Choice Activity

For the last day of the week, allow children to choose an activity from the previous weeks. Some activities they may choose are:
- Line Land: **Steve's New Bike**
- Line Land: **Secret Number Game**
- Picture Land: **Making Equations for Beginners**

Make a note of the activities children select. Do they prefer easy or challenging activities? Continue to provide Challenge opportunities for children who have mastered the basic activities.

Monitoring Student Progress
| If . . . children would benefit from extra practice on specific skills, | Then . . . choose an activity for them. |

204 Number Worlds • Week 15

 **Reflect**

Extended Response REASONING
Ask questions such as the following:
- **What did you like about playing** Linking and Thinking?
- **Was there anything about playing this game you didn't like?**
- **Did this game help you do something you couldn't do before? What did it help you do?**
- **What was easy when you were playing** Making and Writing Equations?
- **What was hard when you were playing** Making and Writing Equations?
- **What do you remember most about this game?**
- **What was your favorite part?**

Using Student Pages
Have each child complete **Workbook,** page 49. For Question 4 say, "Graph how many Counters you see in Questions 1, 2, and 3." Is the child correctly applying the skills learned in Week 15?

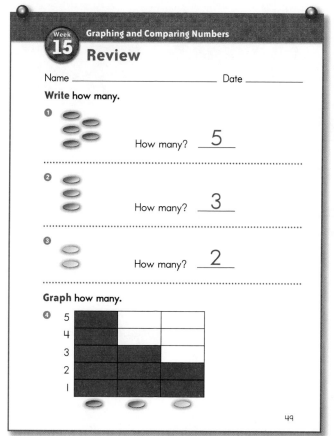

Workbook, p. 49

 Assess

A Gather Evidence
Formal Assessment
Have children complete the weekly test on *Assessment,* page 53. Record formal assessment scores on the Student Assessment Record, *Assessment,* page 100.

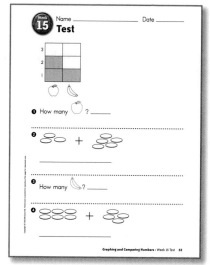

Assessment, p. 53

B Summarize Findings
Review the Student Assessment Records. Determine whether children have Minimal, Basic, or Secure understanding of the concepts presented in Week 15.

C Differentiate Instruction
Based on your observations, use these teaching strategies next week to follow up.

Minimal Understanding
- Repeat the Warm-Up and Engage activities in subsequent weeks to develop concepts of counting objects.
- Repeat the activities from this week. Observe children's understanding of these concepts, and reinforce as necessary.
- Use **Building Blocks** computer activities beginning with Count to develop and reinforce numeration concepts.

Basic Understanding
- Repeat Engage activities in subsequent weeks to reinforce basic concepts.
- Use **Building Blocks** computer activities beginning with Number Compare 2 to reinforce numeration concepts.

Secure Understanding
- Use Challenge variations of this week's activities.
- Use **Building Blocks** computer activities beginning with Barkley's Bones 2.

Graphing and Comparing Numbers • Lesson 5 Review **205**

Week 16: More Counting and Adding

Week at a Glance

This week, children begin **Number Worlds,** Week 16, and continue to explore the relationships between Sky Land, Line Land, and Picture Land.

Skills Focus
- Count to 20
- Add whole numbers
- Use math symbols
- Understand the composition of numbers

Background

In Line Land, an increase in quantity is shown as forward movement along a horizontal number line. In Sky Land, an increase in quantity corresponds to upward movement on a vertical scale, and in Picture Land, it corresponds to a larger dot set.

Teaching for Understanding

As children engage in these activities, they will gradually come to understand that forward movement on a number line, upward movement on a vertical scale, and an increase in set size are ways of representing an increase in quantity, which can be described by an addition equation.

Observe closely while evaluating the Engage activities for this week.

- Can children count on fluently from a given number?
- Are children correctly using math symbols?
- Can children correctly add two whole numbers?

How Children Learn

As children begin this week's lesson, they should understand the concept of plus and equals and should have experience viewing and writing equations using formal notation.

At the end of this week, children should associate addition with counting on and counting up, and should be able to write two-addend equations using formal notation.

Math at Home

Give one copy of the Letter to Home, page A16, to each child. Complete the activity in class, and then encourage children to share it with their caregivers.

Letter to Home, Teacher Edition, p. A16

Math Vocabulary

add To put together
equation Number story
plus sign Symbol that means "add"
equal sign Symbol that means "having the same amount"

English Learners

SPANISH COGNATES

English	Spanish
concepts	conceptos
explore	explorar
result	resultado
scale	escala
thermometer	termómetro

ALTERNATE VOCABULARY

secret Not known

Week 16 Planner: More Counting and Adding

PACING	LESSON	LEARNING GOALS	MATERIALS	TECHNOLOGY
DAY 1	Warm Up 4 **Sky Land:** Thermometer Counting	Children associate counting up and down with movement up and down a vertical scale.	**Program Materials** Thermometer Picture **Additional Materials** Red erasable marker	**Building Blocks** Counting Activity 3 Road Race: Shape Counting
	Activity 28 **Line Land:** Secret Shopping Game*	Children associate addition with counting on and write two-addend equations.	**Program Materials** • Magnetic Number Line, 1 per child • Magnetic Chips • Number 1–6 Cube **Additional Materials** 5 envelopes labeled with a number 1–20, containing pictures of a shopping item	
DAY 2	Warm Up 4 **Sky Land:** Thermometer Counting	Children associate counting up and down with movement on a vertical scale.	**Program Materials** Thermometer Picture **Additional Materials** Red erasable marker	**Building Blocks** Barkley's Bones 2
	Activity 29 **Picture Land:** Changing Addends*	Children use formal notation to write two-addend equations.	**Program Materials** • Counters • Picture Land Activity Sheet 3, 1 per child **Additional Materials** • Two-color box top • Red and blue crayons, pencils	**e MathTools** Set Tool
DAY 3	Warm Up 4 **Sky Land:** Thermometer Counting	Children associate counting up and down with movement on a vertical scale.	**Program Materials** Thermometer Picture **Additional Materials** Red erasable marker	**Building Blocks** Road Race: Shape Counting
	Activity 28 **Line Land:** Secret Shopping Game*	Children associate addition with counting on and write two-addend equations.	Same as Lesson 1	
DAY 4	Warm Up 4 **Sky Land:** Thermometer Counting	Children associate counting up and down with movement on a vertical scale.	**Program Materials** Thermometer Picture **Additional Materials** Red erasable marker	**Building Blocks** Barkley's Bones 2
	Activity 29 **Picture Land:** Changing Addends*	Children use formal notation to write two-addend equations.	Same as Lesson 2	
DAY 5	Warm Up 4 **Sky Land:** Thermometer Counting	Children associate counting up and down with movement on a vertical scale.	**Program Materials** Thermometer Picture **Additional Materials** Red erasable marker	**Building Blocks** Review previous activities
	Review and Assess	Children review favorite activities.	Materials will be selected from those used in previous weeks.	

* Includes Challenge Variations

Week 16 — More Counting and Adding
Lesson 1

Objective
Children practice counting up and down a vertical scale and associate addition with counting on.

Program Materials
Warm Up
Thermometer Picture

Engage
- Magnetic Number Line, 1 per child
- Magnetic Chips
- Number 1–6 Cube

Additional Materials
Warm Up
Red erasable marker

Engage
5 envelopes containing pictures of a shopping item, each labeled with a number between 1 and 20

Access Vocabulary
thermometer Tool used to measure how hot or cold something is
secret Not known

Creating Context
Many children have experience shopping with adults, but English Learners may not know the vocabulary for the products. Using pictures or props, review the names of some common items. For beginning English Learners, provide a sentence frame such as "I want _____." For more proficient English Learners, model the conditional tense: "I would pick _____ and _____."

1 Warm Up 5

Skill Building COMPUTING
Thermometer Counting
Before beginning the small-group activity **Secret Shopping Game**, use the **Thermometer Counting** activity with the whole group.

Purpose Children associate counting up and down with movement up and down a vertical scale.

Warm-Up Card 4

Monitoring Student Progress
If . . . children have trouble counting down from 10,
Then . . . practice counting down with them throughout the day.

Teacher's Note You can use the **Thermometer Counting** activity to record attendance at the beginning of each day.

2 Engage 30

Concept Building UNDERSTANDING
Secret Shopping Game
"Today we will see who can collect the most items from a secret shopping list."

Follow the instructions on the Activity Card to play **Secret Shopping Game**. As children play, ask questions about what is happening in the activity.

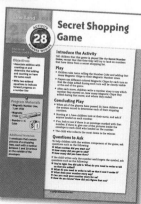
Activity Card 28

Purpose Children associate addition with counting on and learn that adding and counting on have the same result.

Monitoring Student Progress

| If... children make mistakes when writing equations, | Then... remind them that the numbers on both sides of the equal sign must make the same amount. Encourage children to recount the Magnetic Chips on the Magnetic Number Line to check their answers. |

 Teacher's Note Instead of using a shopping theme, you can create a collection game using information from another curriculum. Children might collect pictures of healthful foods, the planets of the solar system, parts of a plant, and so on.

 Building Blocks For further practice with addition, children should complete **Building Blocks** Road Race: Shape Counting.

3 Reflect 10

Extended Response REASONING

Ask questions such as the following:

- **What is this game about?** Accept all reasonable answers. Encourage children to note that the game involves both adding Magnetic Chips on the Magnetic Number Line and writing equations to describe what happened.
- **Who went the farthest?** The child that ended on the highest number went the farthest.
- **Who went the shortest distance?** The child that ended on the lowest number went the shortest distance. **How do you know?** Higher numbers are farther from zero than lower numbers.

Using Student Pages

Have each child complete **Workbook,** page 50. Say, "Write the number story to describe what happens when the Counter moves from where it is to the spot shown by the arrow." Did the child write the correct number story?

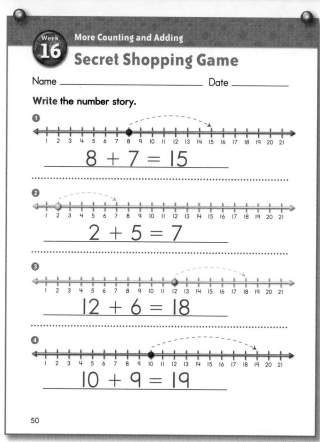

Workbook, p. 50

Assess

Informal Assessment

Use the Student Assessment Record, **Assessment,** page 100, to record informal observations.

COMPUTING	UNDERSTANDING
Thermometer Counting	**Secret Shopping Game**
Did the child	Did the child
❏ respond accurately?	❏ make important observations?
❏ respond quickly?	❏ extend or generalize learning?
❏ respond with confidence?	❏ provide insightful answers?
❏ self-correct?	❏ pose insightful questions?

More Counting and Adding • Lesson 1

Week 16 — More Counting and Adding
Lesson 2

Objective
Children continue to practice counting up and down on a vertical scale and writing two-addend equations.

Program Materials
Warm Up
Thermometer Picture

Engage
- Counters
- Picture Land Activity Sheet 3, 1 per child
- Picture Land Activity Sheets 4–8 for a Challenge

Additional Materials
Warm Up
Red erasable marker

Engage
- Two-color box top
- Red and blue crayons for each child
- Pencils, 1 per child

Access Vocabulary
altogether In all; total
figure out To think about a problem and solve it; find the answer

Creating Context
Some English Learners may know what the plus and equal signs mean, but may not know how to articulate this understanding. Be sure to include questions that let children show comprehension by pointing or by manipulating objects to demonstrate meaning. Model a sentence frame for children to use as they learn to talk about equations: "____ + ____ = ____."

1 Warm Up

Concept Building — COMPUTING
Thermometer Counting
Before beginning the small-group activity **Changing Addends**, use the **Thermometer Counting** activity with the whole group.

Purpose Children associate counting up and down with movement up and down a vertical scale.

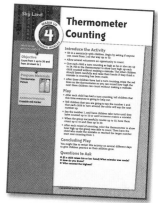
Warm-Up Card 4

Monitoring Student Progress
If . . . children have trouble counting down from 10,
Then . . . practice counting down with them throughout the day.

2 Engage

Concept Building — ENGAGING
Changing Addends
"Today we are going to talk about the number 5 and all of the numbers that can be added together to make 5."

Follow the instructions on the Activity Card to play **Changing Addends**. As children play, ask questions about what is happening in the activity.

Purpose Children practice using formal notation to write two-addend equations.

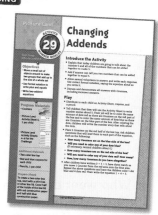
Activity Card 29

Monitoring Student Progress

If . . . children have trouble remembering to use the plus and equal signs,

Then . . . post the signs in the classroom for children to reference. Do some equations together as a class, emphasizing the use of the plus and equal signs.

 MathTools Use the Set Tool to demonstrate and explore counting.

Teacher's Note This activity makes a good introduction to **Three Area Counter Drop** (Activity Card 30) and **Counter Dropping** (Activity Card 31).

Building Blocks For further practice with addition, children should complete **Building Blocks** Barkley's Bones 2.

3 Reflect   10

Extended Response   REASONING

Ask questions such as the following:

- **How many blue dots did you have? How many red dots did you need to make 5?** Answers will vary.
- **How did you figure this out?** Children may count up from the number of blue Counters to 5 to find the missing addend.

Using Student Pages

Have each child complete **Workbook,** page 51. Say, "Write the number story to show how many Counters you have in each group and how many you have altogether when you add the Counters together." Did the child write the correct number story?

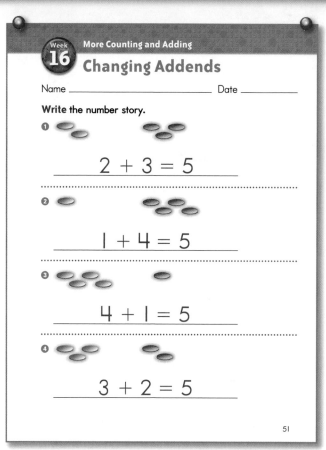

Workbook, p. 51

4 Assess

Informal Assessment

Use the Student Assessment Record, **Assessment,** page 100, to record informal observations.

COMPUTING	ENGAGING
Thermometer Counting	**Changing Addends**
Did the child	Did the child
❏ respond accurately?	❏ pay attention to the contributions of others?
❏ respond quickly?	
❏ respond with confidence?	❏ contribute information and ideas?
❏ self-correct?	
	❏ improve on a strategy?
	❏ reflect on and check accuracy of work?

More Counting and Adding • Lesson 2   **211**

Week 16 — More Counting and Adding
Lesson 3

Objective
Children will continue to practice counting up and down a vertical scale and will associate addition with counting on.

Program Materials
Warm Up
Thermometer Picture

Engage
- Magnetic Number Line, 1 per child
- Magnetic Chips
- Number 1–6 Cube

Additional Materials
Warm Up
Red erasable marker

Engage
5 envelopes containing pictures of a shopping item, each labeled with a number between 1 and 20

Access Vocabulary
add To put together
plus sign Symbol that means "add"

Creating Context
In the **Secret Shopping Game,** the player with the most items wins. *Most* means "more than anyone else." With the class, brainstorm other ways to describe *most* (the top number, the greatest amount, and so on). Can children think of a game in which the person with the least or fewest wins?

1 Warm Up 5

Skill Building COMPUTING
Thermometer Counting
Before beginning the small-group activity **Secret Shopping Game,** use the **Thermometer Counting** activity with the whole group.

Purpose Children associate counting up and down with movement up and down a vertical scale.

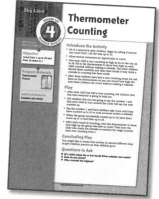
Warm-Up Card 4

Monitoring Student Progress
If . . . children are counting up and down fluently to 10,
Then . . . challenge them to count to 15.

2 Engage 30

Concept Building REASONING
Secret Shopping Game
"Today we will see who can collect the most items from a secret shopping list."

Follow the instructions on the Activity Card to play **Secret Shopping Game.** As children play, ask questions about what is happening in the activity.

Purpose Children associate addition with counting on and learn that adding and counting have the same result.

Activity Card 28

212 Number Worlds • Week 16

Monitoring Student Progress

If . . . children can fluently write their equations,

Then . . . have them do the Challenge variation by writing the equation before they add the chips to the Magnetic Number Line.

Teacher's Note Encourage children to use this week's vocabulary words as they engage in activities, discuss math concepts, and make predictions.

 For further practice with addition, children should complete **Building Blocks** Road Race: Shape Counting.

3 Reflect — 10

Extended Response REASONING

Ask questions such as the following:

- **How would you explain this to someone who had never done it before?** Accept all reasonable answers. Children may indicate that you count the new Magnetic Chips after each turn to find the total number of chips.

- **How did you solve this problem? Could you have solved it another way?** Children may say that they just know the answer, that they counted on from their last Magnetic Chip, or that they counted all the Magnetic Chips from 1 to get the answer.

4 Assess

Informal Assessment

Use the Student Assessment Record, **Assessment,** page 100, to record informal observations.

COMPUTING	REASONING
Thermometer Counting	**Secret Shopping Game**
Did the child	Did the child
❏ respond accurately?	❏ provide a clear explanation?
❏ respond quickly?	❏ communicate reasons and strategies?
❏ respond with confidence?	
❏ self-correct?	❏ choose appropriate strategies?
	❏ argue logically?

More Counting and Adding • Lesson 3 213

Week 16 — More Counting and Adding

Objective
Children continue to practice counting up and down on a vertical scale and writing two-addend equations.

Program Materials
Warm Up
Thermometer Picture

Engage
- Counters
- Picture Land Activity Sheet 3, 1 per child
- Picture Land Activity Sheets 4–8 for a Challenge

Additional Materials
Warm Up
Red erasable marker

Engage
- Two-color box top
- Red and blue crayons for each child
- Pencils, 1 per child

Access Vocabulary
equal sign Symbol that means "having the same amount"
equation Number story

Creating Context
Invite children to use classroom objects to practice counting fluently. For example, they can link together paper clips and count them, or they can collect pencils or crayons in order to have many items to count. Before children begin counting, have each child say the name of the object he or she is going to count.

1 Warm Up

Concept Building — COMPUTING
Thermometer Counting
Before beginning the small-group activity **Changing Addends**, use the **Thermometer Counting** activity with the whole group.

Purpose Children associate counting up and down with movement up and down a vertical scale.

Warm-Up Card 4

Monitoring Student Progress
If . . . children are fluently counting up and down to 10 or 15,

Then . . . challenge them to count to 20. Invite volunteers to color the thermometer.

2 Engage

Concept Building — APPLYING
Changing Addends
"Today we are going to talk about the number 5 and all of the numbers that can be added together to make 5."

Follow the instructions on the Activity Card to play **Changing Addends**. As children play, ask questions about what is happening in the activity.

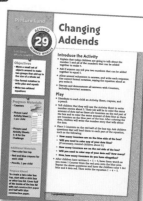

Activity Card 29

Purpose Children practice using formal notation to write two-addend equations.

214 Number Worlds • Week 16

Monitoring Student Progress

If . . . children can write their equations fluently,

Then . . . introduce other numbers by using the Challenge variation of the activity.

Teacher's Note Children will notice the pattern that is created if they have consistently filled in the same color first. You might want to ask children to discuss the pattern they have created. Give each child a chance to describe the pattern in his or her own words.

Building Blocks For further practice with addition, children should complete **Building Blocks** Barkley's Bones 2.

3 Reflect 10

Extended Response REASONING

Ask questions such as the following:

- **What patterns do you see?** As the set of blue dots gets bigger by 1, the set of red dots gets smaller by 1.
- **What would happen if we found all of the numbers that can make 6? Would the pattern change?** Children may note that although the sets would be bigger, the pattern would be the same (as blue dots increase by 1, red dots decrease by 1). If children cannot answer this question, have them complete Activity Sheet 4 to discover the answer.

4 Assess

Informal Assessment

Use the Student Assessment Record, **Assessment,** page 100, to record informal observations.

COMPUTING	APPLYING
Thermometer Counting	**Changing Addends**
Did the child	Did the child
❑ respond accurately?	❑ apply learning to new situations?
❑ respond quickly?	❑ contribute concepts?
❑ respond with confidence?	❑ contribute answers?
❑ self-correct?	❑ connect mathematics to the real world?

More Counting and Adding • Lesson 4

Week 16 — More Counting and Adding

Lesson 5: Review

Objective
Children review the material presented in Week 16 of *Number Worlds*.

Program Materials
Warm Up
Thermometer Picture

Additional Materials
Warm Up
Red erasable marker

Creating Context
In the Thermometer Counting activity, children are asked to raise their hands if they think another child has made a mistake. Some children may think this is unkind and may not want to participate. Provide alternative ways for children to show they have detected a mistake, such as writing down a skipped or repeated number.

1 Warm Up — 5

Skill Practice COMPUTING
Thermometer Counting
Before beginning the **Free-Choice** activity, use the **Thermometer Counting** activity with the whole group.

Purpose Children associate counting up and down with movement up and down a vertical scale.

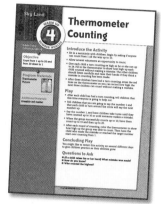

Warm-Up Card 4

Monitoring Student Progress

| If . . . children can count using the Thermometer Picture, | Then . . . challenge them to count on other vertical scales. For example, children could determine how tall they are using either standard or nonstandard units of measure. |

2 Engage — 20

Concept Building APPLYING
Free-Choice Activity

For the last day of the week, allow children to choose an activity from the previous weeks. Some activities they may choose include the following:
- Object Land: **Add Them Up**
- Picture Land: **Making Equations for Beginners**
- Picture Land: **Making and Writing Equations**

Do they prefer easy or challenging activities? Continue to provide Challenge opportunities for children who have mastered the basic activities.

Monitoring Student Progress

| If . . . children would benefit from extra practice on specific skills, | Then . . . choose an activity for them. |

Reflect   10

Extended Response REASONING
Ask questions such as the following:
- What did you like about playing Secret Shopping Game?
- Did this game help you do something you couldn't do before? What did it help you do?
- What was easy when you were playing Changing Addends?
- What was your favorite part?

Using Student Pages
Have each child complete **Workbook,** page 52. For Questions 1 through 3 say, "Write the number story to describe what happens when the Counter moves from where it is to the spot shown by the arrow." For Question 4 say, "Write the number story to show how many Counters you have in each group and how many you have when you add them together." Is the child correctly applying the concepts learned in Week 16?

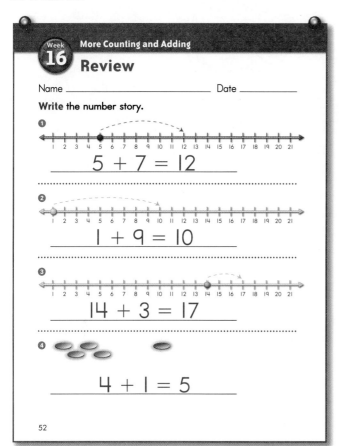

Workbook, p. 52

 ## Assess   10

A Gather Evidence
Formal Assessment
Have students complete the weekly test on **Assessment,** page 55. Record formal assessment scores on the Student Assessment Record, **Assessment,** page 100.

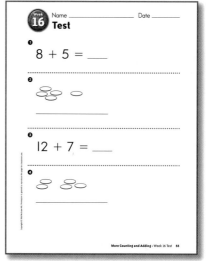

Assessment, p. 55

B Summarize Findings
Review the Student Assessment Records. Determine whether students have Minimal, Basic, or Secure understanding of the concepts presented in Week 16.

C Differentiate Instruction
Based on your observations, use these teaching strategies next week to follow up.

Minimal Understanding
- Repeat the Warm-Up and Engage activities in subsequent weeks to develop number and addition concepts.
- Repeat the activities from this week. Observe children's understanding of these concepts and reinforce as necessary.
- Use **Building Blocks** computer activities beginning with Counting Activity 1 to develop and reinforce numeration and addition concepts.

Basic Understanding
- Repeat Engage activities in subsequent weeks to reinforce basic concepts.
- Use **Building Blocks** computer activities beginning with Counting Activity 3 to reinforce numeration and addition concepts.

Secure Understanding
- Use Challenge variations of this week's activities.
- Use **Building Blocks** computer activities beginning with Barkley's Bones 2.

More Counting and Adding • Lesson 5 Review

Week 17: Solving Equations

Week at a Glance

This week, children begin **Number Worlds,** Week 17, and continue to explore Object Land and Picture Land.

Background

In Picture Land, numbers are represented by numerals, which enables children to forge links between the world of real quantities (objects) and the world of formal symbols.

How Children Learn

As children begin this week's lesson, they should associate addition with counting on and counting up and should be able to write two-addend equations using formal notation.

At the end of this week, children should be able to use formal notation to write equations with two or more addends.

Skills Focus

- Compare and order numbers
- Add whole numbers
- Use math symbols
- Understand equality

Teaching for Understanding

As children engage in these activities, they will come to realize the correspondence between set size and addition.

Observe closely while evaluating the Engage activities assigned for this week.

- Are children correctly adding whole numbers?
- Do children use formal notation when writing addition equations?
- Do children understand the concept of equality?

Math at Home

Give one copy of the Letter to Home, page A17 to each child. Complete the activity in class, and then encourage children to share it with their caregivers.

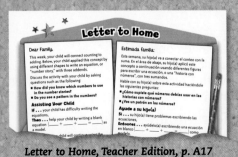

Letter to Home, Teacher Edition, p. A17

Math Vocabulary

plus sign Symbol that means "add"

equation Number story

pattern Something that repeats the same way each time

figure out To think about a problem and solve it; to discover the answer

English Learners

SPANISH COGNATES

English	Spanish
area	área
equations	equaciones
practice	práctica

ALTERNATE VOCABULARY

compare To think about how things are the same and how they are different

Week 17 Planner: Solving Equations

PACING	LESSON	LEARNING GOALS	MATERIALS	TECHNOLOGY
DAY 1	Warm Up 5 Object Land: What Number Am I?*	Children use the spatial terms *before*, *after*, and *between* with the number sequence.	**Program Materials** Neighborhood Number Line	**Building Blocks** Road Race: Shape Counting
	Activity 30 Object Land: Three Area Counter Drop*	Children create and solve equations with more than two addends.	**Program Materials** Counters, 10 per team **Additional Materials** • Three-color box top, 1 per team • Paper and pencils	
DAY 2	Warm Up 5 Object Land: What Number Am I?*	Children use the spatial terms *before*, *after*, and *between* with the number sequence.	**Program Materials** Neighborhood Number Line	**Building Blocks** Barkley's Bones 2
	Activity 31 Picture Land: Counter Dropping*	Children write equations with two addends.	**Program Materials** • Counters, 5 per team • Picture Land Activity Sheet 9, 1 per team **Additional Materials** • Two-color box top • Pencils	
DAY 3	Warm Up 5 Object Land: What Number Am I?*	Children use the spatial terms *before*, *after*, and *between* with the number sequence.	**Program Materials** Neighborhood Number Line	**Building Blocks** Road Race: Shape Counting
	Activity 30 Object Land: Three Area Counter Drop*	Children create and solve equations with more than two addends.	Same as Lesson 1	
DAY 4	Warm Up 5 Object Land: What Number Am I?*	Children use the spatial terms *before*, *after*, and *between* with the number sequence.	**Program Materials** Neighborhood Number Line	**Building Blocks** Barkley's Bones 2
	Activity 31 Picture Land: Counter Dropping*	Children write equations with two addends.	Same as Lesson 2	
DAY 5	Warm Up 5 Object Land: What Number Am I?*	Children use the spatial terms *before*, *after*, and *between* with the number sequence.	**Program Materials** Neighborhood Number Line	**Building Blocks** Review previous activities
	Review and Assess	Children review their favorite activities to improve understanding.	Materials will be selected from those used in previous weeks.	

*Includes Challenge Variations

Week 17 — Solving Equations
Lesson 1

Objective
Children use spatial terms with the number sequence and create and solve equations with more than two addends.

Program Materials
Warm Up
Neighborhood Number Line

Engage
Counters, 10 per team

Additional Materials
Engage
- Three-color box top, 1 per team
- Additional three-color box tops, if needed
- Paper and pencils for each team

Access Vocabulary
before In front of; to come first
after Behind; to come later

Creating Context
Sentence frames are useful tools to help English Learners structure a response to a question. In this lesson, children work on equations with more than two addends. Provide the equation frame _____ + _____ + _____ = _____ for children to complete.

---

1 Warm Up 5

Skill Building COMPUTING
What Number Am I?
Before beginning the activity **Three Area Counter Drop,** use the **What Number Am I?** activity with the whole group.

Purpose Children will use the spatial terms *before*, *after*, and *between* with the number sequence.

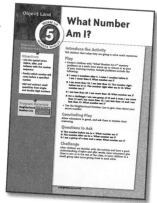

Warm-Up Card 5

Monitoring Student Progress
If . . . children have trouble remembering what clues have been given,

Then . . . give them problems that can be solved with only one clue (for example, "When you're counting by twos, what number comes after 12?").

 Encourage children and families to play this game at home.

2 Engage 30

Concept Building UNDERSTANDING
Three Area Counter Drop
"Today we're going to write and solve number stories."

Follow the instructions on the activity card to play **Three Area Counter Drop.** As children play, ask questions about what is happening in the activity.

Purpose Children will create and solve equations with more than two addends.

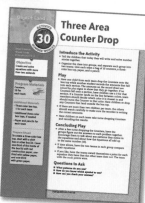

Activity Card 30

220 Number Worlds • Week 17

Monitoring Student Progress

If . . . children have trouble remembering to use a 0 when no Counters fall onto a section,

Then . . . write a blank equation on the board or on chart paper (_____ + _____ + _____ = _____) that children can refer to as they do the activity.

 MathTools Use the Number Line Tool to demonstrate and explore number sequence.

 Teacher's Note This activity can be used with one small group or can be used by organizing the whole class into small groups for simultaneous play.

Building Blocks For further practice with addition, children should complete **Building Blocks** Road Race: Shape Counting.

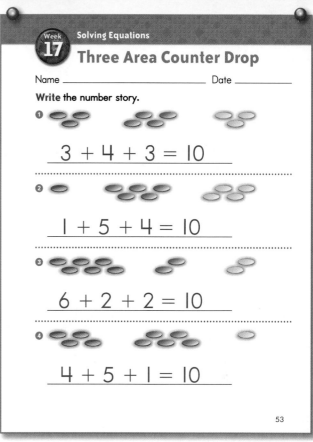

Workbook, p. 53

3 Reflect 10

Extended Response REASONING
Ask questions such as the following:
- **Which number is smaller, the number after 8 or the number before 11?** 9 is smaller than 10. **How did you figure that out?** It comes before 10 when we count.
- **How many Counters landed in the red section? How many more did you need to make 10? How did you figure that out?** Accept all reasonable answers. Children may count on to determine the number of Counters needed to make 10.

Using Student Pages
Have each child complete **Workbook,** page 53. Say, "Write the number story to show how many Counters you have in each group and how many you have when you add them together." Did the child write the correct number story?

4 Assess

Informal Assessment
Use the Student Assessment Record, **Assessment,** page 100, to record informal observations.

COMPUTING	UNDERSTANDING
What Number Am I? Did the child	**Three Area Counter Drop** Did the child
❑ respond accurately?	❑ make important observations?
❑ respond quickly?	❑ extend or generalize learning?
❑ respond with confidence?	❑ provide insightful answers?
❑ self-correct?	❑ pose insightful questions?

Solving Equations • Lesson 1 **221**

Week 17 — Solving Equations
Lesson 2

Objective
Children continue to use spatial terms with the number sequence and write equations with two addends.

Program Materials
Warm Up
Neighborhood Number Line

Engage
- Counters, 5 for each team
- Picture Land Activity Sheet 9, per team

Additional Materials
Engage
- Two-color box top
- Pencils for each team

Access Vocabulary
between In the middle of two things
compare To think about how things are the same and how they are different

Creating Context
Although it is important to ask beginning English Learners questions that have a lower language demand, this does not mean that questions should have a lower cognitive load. Ask English Learners to demonstrate their understanding with manipulatives, models, and illustrations. Check understanding by asking children to show, demonstrate, or point to the answer.

1 Warm Up — 5

Concept Building — COMPUTING
What Number Am I?
Before beginning the small-group activity **Counter Dropping**, use the **What Number Am I?** activity with the whole group.

Purpose Children use the spatial terms *before, after,* and *between* with the number sequence and add and subtract small quantities from one- and two-digit numbers.

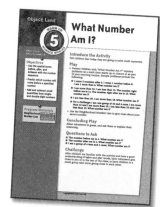

Warm-Up Card 5

Monitoring Student Progress
If . . . children have trouble remembering the clues or solving the problem,

Then . . . give them a paper number line and a pencil so they can mark off eliminated numbers.

2 Engage — 30

Concept Building — ENGAGING
Counter Dropping
"Today we are going to see which team can find the most combinations for the number 5."

Follow the instructions on the Activity Card to play **Counter Dropping**. As children play, ask questions about what is happening in the activity.

Purpose Children practice writing equations with two addends.

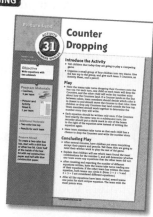

Activity Card 31

222 Number Worlds • Week 17

Monitoring Student Progress

If . . . children have trouble writing correct equations,

Then . . . determine the cause of the problem. Do children have trouble writing numerals or adding numbers to find the sum, or are they simply careless when employing addition strategies or counting the Counters that landed in each area? When you know where the problem lies, give children additional practice to improve skills in this area.

Teacher's Note This activity can easily be used with several small groups at once.

 Building Blocks For further practice with addition, children should complete **Building Blocks** Barkley's Bones 2.

3 Reflect — 10

Extended Response APPLYING

Ask questions such as the following:

- **How many Counters landed on the red area? How many more did you need to make 5? How did you figure this out?** Answers will vary. Children may count on from the number of red Counters to determine how many are needed to make 5.

- **How can you check your answer?** Accept all reasonable answers. Children may suggest counting the blue Counters.

Using Student Pages

Have each child complete **Workbook,** page 54. Say, "Write the number story to show how many Counters you have in each group and how many you have when you add them together." Did the child write the correct number story?

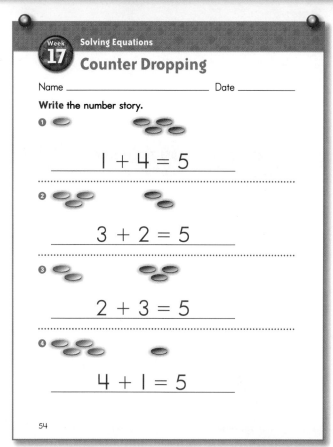

Workbook, p. 54

4 Assess

Informal Assessment

Use the Student Assessment Record, **Assessment,** page 100, to record informal observations.

COMPUTING	ENGAGING
What Number Am I?	**Counter Dropping**
Did the child	Did the child
❏ respond accurately?	❏ pay attention to the contributions of others?
❏ respond quickly?	
❏ respond with confidence?	❏ contribute information and ideas?
❏ self-correct?	
	❏ improve on a strategy?
	❏ reflect on and check accuracy of work?

Solving Equations • Lesson 2   223

Week 17: Solving Equations

Objective
Children gain additional practice creating and solving equations with more than two addends.

Program Materials
Warm Up
Neighborhood Number Line

Engage
Counters, 10 per team

Additional Materials
Engage
- Three-color box top, 1 per team
- Paper and pencils for each team
- Additional three-color box tops, if needed

Access Vocabulary
plus sign Symbol that means "add"
figure out To think about a problem and solve it; discover the answer

Creating Context
One of the Reflect questions for this lesson is "What would happen if there were four areas on the box top?" This grammatical construction will be new for some English Learners, but it is important for them to learn as they consider alternatives and practice critical thinking. Explain that the word *were* lets us talk about possibilities. "What if" questions like this one can be modeled in a sentence frame: "What would happen if _____ were _____?"

1 Warm Up

Skill Building — COMPUTING
What Number Am I?
Before beginning the activity **Three Area Counter Drop,** use the **What Number Am I?** activity with the whole group.

Purpose Children use the spatial terms *before,* *after,* and *between* with the number sequence and add and subtract small quantities from one- and two-digit numbers.

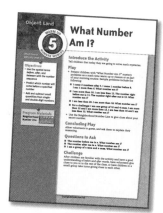

Warm-Up Card 5

Monitoring Student Progress
If . . . children have a good understanding of *before* and *after* concepts,

Then . . . organize the class into groups, and have children take turns giving number clues to each other.

2 Engage

Concept Building — REASONING
Three Area Counter Drop
"Today we're going to write and solve number stories."

Follow the instructions on the Activity Card to play **Three Area Counter Drop.** As children play, ask questions about what is happening in the activity.

Purpose Children will create and solve equations with more than two addends.

Activity Card 30

224 Number Worlds • Week 17

Monitoring Student Progress

| If . . . children can fluently work to 10 with multiple addends, | Then . . . increase the number of Counters to 15. |

 Teacher's Note Encourage children to use this week's vocabulary words as they engage in the activities, discuss math concepts, and make predictions.

 For further practice with addition, children should complete **Building Blocks** Road Race: Shape Counting.

3 Reflect 10

Extended Response REASONING
Ask questions such as the following:
- **What patterns do you see?** Accept all reasonable answers.
- **What would happen if there were four areas on the box top?** Accept all reasonable answers. Children may realize that there would be four numbers (addends) in the equation instead of three or that the numbers would still add up to 10.

4 Assess

Informal Assessment
Use the Student Assessment Record, **Assessment**, page 100, to record informal observations.

COMPUTING	REASONING
What Number Am I? Did the child ❑ respond accurately? ❑ respond quickly? ❑ respond with confidence? ❑ self-correct?	**Three Area Counter Drop** Did the child ❑ provide a clear explanation? ❑ communicate reasons and strategies? ❑ choose appropriate strategies? ❑ argue logically?

Solving Equations • Lesson 3   **225**

Week 17 — Solving Equations
Lesson 4

Objective
Children continue to use spatial terms with the number sequence and to write equations with two addends.

Program Materials
Warm Up
Neighborhood Number Line

Engage
- Counters, 5 for each team
- Picture Land Activity Sheet 9, 1 per team

Additional Materials
Engage
- Two-color box top
- Pencils for each team

Access Vocabulary
pattern Something that repeats the same way each time
equation Number story

Creating Context
English Learners may benefit from modeling the meaning of *pattern* and how to express their findings orally. An excellent strategy for English Learners is to model an example and provide an alternative way for children to respond, such as drawing the pattern.

1 Warm Up — 5

Skill Building — COMPUTING
What Number Am I?
Before beginning the small-group activity **Counter Dropping**, use the **What Number Am I?** activity with the whole group.

Purpose Students will use the spatial terms *before*, *after*, and *between* with the number sequence and will add small quantities using formal notation.

Warm-Up Card 5

Monitoring Student Progress
If . . . children have a good understanding of *before* and *after* concepts,

Then . . . organize the class into groups, and have children take turns giving number clues to each other.

2 Engage — 30

Concept Building — APPLYING
Counter Dropping
"Today we are going to see which team can find the most combinations for the number 5."

Follow the instructions on the Activity Card to play **Counter Dropping**. As children play, ask questions about what is happening in the activity.

Purpose Children practice writing equations with two addends.

Activity Card 31

226 Number Worlds • Week 17

Monitoring Student Progress

| If . . . children are comfortable playing this activity, | Then . . . have them determine the most "popular" drops by completing the Challenge 1 variation. |

 For further practice with addition, children should complete **Building Blocks** Barkley's Bones 2.

3 Reflect 10

Extended Response REASONING

Ask questions such as the following:

- **What patterns do you see?** Answers will vary. Children may note that as one addend gets bigger, the other addend gets smaller.

- **What do you notice about this activity?** Accept all reasonable answers. Children may note that the order of the addends does not change the answer: 1 + 4 = 5 and 4 + 1 = 5.

4 Assess

Informal Assessment

Use the Student Assessment Record, **Assessment**, page 100, to record informal observations.

COMPUTING	APPLYING
What Number Am I?	**Counter Dropping**
Did the child	Did the child
❏ respond accurately?	❏ apply learning to new situations?
❏ respond quickly?	❏ contribute concepts?
❏ respond with confidence?	❏ contribute answers?
❏ self-correct?	❏ connect mathematics to the real world?

Solving Equations • Lesson 4 227

Week 17 — Solving Equations

Lesson 5

Review

Objective
Children review the material presented in Week 17 of *Number Worlds*.

Program Materials
Warm Up
Neighborhood Number Line

Creating Context
Discuss with children multiple meanings of the word *free*. Children may know that *free* means "no cost," but may not know that *free choice* means they can choose which activity they would like to do.

1 Warm Up — 5

Skills Practice COMPUTING
What Number Am I?
Before beginning the **Free-Choice** activity, use the **What Number Am I?** activity with the whole group.

Purpose Children continue to use the spatial terms *before*, *after*, and *between* with the number sequence and to add and subtract small quantities from one- and two-digit numbers.

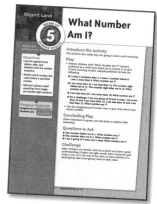

Warm-Up Card 5

Monitoring Student Progress
| If . . . children are comfortable playing the activity, | Then . . . invite volunteers to make up number riddles for the class. |

2 Engage — 20

Concept Building APPLYING
Free-Choice Activity

For the last day of the week, allow children to choose an activity from the previous weeks. Some activities they may choose include the following:
- Object Land: **Add Them Up**
- Picture Land: **Making and Writing Equations**
- Picture Land: **Changing Addends**

Make a note of the activities children select. Do they prefer easy or challenging activities? Continue to provide Challenge opportunities for children who have mastered the basic activities.

Monitoring Student Progress
| If . . . children would benefit from extra practice on specific skills, | Then . . . choose an activity for them. |

228 Number Worlds • Week 17

3 Reflect

Extended Response REASONING
Ask the following questions:
- **What did you like about playing** Three Area Counter Drop?
- **Was there anything about playing this game you didn't like?**
- **Did this game help you do something you couldn't do before? What did it help you do?**
- **What was easy when you were playing** Counter Dropping?
- **What was hard when you were playing** Counter Dropping?
- **What do you remember most about this game?**
- **What was your favorite part?**

Using Student Pages
Have each child complete **Workbook,** page 55. Say, "Write the number story to show how many Counters you have in each group and how many you have when you add them together." Is the child correctly applying concepts learned in Week 17?

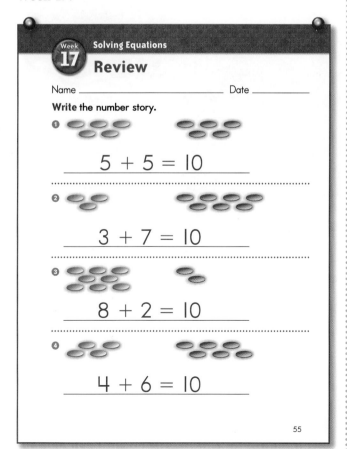

Workbook, p. 55

4 Assess

A Gather Evidence
Formal Assessment
Have students complete the weekly test on **Assessment,** page 57. Record formal assessment scores on the Student Assessment Record, **Assessment,** page 100.

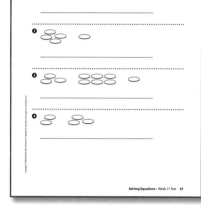

Assessment, p. 57

B Summarize Findings
Review the Student Assessment Records. Determine whether children have Minimal, Basic, or Secure understanding of the concepts presented in Week 17.

C Differentiate Instruction
Based on your observations, use these teaching strategies next week to follow up.

Minimal Understanding
- Repeat the Warm-Up and Engage activities in subsequent weeks to develop number and addition concepts.
- Repeat the activities from this week. Observe children's understanding of these concepts and reinforce as necessary.
- Use **Building Blocks** computer activities beginning with Counting Activity 1 to develop and reinforce numeration and addition concepts.

Basic Understanding
- Repeat Engage activities in subsequent weeks to reinforce basic concepts.
- Use **Building Blocks** computer activities beginning with Counting Activity 3 to reinforce numeration and addition concepts.

Secure Understanding
- Use Challenge variations of this week's activities.
- Use **Building Blocks** computer activities beginning with Barkley's Bones 2.

Solving Equations • Lesson 5 Review 229

Week 18: Adding and Subtracting

Week at a Glance

This week, children begin **Number Worlds**, Week 18, and continue to explore Object Land and Picture Land.

Background

In Picture Land, numbers are represented by numerals, which enable children to forge links between the world of real quantities (objects) and the world of formal symbols.

How Children Learn

As children begin this week's lesson, they should be able to write addition equations with two or more addends using formal notation.

At the end of this week, children should be able to write and solve both addition and subtraction problems using formal notation.

Skills Focus
- Compare and order numbers
- Add whole numbers
- Subtract whole numbers

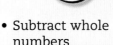

Teaching for Understanding

As children engage in these activities, they will associate subtraction with a decrease in quantity and will gradually become comfortable writing subtraction equations.

Observe closely while evaluating the Engage activities assigned for this week.

- Are children correctly adding and subtracting whole numbers?
- Do children use formal notation when writing addition and subtraction equations?
- Do children understand the concept of equality?

Math at Home

Give one copy of the Letter to Home, page A18, to each child. Complete the activity in class, and then encourage children to share it with their caregivers.

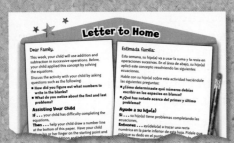

Letter to Home, Teacher Edition, p. A18

Math Vocabulary
add To put together
subtract To take away
plus sign Symbol that means "add"
minus sign Symbol that means "take away"

English Learners

SPANISH COGNATES

English	Spanish
compare	comparar
equations	ecuaciones
group	grupo
numbers	número
subtract	substraer

ALTERNATE VOCABULARY
quest A search or adventure

230 Number Worlds • Week 18

Week 18 Planner: Adding and Subtracting

PACING	LESSON	LEARNING GOALS	MATERIALS	TECHNOLOGY
DAY 1	**Warm Up 5** **Object Land:** What Number Am I?*	Children use the spatial terms *before*, *after*, and *between* with the number sequence.	**Program Materials** Neighborhood Number Line	**Building Blocks** Function Machine 1
	Activity 32 **Picture Land:** Dragon Quest 2*	Children use addition and subtraction in successive operations.	**Program Materials** • Dragon Quest Game Board • Pawns, Spinner • Dragon Quest Cards +1 through +5, −1, and −2 • Picture Land Activity Sheet 10, 1 per child **Additional Materials** Paper and pencils	
DAY 2	**Warm Up 5** **Object Land:** What Number Am I?*	Children use spatial terms with the number sequence.	**Program Materials** Neighborhood Number Line	**Building Blocks** Function Machine 1
	Activity 33 **Object Land:** Shopping Trip	Children add a series of numbers and subtract that amount from a specified quantity.	**Program Materials** • Counters, 10 per child • Multi-Land Activity Sheet 1, 1 per child	
DAY 3	**Warm Up 5** **Object Land:** What Number Am I?*	Children use spatial terms with the number sequence.	**Program Materials** Neighborhood Number Line	**Building Blocks** Function Machine 1
	Activity 32 **Picture Land:** Dragon Quest 2*	Children use addition and subtraction in successive operations.	**Program Materials** Same as Day 1	
DAY 4	**Warm Up 5** **Object Land:** What Number Am I?*	Children use spatial terms with the number sequence.	**Program Materials** Neighborhood Number Line	**Building Blocks** Function Machine 1
	Activity 33 **Object Land:** Shopping Trip	Children add numbers and subtract that amount from a specified quantity.	**Program Materials** • Counters, 10 per child • Multi-Land Activity Sheet 1, for each child	
DAY 5	**Warm Up 5** **Object Land:** What Number Am I?*	Children use spatial terms with the number sequence.	**Program Materials** Neighborhood Number Line	**Building Blocks** Review previous activities
	Review and Assess	Children review their favorite activities to improve understanding.	Materials will be selected from those used in previous weeks.	

Week 18 Adding and Subtracting

Lesson 1

Objective
Children use spatial terms with the number sequence and use addition and subtraction in successive operations.

Program Materials
Warm Up
Neighborhood Number Line

Engage
- Dragon Quest Game Board
- Pawns, 1 per student
- Spinner (1–4)
- Dragon Quest Cards +1 through +5, −1, and −2
- Picture Land Activity Sheet 10, 1 per child

Additional Materials
Engage
Paper and pencils

Prepare Ahead
Make sure the pile of Dragon Quest Cards contains only 1–5 Addition Cards and 1–2 Subtraction Cards. Make sure the top six cards in the facedown pile are addition cards so children will not have to subtract from 0 until they are ready to handle this challenge.

Access Vocabulary
quest A search or adventure
pawn Game piece

Creating Context
Although it is important to ask beginning English Learners questions that have a lower language demand, this does not mean that questions should have a lower cognitive load. Ask English Learners to demonstrate their understanding with manipulatives, models, and illustrations.

1 Warm Up

Skill Building — COMPUTING
What Number Am I?
Before beginning the small-group activity **Dragon Quest 2,** use the **What Number Am I?** activity with the whole group.

Purpose Children will use the spatial terms *before, after,* and *between* with the number sequence.

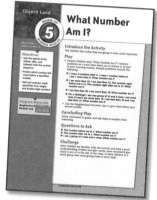

Warm-Up Card 5

Monitoring Student Progress
If . . . children have trouble remembering what clues have been given,

Then . . . give them paper number lines and pencils that they can use to mark off eliminated numbers.

 Encourage children and families to play this game at home.

2 Engage

Concept Building — UNDERSTANDING
Dragon Quest 2: The Dragon's Sister
"In today's game, you will collect buckets of water that you can use to put out a dragon's fire."

Follow the instructions on the Activity Card to play **Dragon Quest 2: The Dragon's Sister.** As children play, ask questions about what is happening in the activity.

Activity Card 32

232 Number Worlds • Week 18

Purpose Children will use addition and subtraction in successive operations.

> **Monitoring Student Progress**
>
> **If . . .** children have trouble doing the subtraction problems,
>
> **Then . . .** allow them to use a number line for reference.

 Building Blocks For further practice with addition and subtraction, children should complete **Building Blocks** Function Machine 1.

3 Reflect — 10

Extended Response REASONING
Ask questions such as the following:

- **What symbol do we use to show we are adding (or subtracting) something?** Plus sign (minus sign).
- **After each turn, how did you figure out how many more buckets of water you needed to put out the dragon's fire?** Children may count on to determine the number of buckets still needed.

Using Student Pages
Have each child complete **Workbook,** page 56. Did the child write the correct answer?

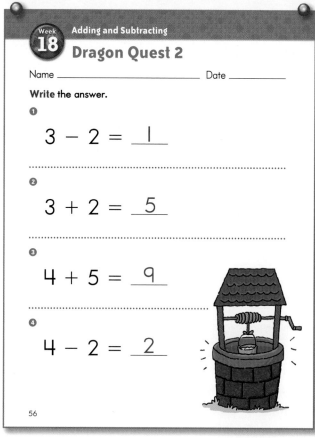

Workbook, p. 56

4 Assess

Informal Assessment
Use the Student Assessment Record, **Assessment,** page 100, to record informal observations.

COMPUTING	UNDERSTANDING
What Number Am I?	**Dragon Quest 2**
Did the child	Did the child
❏ respond accurately?	❏ make important observations?
❏ respond quickly?	❏ extend or generalize learning?
❏ respond with confidence?	❏ provide insightful answers?
❏ self-correct?	❏ pose insightful questions?

Adding and Subtracting • Lesson 1 233

Week 18 — Adding and Subtracting

Lesson 2

Objective
Children continue to use spatial terms with the number sequence and to add and subtract whole numbers.

Program Materials

Warm Up
Neighborhood Number Line

Engage
- Counters, 10 per child
- Multi-Land Activity Sheet 1, 1 per child

Access Vocabulary
before In front of; coming first
after Behind; coming later

Creating Context
Prepositions are not in the beginning level of language proficiency. Help English Learners practice prepositions such as *before*, *after*, and *between* when they line up for recess. Have children identify who is before them in line, who is after them, and who is between two other children.

1 Warm Up   5

Concept Building — COMPUTING

What Number Am I?
Before beginning the small-group activity **Shopping Trip**, use the **What Number Am I?** activity with the whole group.

Purpose Children will use the spatial terms *before*, *after*, and *between* with the number sequence and will add and subtract small quantities from one- and two-digit numbers.

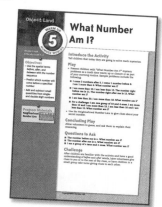

Warm-Up Card 5

Monitoring Student Progress

If . . . children have trouble remembering the clues,

Then . . . give them problems with fewer clues (for example, "I come 2 numbers before 19").

eMathTools Use the Number Line Tool to demonstrate and explore number sequence.

2 Engage   30

Concept Building — ENGAGING

Shopping Trip
"Today we are going to take a pretend shopping trip."

Follow the instructions on the Activity Card to play **Shopping Trip**. As children play, ask questions about what is happening in the activity.

Purpose Children will add together a series of numbers and then subtract that amount from a specified quantity.

Activity Card 33

234 Number Worlds • Week 18

Monitoring Student Progress

If . . . children have trouble remembering how many Counters they have spent,

Then . . . allow them to keep a running tally on a piece of paper.

Teacher's Note This activity can be used again with different shopping "goals." For example, you can ask children to spend all ten Counters, or to spend them as quickly (with the fewest purchases) or as slowly (with the most purchases) as possible. Note that you can and should adjust the amount of Counters each child starts with to meet the needs of your group.

 Building Blocks For further practice with addition, children should complete **Building Blocks** Function Machine 1.

3 Reflect — 10

Extended Response — REASONING

Ask questions such as the following:

- **How did you figure out which items to buy?** Answers will vary.
- **How did you keep track of how many Counters you had spent when you were deciding what to buy next?** Accept all reasonable answers.

Using Student Pages

Have each child complete **Workbook,** page 57. Did the child write the correct number story?

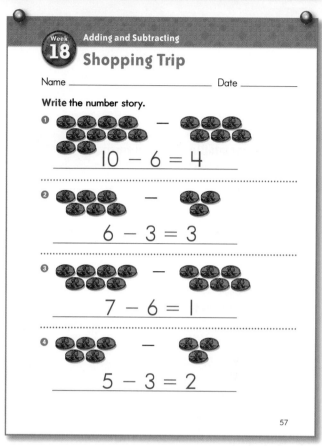

Workbook, p. 57

4 Assess

Informal Assessment

Use the Student Assessment Record, **Assessment,** page 100, to record informal observations.

COMPUTING	ENGAGING
What Number Am I?	**Shopping Trip**
Did the child	Did the child
❏ respond accurately?	❏ pay attention to the contributions of others?
❏ respond quickly?	
❏ respond with confidence?	❏ contribute information and ideas?
❏ self-correct?	
	❏ improve on a strategy?
	❏ reflect on and check accuracy of work?

Adding and Subtracting • Lesson 2 235

Week 18 — Adding and Subtracting

Objective

Children gain additional practice using addition and subtraction in successive operations.

Program Materials

Warm Up
Neighborhood Number Line

Engage
- Dragon Quest Game Board
- Pawns, 1 per student
- Spinner (1–4)
- Dragon Quest Cards +1 through +5, −1, and −2
- Picture Land Activity Sheet 10, 1 per child

Additional Materials

Engage
Paper and pencils

Prepare Ahead

Make sure the pile of Dragon Quest Cards contains only 1–5 Addition Cards and 1–2 Subtraction Cards. Make sure the top six cards in the facedown pile are addition cards so children will not have to subtract from 0 until they are ready to handle this challenge.

Access Vocabulary

add To put together
subtract To take away

Creating Context

English Learners often understand math concepts better when they can interact with other children who speak the same primary language. Allow children to work with partners or cross-age helpers who speak the same primary language so they can check understanding with one another in team situations.

1 Warm Up 5

Skill Building — COMPUTING

What Number Am I?
Before beginning the small-group activity **Dragon Quest 2,** use the **What Number Am I?** activity with the whole group.

Purpose Children use the spatial terms *before, after,* and *between* with the number sequence and add and subtract small quantities from one- and two-digit numbers.

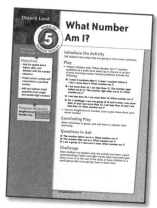

Warm-Up Card 5

Monitoring Student Progress

If . . . children have a good understanding of *before* and *after* concepts,

Then . . . divide the class into groups and have children take turns giving number clues to each other.

2 Engage 30

Concept Building — REASONING

Dragon Quest 2: The Dragon's Sister

"In today's game, you will collect buckets of water you can use to put out a dragon's fire."

Follow the instructions on the Activity Card to play **Dragon Quest 2: The Dragon's Sister.** As children play, ask questions about what is happening in the activity.

Activity Card 32

Purpose Children will use addition and subtraction in successive operations.

Monitoring Student Progress

If . . . children can fluently perform the addition and subtraction required for the activity,

Then . . . have them play one of the Challenge variations.

 Teacher's Note Encourage children to use this week's vocabulary words as they engage in the activities, discuss math concepts, and make predictions.

Building Blocks For further practice with addition, children should complete **Building Blocks** Function Machine 1.

3 Reflect 10

Extended Response REASONING

Ask questions such as the following:

- **How do you know if you've won the game? How can you prove you're the winner?** Children have won when they get at least fifteen buckets of water and land in the dragon's lair. They may check their answers by adding and subtracting their cards a second time or by reviewing the Activity Sheet.

- **Did anyone get a lot of big numbers? Did this help you get a lot of buckets or did it slow you down? Why?** Encourage children to refer to their completed Activity Sheets to explain their answers.

4 Assess

Informal Assessment

Use the Student Assessment Record, **Assessment,** page 100, to record informal observations.

COMPUTING	REASONING
What Number Am I?	**Dragon Quest 2**
Did the child	Did the child
❑ respond accurately?	❑ provide a clear explanation?
❑ respond quickly?	❑ communicate reasons and strategies?
❑ respond with confidence?	
❑ self-correct?	❑ choose appropriate strategies?
	❑ argue logically?

Adding and Subtracting • Lesson 3 237

Week 18: Adding and Subtracting

Lesson 4

Objective

Children continue to use spatial terms with the number sequence and to add and subtract whole numbers.

Program Materials

Warm Up
Neighborhood Number Line

Engage
- Counters, 10 for each child
- Multi-Land Activity Sheet 1, 1 per child

Access Vocabulary

between In the middle of two things
plus sign Symbol that means "add"

Creating Context

Some English Learners may have lived outside the United States and may have different shopping experiences, such as shopping in an open-air market. Ask children to describe a recent shopping trip. Did they have a list of items to buy? Have children create a shopping list by drawing pictures and by writing a numeral to indicate the quantity of each item they would buy.

1. Warm Up

Skill Building — COMPUTING

What Number Am I?

Before beginning the small-group activity **Shopping Trip**, use the **What Number Am I?** activity with the whole group.

Purpose Children will use the spatial terms *before, after,* and *between* with the number sequence and will add and subtract small quantities from one- and two-digit numbers.

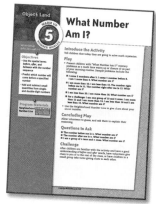

Warm-Up Card 5

Monitoring Student Progress

If . . . children have a good understanding of *before* and *after* concepts,

Then . . . organize the class into groups, and have children take turns giving number clues to each other.

2. Engage

Concept Building — APPLYING

Shopping Trip

"Today we are going to take a pretend shopping trip."

Follow the instructions on the Activity Card to play **Shopping Trip**. As children play, ask questions about what is happening in the activity.

Activity Card 33

Purpose Children add together a series of numbers and subtract that amount from a specified quantity.

238 Number Worlds • Week 18

Monitoring Student Progress

If . . . children can play the game fluently,

Then . . . challenge them to play with fifteen Counters.

 For further practice with addition, children should complete **Building Blocks** Function Machine 1.

 Reflect 10

Extended Response REASONING

Ask questions such as the following:

- **How did you solve this problem? Why did you choose this strategy?** Accept all reasonable answers.
- **Can you think of other times at school or at home when you need to count money?** Accept all reasonable answers.

 Assess

Informal Assessment

Use the Student Assessment Record, **Assessment**, page 100, to record informal observations.

COMPUTING	APPLYING
What Number Am I? Did the child	**Shopping Trip** Did the child
❏ respond accurately?	❏ apply learning to new situations?
❏ respond quickly?	❏ contribute concepts?
❏ respond with confidence?	❏ contribute answers?
❏ self-correct?	❏ connect mathematics to the real world?

Adding and Subtracting • Lesson 4 239

Week 18 — Adding and Subtracting

Lesson 5 Review

Objective
Children review the material presented in Week 18 of *Number Worlds*.

Program Materials
Warm Up
Neighborhood Number Line

Creating Context
Playing games is an excellent way to give English Learners practice listening to and speaking English. Wait time is built into the process, and the low-stakes environment makes the use of games an enjoyable learning tool.

✓ Cumulative Assessment
After Lesson 5 is complete, children should complete the Cumulative Assessment, **Assessment,** pages 91–92. Using the key on **Assessment,** page 90, identify incorrect responses. Reteach and review the suggested activities to reinforce concept understanding.

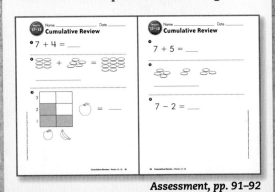

Assessment, pp. 91–92

---

1 Warm Up 5

Skill Practice — COMPUTING
What Number Am I?
Before beginning the **Free-Choice** activity, use the **What Number Am I?** activity with the whole group.

Purpose Children continue to use the spatial terms *before, after,* and *between* with the number sequence and to add and subtract small quantities from one- and two-digit numbers.

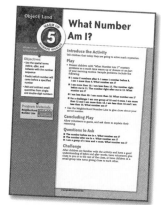

Warm-Up Card 5

Monitoring Student Progress
If . . . children are comfortable playing the activity,

Then . . . invite volunteers to make up number riddles for the class.

---

2 Engage 20

Concept Building — APPLYING
Free-Choice Activity
For the last day of the week, allow children to choose an activity from the previous weeks. Some activities they may choose include the following:
- Line Land: **Secret Shopping Game**
- Object Land: **Three Area Counter Drop**
- Picture Land: **Counter Dropping**

Make a note of the activities children select. Do they prefer easy or challenging activities? Continue to provide Challenge opportunities for children who have mastered the basic activities.

Monitoring Student Progress
If . . . children would benefit from extra practice on specific skills,

Then . . . choose an activity for them.

Reflect 10

Extended Response   REASONING
Ask questions such as the following:
- **What did you like about playing** Dragon Quest 2: The Dragon's Sister?
- **Was there anything about playing this game you didn't like?**
- **Did this game help you do something you couldn't do before? What did it help you do?**
- **What was easy when you were playing** Shopping Trip?
- **What was hard when you were playing** Shopping Trip?
- **What do you remember most about this game?**
- **What was your favorite part?**

Using Student Pages
Have each child complete **Workbook,** page 58. Is the child correctly applying the concepts learned in Week 18?

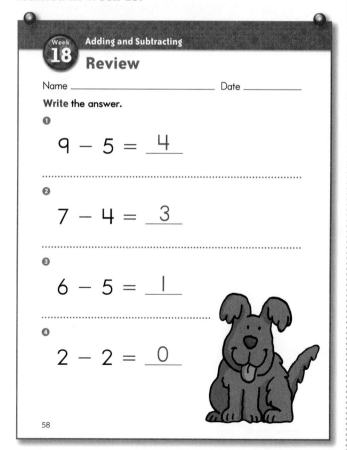

Workbook, p. 58

 ## Assess 10

A Gather Evidence
Formal Assessment
Have children complete the weekly test on *Assessment,* page 59. Record formal assessment scores on the Student Assessment Record, *Assessment,* page 100.

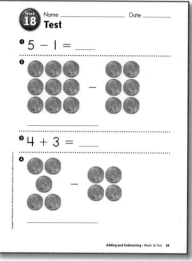

Assessment, p. 59

B Summarize Findings
Review the Student Assessment Records. Determine whether children have Minimal, Basic, or Secure understanding of the concepts presented in Week 18.

C Differentiate Instruction
Based on your observations, use these teaching strategies next week to follow up.

Minimal Understanding
- Repeat the Warm-Up and Engage activities in subsequent weeks to develop concepts of counting objects.
- Repeat the activities from this week. Observe children's understanding of these concepts, and reinforce as necessary.
- Use **Building Blocks** computer activities beginning with Counting Activity 1 to develop and reinforce numeration and addition concepts.

Basic Understanding
- Repeat Engage activities in subsequent weeks to reinforce basic concepts.
- Use **Building Blocks** computer activities beginning with Function Machine 1 to reinforce addition and subtraction concepts.

Secure Understanding
- Use Challenge variations of this week's activities.
- Use **Building Blocks** computer activities beginning with Function Machine 1.

Adding and Subtracting • Lesson 5 Review   **241**

Week 19 Subtracting

Week at a Glance

This week, children begin **Number Worlds**, Week 19, and explore Circle Land and Line Land.

Background

In Circle Land, numbers are represented as points on a dial. When a child is asked "How far around are you now?" he or she might think of the point on the dial, the size of the set that corresponds to that point, or the number that shows the point on the dial and indicates the set size it marks.

How Children Learn

As children begin this week's lesson, they should be able to write and solve both addition and subtraction problems using formal notation.

At the end of the week, children should become more fluent solving subtraction problems involving both single- and double-digit numbers.

Skills Focus
- Compare and order numbers
- Subtract whole numbers
- Use math symbols
- Understand equality

Teaching for Understanding

As children engage in these activities, they will associate subtraction with a decrease in quantity and will gradually become as comfortable writing subtraction equations as they are writing addition equations.

Observe closely while evaluating the Engage activities assigned for this week.

- Are children correctly adding and subtracting whole numbers?
- Do children use formal notation when writing addition and subtraction equations?
- Do children understand the concept of equality?

Math at Home

Give one copy of the Letter to Home, page A19, to each child. Complete the activity in class, and then encourage children to share it with their caregivers.

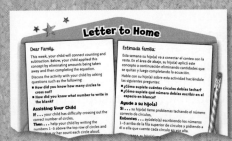

Letter to Home, Teacher Edition, p. A19

Math Vocabulary

count back To begin with a number and count backward, subtracting one each time

subtract To take away

minus sign Symbol that means "take away"

English Learners

SPANISH COGNATES

English	Spanish
number	número
compare	comparar
symbols	símbolos

ALTERNATE VOCABULARY

predict To say what you think will happen

Week 19 Planner Subtracting

PACING	LESSON	LEARNING GOALS	MATERIALS	TECHNOLOGY
DAY 1	Warm Up 6 **Circle Land:** Count Up	Children associate counting up with movement around a dial.	**Program Materials** Circle Land Classroom Dial	**Building Blocks** Function Machine 1
	Activity 34 **Line Land:** Tear It Up and Throw It Away	Children subtract small quantities from single- and double-digit numbers.	**Program Materials** • Dot Cube • Line Land Activity Sheet 2, 1 per child **Prepare Ahead** Cut Activity Sheet 2 into thirds.	
DAY 2	Warm Up 6 **Circle Land:** Count Up	Children associate counting up with movement around a dial.	**Program Materials** Circle Land Classroom Dial	**Building Blocks** Function Machine 1
	Activity 35 **Line Land:** Subtracting Counters	Children subtract small quantities from single- and double-digit numbers and write corresponding equations.	**Program Materials** • Counters, 20 per child • Line Land Activity Sheet 2, 1 per child **Additional Materials** • Cup for each child • Chart paper and marker	
DAY 3	Warm Up 6 **Circle Land:** Count Up	Children associate counting up with movement around a dial.	**Program Materials** Circle Land Classroom Dial	**Building Blocks** Function Machine 1
	Activity 34 **Line Land:** Tear It Up and Throw It Away	Children subtract small quantities from single- and double-digit numbers.	Same as Lesson 1	
DAY 4	Warm Up 6 **Circle Land:** Count Up	Children associate counting up with movement around a dial.	**Program Materials** Circle Land Classroom Dial	**Building Blocks** Function Machine 1
	Activity 35 **Line Land:** Subtracting Counters	Children subtract small quantities and write equations.	Same as Lesson 2	
DAY 5	Warm Up 6 **Circle Land:** Count Up	Children associate counting up with movement around a dial.	**Program Materials** Circle Land Classroom Dial	**Building Blocks** Review previous activities
	Review and Assess	Children review their favorite activities to improve understanding.	Materials will be selected from those used in previous weeks.	

Week 19 Subtracting
Lesson 1

Objective
Children expand number sequence skills and practice subtracting small quantities from single- and double-digit numbers.

Program Materials
Warm Up
Circle Land Classroom Dial

Engage
- Dot Cube
- Line Land Activity Sheet 2, 1 per child

Prepare Ahead
Engage
Cut Activity Sheet 2 into thirds so that each child has three separate number lines to use on successive rounds of this game.

Access Vocabulary
add To put together
subtract To take away

Creating Context
English Learners may find it puzzling that the word *up* is used in many different idiomatic expressions, such as *tear it up*. Explain that *tear it up* means "to rip something into pieces." Create a word bank to list idioms and expressions that use the word *up,* and have children illustrate their meanings.

1 Warm Up   5

Skill Building   COMPUTING
Count Up
Before beginning the small-group activity **Tear It Up and Throw It Away,** use the **Count Up** activity with the whole class.

Purpose Children associate counting up with movement around a dial.

Warm-Up Card 6

Monitoring Student Progress
If . . . children have trouble remembering how many revolutions they have made around the dial,

Then . . . make tally marks on the board to help them keep track.

2 Engage   30

Concept Building   UNDERSTANDING
Tear It Up and Throw It Away

"Today you will use number lines to help you with subtraction."

Follow the instructions on the Activity Card to play **Tear It Up and Throw It Away.** As children play, ask questions about what is happening in the activity.

Purpose Children practice subtracting small quantities from single- and double-digit numbers.

Activity Card 34

244 Number Worlds • Week 19

Monitoring Student Progress

| If . . . children have trouble tearing the numbers, | Then . . . allow them to use safety scissors. If children are still struggling, have them use a pencil to mark the line on which they will cut, and have them discuss their choice before they cut. |

 Teacher's Note Before starting the game, tell children they get to make up a rule about what to do when they get to the end of the number line. Must a child roll the exact amount needed to get to 0, or will any number that is the same or larger than the remaining numbers end the game?

Building Blocks For additional practice with subtraction, children should complete **Building Blocks** Function Machine 1.

3 Reflect — 10

Extended Response REASONING
Ask questions such as the following:
- **How many did you start with? How many did you take away? How many did you have remaining?** Answers will vary.
- **How can you check your answer?** Possible answers: Count the remaining numbers; the highest number remaining on the number line shows how many are remaining.

Using Student Pages
Have each child complete **Workbook**, page 59. Did children write the correct answer?

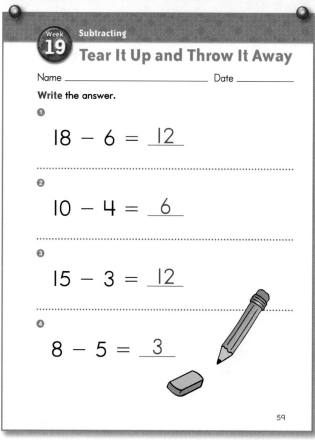

Workbook, p. 59

4 Assess

Informal Assessment
Use the Student Assessment Record, **Assessment**, page 100, to record informal observations.

COMPUTING	UNDERSTANDING
Count Up Did the child ❏ respond accurately? ❏ respond quickly? ❏ respond with confidence? ❏ self-correct?	**Tear It Up and Throw It Away** Did the child ❏ make important observations? ❏ extend or generalize learning? ❏ provide insightful answers? ❏ pose insightful questions?

Subtracting • Lesson 1 245

Week 19 Subtracting
Lesson 2

Objective
Children reinforce counting and number sequence skills, subtract small quantities from single- and double-digit numbers, and write subtraction equations.

Program Materials
Warm Up
Circle Land Classroom Dial
Engage
- Counters, 20 per child
- Line Land Activity Sheet 2, 1 per child

Additional Materials
Engage
- Cup for each child
- Chart paper and marker

Access Vocabulary
count back To begin with a number then count backward, saying a smaller number each time
predict To say what you think will happen

Creating Context
Prepositions are used in many idiomatic expressions in English that may actually have conflicting meanings. In the previous lesson, children heard the phrase *throw it away*, and in this lesson they work with the phrase *take away*. Have children demonstrate each action to see if there is a difference.

1 Warm Up

Skill Building COMPUTING
Count Up
Before beginning the activity **Subtracting Counters**, use the **Count Up** activity with the whole class.

Purpose Children associate counting up with movement around a dial.

Warm-Up Card 6

Monitoring Student Progress
If . . . children have trouble remembering how many revolutions they have made around the dial,

Then . . . make tally marks on the board to help them keep track.

2 Engage

Concept Building ENGAGING
Subtracting Counters
"Today you will use the Counters to help with subtraction."

Follow the instructions on the Activity Card to play **Subtracting Counters**. As children play, ask questions about what is happening in the activity.

Purpose Children subtract small quantities from single- and double-digit numbers and write subtraction equations.

Activity Card 35

Monitoring Student Progress

| If . . . children have trouble subtracting from larger numbers, | Then . . . have them start with only ten Counters. |

 For additional practice with subtraction, children should complete **Building Blocks** Function Machine 1.

3 Reflect · 10

Extended Response REASONING

Ask questions such as the following:

- **If you started with 20 Counters and took away 5, how many Counters would you have left?** 15
- **How did you figure that out?** Accept all reasonable answers. Children may count back or may look at the highest number on the number line after taking away the 5 Counters.

Using Student Pages

Have each child complete **Workbook,** page 60. Did the child write the correct number story?

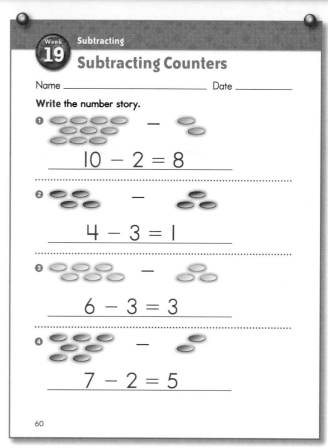

Workbook, p. 60

4 Assess

Informal Assessment

Use the Student Assessment Record, **Assessment,** page 100, to record informal observations.

COMPUTING	ENGAGING
Count Up	**Subtracting Counters**
Did the child	Did the child
❏ respond accurately?	❏ pay attention to the contributions of others?
❏ respond quickly?	
❏ respond with confidence?	❏ contribute information and ideas?
❏ self-correct?	
	❏ improve on a strategy?
	❏ reflect on and check accuracy of work?

Subtracting • Lesson 2

Week 19 Subtracting

Objective
Children reinforce counting and number-sequence skills and practice subtracting small quantities from single- and double-digit numbers.

Program Materials
Warm Up
Circle Land Classroom Dial

Engage
- Dot Cube
- Line Land Activity Sheet 2, 1 per child

Prepare Ahead

Engage
- Cut Activity Sheet 2 into thirds so that each child has three separate number lines to use on successive rounds of this game.

Access Vocabulary
number line A line with the number sequence on it that is used to solve math problems

numeral The written symbol that shows how many

Creating Context
Some English Learners might be confused by numbers described as *higher* or *lower* and *counting up* and *counting down* because these words commonly refer to position. Discuss with children that *higher* and *up* refer to numbers that come later in the counting sequence, and *lower* and *down* refer to numbers that come earlier in the sequence.

1 Warm Up

Skill Building — COMPUTING
Count Up
Before beginning the small-group activity **Tear It Up and Throw It Away**, use the **Count Up** activity with the whole class.

Purpose Children associate counting up with movement around a dial.

Warm-Up Card 6

Monitoring Student Progress
If . . . children have trouble remembering the value of the 0 space when they complete a revolution around the dial,

Then . . . ask them what number comes next when they're counting. Remind them to look at the tally marks on the board to see how many times they have gone around the dial. Because each tally mark indicates 10 spaces, children can count the tally marks by tens to determine the number they should say when they reach 0 again.

2 Engage

Concept Building — REASONING
Tear It Up and Throw It Away

"Today you will use number lines to help you with subtraction."

Follow the instructions on the Activity Card to play **Tear It Up and Throw It Away.** As children play, ask questions about what is happening in the activity.

Activity Card 34

248 Number Worlds • Week 19

Purpose Children practice subtracting small quantities from single- and double-digit numbers.

Monitoring Student Progress

If . . . children are working fluently with the number line to 20,

Then . . . challenge them to write the equation that corresponds with each roll of the Number Cube.

 MathTools Use the Number Line Tool to demonstrate and explore number sequence.

Teacher's Note Encourage children to use this week's vocabulary words as they engage in the activities, discuss math concepts, and make predictions.

Building Blocks For additional practice with subtraction, children should complete **Building Blocks** Function Machine 1.

3 Reflect   10

Extended Response REASONING

Ask questions such as the following:

- **How would you explain this to someone who had never done it before?** Possible answer: You roll the Dot Cube and take away that many from the number line.
- **Can you think of other times at school or at home when you subtract?** Accept all reasonable answers.

4 Assess

Informal Assessment

Use the Student Assessment Record, **Assessment,** page 100, to record informal observations.

COMPUTING	REASONING
Count Up	**Tear It Up and Throw It Away**
Did the child	Did the child
❏ respond accurately?	❏ provide a clear explanation?
❏ respond quickly?	❏ communicate reasons and strategies?
❏ respond with confidence?	❏ choose appropriate strategies?
❏ self-correct?	❏ argue logically?

Subtracting • Lesson 3   **249**

Week 19 Subtracting
Lesson 4

Objective
Children reinforce counting and number-sequence skills, subtract small quantities from single- and double-digit numbers, and write subtraction equations.

Program Materials
Warm Up
Circle Land Classroom Dial

Engage
- Counters, 20 per child
- Line Land Activity Sheet 2, 1 per child

Additional Materials
Engage
- Cup for each child
- Chart paper and marker

Access Vocabulary
minus sign Symbol that means "take away"
solve To figure out; to find the answer

Creating Context
As children work with the number line, they will be asked to work with bigger numbers. Keep in mind that some English Learners may think of larger-sized numerals rather than an increased quantity. When you use a term such as *bigger numbers*, repeat the idea with an alternate term that helps define the first term. For example, say, "Now write a bigger number—a number that has more."

1 Warm Up   5

Skill Building COMPUTING
Count Up
Before beginning the activity **Subtracting Counters**, use the **Count Up** activity with the whole class.

Purpose Children associate counting up with movement around a dial.

Warm-Up Card 6

Monitoring Student Progress
| If . . . children can fluently count on the dial to 20, | Then . . . challenge them to complete revolutions to 50. |

2 Engage   30

Concept Building APPLYING
Subtracting Counters
"Today you will use the Counters to help with subtraction."

Follow the instructions on the Activity Card to play **Subtracting Counters**. As children play, ask questions about what is happening in the activity.

Activity Card 35

Purpose Children subtract small quantities from single- and double-digit numbers and write subtraction equations.

250 Number Worlds • Week 19

Monitoring Student Progress

If . . . children are comfortable subtracting quantities from 20,

Then . . . challenge them to write all the corresponding equations.

 For additional practice with subtraction, children should complete **Building Blocks** Function Machine 1.

3 Reflect 10

Extended Response REASONING

Ask questions such as the following:

- **What does this remind you of?** Accept all reasonable answers. Children may mention the **Food Fun Teaches "Zero"** and **They're Gone** activities. They may mention real-life situations at home or at school.

- **If you start with 20 Counters, how many would you need to take away to get to 17? How did you figure that out?** 3; Children may count down from 20 to find the correct answer, or they may start at 17 and count up to 20.

4 Assess

Informal Assessment

Use the Student Assessment Record, **Assessment,** page 100, to record informal observations.

COMPUTING	APPLYING
Count Up	**Subtracting Counters**
Did the child	Did the child
❑ respond accurately?	❑ apply learning to new situations?
❑ respond quickly?	❑ contribute concepts?
❑ respond with confidence?	❑ contribute answers?
❑ self-correct?	❑ connect mathematics to the real world?

Subtracting • Lesson 4   251

Week 19 Subtracting

Lesson 5

Review

Objective
Children review the material presented in Week 19 of *Number Worlds*.

Program Materials
Warm Up
Circle Land Classroom Dial

Additional Materials
See materials from previous weeks.

Creating Context
The variety of words used to describe subtraction (*take away, minus, less, subtracted from,* and so on) may be puzzling for English Learners. Help children brainstorm a list of words and phrases used to describe subtraction.

1 Warm Up — 5

Skill Building COMPUTING
Count Up
Before beginning the **Free-Choice** activity, use the **Count Up** activity with the whole class.

Purpose Children associate counting up with movement around a dial.

Warm-Up Card 6

Monitoring Student Progress
| If . . . children can fluently count on the dial to 20, | Then . . . challenge them to complete revolutions to 50 and eventually to 100. |

2 Engage — 20

Concept Building APPLYING
Free-Choice Activity
For the last day of the week, allow children to choose an activity from the previous weeks. Some activities they may choose are the following:
- Object Land: **They're Gone**
- Picture Land: **Dragon Quest 2**
- Object Land: **Shopping Trip**

Make a note of the activities children select. Do they prefer easy or challenging activities? Continue to provide Challenge opportunities for children who have mastered the basic activities.

Monitoring Student Progress
| If . . . children would benefit from extra practice on specific skills, | Then . . . choose an activity for them. |

Reflect

Extended Response REASONING

Ask questions such as the following:
- **What did you like about playing** Tear It Up and Throw It Away?
- **Was there anything about playing this game you didn't like?**
- **Did this game help you do something you couldn't do before? What did it help you do?**
- **What was easy when you were playing** Subtracting Counters?
- **What was hard when you were playing** Subtracting Counters?
- **What do you remember most about this game?**
- **What was your favorite part?**

Using Student Pages

Have each child complete **Workbook,** page 61. Say, "Write the number story to describe what happens when the Counter moves from where it is to the spot shown by the arrow." Is the child correctly applying the skills learned in Week 19?

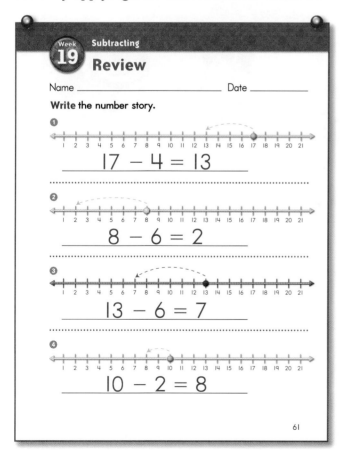

Workbook, p. 61

Assess

A Gather Evidence

Formal Assessment

Have children complete the weekly test on *Assessment,* page 61. Record formal assessment scores on the Student Assessment Record, *Assessment,* page 100.

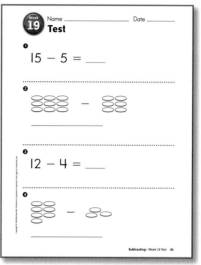

Assessment, p. 61

B Summarize Findings

Review the Student Assessment Records. Determine whether children have Minimal, Basic, or Secure understanding of the concepts presented in Week 19.

C Differentiate Instruction

Based on your observations, use these teaching strategies next week to follow up.

Minimal Understanding
- Repeat the Warm-Up and Engage activities in subsequent weeks to develop concepts of counting objects.
- Repeat the activities from this week. Observe children's understanding of these concepts, and reinforce as necessary.
- Use **Building Blocks** computer activities beginning with Counting Activity 1 to develop and reinforce numeration concepts.

Basic Understanding
- Repeat Engage activities in subsequent weeks to reinforce basic concepts.
- Use **Building Blocks** computer activities beginning with Function Machine 1 to reinforce addition and subtraction concepts.

Secure Understanding
- Use Challenge variations of the free-choice activity.
- Use **Building Blocks** computer activities beginning with Function Machine 1.

Subtracting • Lesson 5 Review 253

Week 20: Subtracting and Predicting

Week at a Glance

This week, children begin **Number Worlds**, Week 20, and continue to work with addition and subtraction in Picture Land.

Background

In Circle Land, numbers are represented as points on a dial. When a child is asked "How far around are you now?" he or she might think of the point on the dial, the size of the set that corresponds to that point, or the number that shows the point on the dial.

How Children Learn

As children begin this week's lesson, they should have experience using formal notation and writing both addition and subtraction problems.

At the end of the week, children should become more comfortable solving and writing subtraction problems involving both single- and double-digit numbers. They should be able to solve missing-addend problems involving small numbers.

Skills Focus

- Add and subtract whole numbers
- Use math symbols
- Understand equality
- Predict

Teaching for Understanding

As children engage in these activities, they will deepen their understanding of number relationships. The **Adding Concentration** activity lays the groundwork for formal missing-addend problems and helps children discover the commutative property of addition (3 + 2 = 2 + 3).

Observe closely while evaluating the Engage activities assigned for this week.

- Are children correctly subtracting from both single- and double-digit numbers?
- Are children correctly writing subtraction equations?
- Can children identify missing addends?

Math at Home

Give one copy of the Letter to Home, page A20, to each child. Complete the activity in class, and then encourage children to share it with their caregivers.

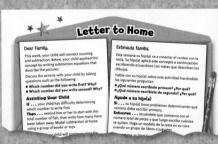

Letter to Home, Teacher Edition, p. A20

Math Vocabulary

predict To say what you think will happen

equal Having the same amount

equation Number story that includes an equal sign

English Learners

SPANISH COGNATES

English	Spanish
number	número
practice	práctica
predict	predecir

ALTERNATE VOCABULARY

concentrate To think carefully about something

fishing Catching fish, usually with a pole

Week 20 Planner: Subtracting and Predicting

PACING	LESSON	LEARNING GOALS	MATERIALS	TECHNOLOGY
DAY 1	Warm Up 7 **Circle Land:** Blastoff!*	Children associate counting down with moving backward on a dial.	**Program Materials** Circle Land Classroom Dial	**Building Blocks** Function Machine 1
	Activity 36 **Picture Land:** Going Fishing	Children subtract small quantities from single- and double-digit numbers and use formal notation.	**Program Materials** • Picture Land Activity Sheets 11 and 12, 1 per team • Spinner for each team **Additional Materials** Pencils, 1 per child	
DAY 2	Warm Up 7 **Circle Land:** Blastoff!*	Children associate counting down with moving backward on a dial.	**Program Materials** Circle Land Classroom Dial	**Building Blocks** Pizza Pizzazz 3
	Activity 37 **Picture Land:** Adding Concentration*	Children add two sets of numbers and identify a missing addend.	**Program Materials** • Number Cards (1–5), 4 sets • Number 0 Cards, 1 set **Additional Materials** Paper clips	
DAY 3	Warm Up 7 **Circle Land:** Blastoff!*	Children associate counting down with moving backward on a dial.	**Program Materials** Circle Land Classroom Dial	**Building Blocks** Function Machine 1
	Activity 36 **Picture Land:** Going Fishing	Children subtract small quantities and use formal notation.	Same as Lesson 1	
DAY 4	Warm Up 7 **Circle Land:** Blastoff!*	Children associate counting down with moving backward on a dial.	**Program Materials** Circle Land Classroom Dial	**Building Blocks** Pizza Pizzazz 3
	Activity 37 **Picture Land:** Adding Concentration*	Children add two sets of numbers and identify a missing addend.	Same as Lesson 2	
DAY 5	Warm Up 7 **Circle Land:** Blastoff!*	Children associate counting down with moving backward on a dial.	**Program Materials** Circle Land Classroom Dial	**Building Blocks** Review previous activities
	Review and Assess	Children review their favorite activities to improve understanding.	Materials will be selected from those used in previous weeks.	

* Includes Challenge Variations

Week 20 — Lesson 1: Subtracting and Predicting

Objective
Children reinforce counting and number sequence skills and subtract small quantities from single- and double-digit numbers.

Program Materials
Warm Up
Circle Land Classroom Dial

Engage
- Picture Land Activity Sheets 11 and 12, 1 per team
- Spinner (or Dot 1–6 Cube or Number 7–12 Cube) for each pair of children

Additional Materials
Engage
Pencils, 1 per child

Access Vocabulary
subtract To take away
minus Symbol that means "take away"

Creating Context
Words that are familiar to English Learners in everyday situations may take on different meanings in the context of math. Help children understand that questions beginning with *how many* are asking for a precise number. Questions that begin with *how much* may sometimes be answered more broadly with words such as *a lot* or *more*.

---

1 Warm Up — 5

Skill Building COMPUTING
Blastoff!
Before beginning the small-group activity **Going Fishing,** use the **Blastoff!** activity with the whole class.

Purpose Children associate counting down with moving backward on a dial.

Warm-Up Card 7

Monitoring Student Progress
If . . . children have trouble counting back from 10,
Then . . . have them count back from 9 to 0.

---

2 Engage — 30

Concept Building UNDERSTANDING
Going Fishing
"Today we are going on a fishing trip."
Follow the instructions on the Activity Card to play **Going Fishing.** As children play, ask questions about what is happening in the activity.

Activity Card 36

Purpose Children subtract small quantities from single- and double-digit numbers and use formal notation to write subtraction problems.

> **Monitoring Student Progress**
>
> **If . . .** children have trouble subtracting from 20,
>
> **Then . . .** have them play the game to 10 fish with a Dot Cube or Number Cube (1–6).

 Building Blocks For additional practice with subtraction, children should complete **Building Blocks** Function Machine 1.

3 Reflect — 10

Extended Response REASONING

Toward the end of the game, ask questions such as the following:

- **How many fish do you have now? How many more do you need to make 20? How did you figure that out?** Answers will vary. Children may say they subtracted the number of fish they have now from 20 to find out how many they still need.

- **How could you check your answer?** Accept all reasonable answers. Possible answers include counting on from the number of fish children currently have; counting the remaining fish on the Activity Sheet.

Using Student Pages

Have each child complete **Workbook,** page 62. Did the child write the correct answer?

Week 20 — Subtracting and Predicting
Going Fishing

Name _____ Date _____

Write the answer.

1. $14 - 7 = \underline{\ 7\ }$

2. $8 - 2 = \underline{\ 6\ }$

3. $11 - 5 = \underline{\ 6\ }$

4. $5 - 1 = \underline{\ 4\ }$

62

Workbook, p. 62

4 Assess

Informal Assessment

Use the Student Assessment Record, **Assessment,** page 100, to record informal observations.

COMPUTING	UNDERSTANDING
Blastoff!	**Going Fishing**
Did the child	Did the child
❏ respond accurately?	❏ make important observations?
❏ respond quickly?	❏ extend or generalize learning?
❏ respond with confidence?	❏ provide insightful answers?
❏ self-correct?	❏ pose insightful questions?

Subtracting and Predicting • Lesson 1   **257**

Week 20 — Subtracting and Predicting

Lesson 2

Objective
Children practice counting and number-sequence skills, add two sets of numbers, and identify a missing addend.

Program Materials
Warm Up
Circle Land Classroom Dial

Engage
- Number Cards (1–5), 4 sets; (1–10 for a Challenge)
- Number 0 Cards, 1 set

Additional Materials
Engage
Paper clips

Prepare Ahead
Add a Number 0 Card to each set of Number Cards.

Access Vocabulary
add To put together
plus sign Symbol that means "add"

Creating Context
Concentration games are excellent for English Learners because they do not rely heavily on oral language to excel and win. When children identify pairs of numbers that add up to 5, have them say the equation for practice (for example, "two plus three equals five").

1. Warm Up

Skill Building — COMPUTING
Blastoff!
Before beginning the small-group activity **Adding Concentration,** use the **Blastoff!** activity with the whole class.

Purpose Children associate counting down with moving backward on a dial.

Warm-Up Card 7

Monitoring Student Progress
If . . . children have trouble counting back from 10,
Then . . . have them count back from 9 to 0.

2. Engage

Concept Building — ENGAGING
Adding Concentration
"Today we will add numbers together to make 5."

Follow the instructions on the Activity Card to introduce and play **Adding Concentration.** As children play, ask questions about what is happening in the activity.

Activity Card 37

Purpose Children add two sets of numbers and predict how many more need to be added to a small set to make 5 altogether.

258 Number Worlds • Week 20

Monitoring Student Progress

| If . . . children have trouble selecting a second card that makes 5, | Then . . . give them Counters or a desktop number line to work with. |

eMathTools Use the Set Tool to demonstrate and explore counting.

 Teacher's Note If you place the Number Cards in an envelope ahead of time, this can be a great activity for an independent learning center or for early finishers.

Building Blocks For additional practice with missing addends, children should complete *Building Blocks* Pizza Pizzazz 3.

3 Reflect   10

Extended Response   REASONING

Ask questions such as the following:

- **If you picked a 2, what number would you need to make 5? How did you figure that out?** Accept all reasonable answers.
- **What patterns do you see?** Accept all reasonable answers. Children may note that as one addend increases by 1 the other addend decreases by 1, or that addends can be written in any order (2 + 3 = 3 + 2).

Using Student Pages

Have each child complete *Workbook,* page 63. Did the child identify the missing number?

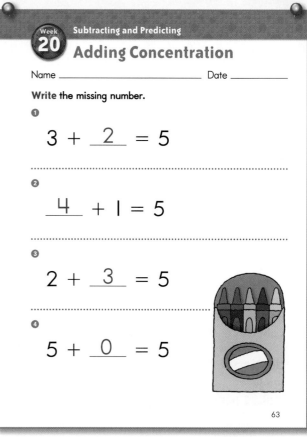

Workbook, p. 63

4 Assess

Informal Assessment

Use the Student Assessment Record, **Assessment,** page 100, to record informal observations.

COMPUTING	ENGAGING
Blastoff!	**Adding Concentration**
Did the child	Did the child
❏ respond accurately?	❏ pay attention to the contributions of others?
❏ respond quickly?	
❏ respond with confidence?	❏ contribute information and ideas?
❏ self-correct?	
	❏ improve on a strategy?
	❏ reflect on and check accuracy of work?

Subtracting and Predicting • Lesson 2 259

Week 20: Subtracting and Predicting

Objective
Children reinforce counting and number sequence skills and use formal notation to record subtraction problems.

Program Materials
Warm Up
Circle Land Classroom Dial

Engage
- Picture Land Activity Sheets 11 and 12, 1 per team
- Spinner (or Dot 1–6 Cube or Number 7–12 Cube) for each pair of children

Additional Materials
Engage
Pencils, 1 per child

Access Vocabulary
equal Having the same amount
equation Number story that includes an equal sign

Creating Context
To help children build vocabulary and practice conversation, ask them to share any experiences they have with fishing. Start the discussion with some visual cues such as pictures of fishing equipment, and have children act out the steps involved in catching a fish. Remind children that the word *fish* can describe one or more fish.

1 Warm Up — 5

Skill Building — COMPUTING
Blastoff!
Before beginning the small-group activity **Going Fishing**, use the **Blastoff!** activity with the whole class.

Purpose Children associate counting down with moving backward on a dial.

Warm-Up Card 7

Monitoring Student Progress
If . . . children can count down fluently from 10,
Then . . . challenge them to count down from 15.

2 Engage — 30

Concept Building — REASONING
Going Fishing
"Today we are going on a fishing trip."

Follow the instructions on the Activity Card to play **Going Fishing**. As children play, ask questions about what is happening in the activity.

Purpose Children subtract small quantities from single- and double-digit numbers and use formal notation to write subtraction problems.

Activity Card 36

260 Number Worlds • Week 20

Monitoring Student Progress

If . . . children are comfortable playing the game,

Then . . . invite them to use tally marks instead of the Fish Pond Activity Sheet to record how many fish they have caught.

Teacher's Note Encourage children to use this week's vocabulary words as they engage in the activities, discuss math concepts, and make predictions.

For additional practice with subtraction, children should complete **Building Blocks** Function Machine 1.

3 Reflect 10

Extended Response REASONING
Ask questions such as the following:
- **What do you notice about this game?** Accept all reasonable answers. Children may note that spinning or rolling bigger numbers helps them reach 20 faster.
- **What does this game remind you of?** Accept all reasonable answers.

Assess

Informal Assessment
Use the Student Assessment Record, **Assessment**, page 100, to record informal observations.

COMPUTING	REASONING
Blastoff!	**Going Fishing**
Did the child	Did the child
❏ respond accurately?	❏ provide a clear explanation?
❏ respond quickly?	❏ communicate reasons and strategies?
❏ respond with confidence?	❏ choose appropriate strategies?
❏ self-correct?	❏ argue logically?

Subtracting and Predicting • Lesson 3

Week 20: Subtracting and Predicting

Lesson 4

Objective
Children practice counting and number-sequence skills, add two sets of numbers, and identify a missing addend.

Program Materials
Warm Up
Circle Land Classroom Dial

Engage
- Number Cards (1–5), 4 sets; (1–10 for a Challenge)
- Number 0 Cards, 1 set

Additional Materials
Engage
Paper clips

Prepare Ahead
Add a Number 0 Card to each set of Number Cards.

Access Vocabulary
faceup Cards on a table with the numerals up (showing)
predict To say what you think will happen

Creating Context
The Reflect section of this lesson includes questions to help children build critical-thinking and problem-solving skills. Some English Learners at early proficiency levels may have trouble describing their thinking processes. You may have children draw step-by-step models of their processes and work with a more proficient partner to label each step.

1 Warm Up — 5

Skill Building — COMPUTING
Blastoff!
Before beginning the small-group activity **Adding Concentration,** use the **Blastoff!** activity with the whole class.

Purpose Children associate counting down with moving backward on a dial.

Warm-Up Card 7

Monitoring Student Progress
| If . . . children can count down fluently from 10, | Then . . . have them play the Challenge variation of the activity. |

2 Engage — 30

Concept Building — APPLYING
Adding Concentration
"Today we will add numbers together to make 5."

Follow the instructions on the Activity Card to play **Adding Concentration.** As children play, ask questions about what is happening in the activity.

Purpose Children add two sets of numbers and predict how many more need to be added to a small set to make 5 altogether.

Activity Card 37

262 Number Worlds • Week 20

Monitoring Student Progress

If . . . children can fluently choose card combinations that equal 5,

Then . . . challenge them to play the game with the cards facedown.

 For additional practice with missing addends, children should complete *Building Blocks* Pizza Pizzazz 3.

3 Reflect

Extended Response REASONING

Ask questions such as the following:

- **What strategy did you use to solve this problem? Why?** Possible answers include guess and check, look for a pattern, model the situation, and so on.

- **How do you know when you're done?** Accept all reasonable answers. Encourage children to discuss how they know whether they have found all possible combinations of addends that equal 5.

4 Assess

Informal Assessment

Use the Student Assessment Record, **Assessment**, page 100, to record informal observations.

COMPUTING	APPLYING
Blastoff!	**Adding Concentration**
Did the child	Did the child
❏ respond accurately?	❏ apply learning to new situations?
❏ respond quickly?	❏ contribute concepts?
❏ respond with confidence?	❏ contribute answers?
❏ self-correct?	❏ connect mathematics to the real world?

Subtracting and Predicting • Lesson 4

Week 20: Subtracting and Predicting

Lesson 5
Review

Objective
Children review the material presented in Week 20 of **Number Worlds**.

Program Materials
Warm Up
Circle Land Classroom Dial

Additional Materials
See materials from previous weeks.

Creating Context
Help children understand the concept of predicting by asking, "How many of you watch the weather report on television? The reporters predict what the weather will be for the next few days. Their predictions are based on information they have. What do you predict the weather will be tomorrow?" Save the predictions until tomorrow to check the outcome.

1 Warm Up 5

Skill Building COMPUTING
Blastoff!
Before beginning the **Free-Choice** activity, use the **Blastoff!** activity with the whole class.

Purpose Children associate counting down with movement backward on a dial.

Warm-Up Card 7

Monitoring Student Progress
| If . . . children can count down fluently from 10, | Then . . . have them play the Challenge variation of the activity. |

2 Engage 20

Concept Building APPLYING
Free-Choice Activity

For the last day of the week, allow children to choose an activity from the previous weeks. Some activities they may choose include the following:
- Picture Land: **Concentration**
- Picture Land: **Making Equations for Beginners**
- Line Land: **Subtracting Counters**

Make a note of the activities children select. Do they prefer easy or challenging activities? Continue to provide Challenge opportunities for children who have mastered the basic activities.

Monitoring Student Progress
| If . . . children would benefit from extra practice on specific skills, | Then . . . choose an activity for them. |

264 Number Worlds • Week 20

Reflect 10

Extended Response REASONING
Ask questions such as the following:
- What did you like about playing Going Fishing?
- Was there anything about playing this game you didn't like?
- Did this game help you do something you couldn't do before? What did it help you do?
- What was easy when you were playing Adding Concentration?
- What was hard when you were playing Adding Concentration?
- What do you remember most about this game?
- What was your favorite part?

Using Student Pages
Have each child complete **Workbook,** page 64. Is the child correctly applying the skills learned in Week 20?

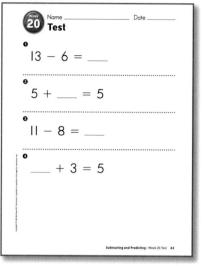

Workbook, p. 64

 Assess 10

A Gather Evidence
Formal Assessment
Have children complete the weekly test on *Assessment,* page 63. Record formal assessment scores on the Student Assessment Record, *Assessment,* page 100.

Assessment, p. 63

B Summarize Findings
Review the Student Assessment Records. Determine whether children have Minimal, Basic, or Secure understanding of the concepts presented in Week 20.

C Differentiate Instruction
Based on your observations, use these teaching strategies next week to follow up.

Minimal Understanding
- Repeat the Warm-Up and Engage activities in subsequent weeks to develop concepts of counting objects.
- Repeat the activities from this week. Observe children's understanding of these concepts, and reinforce as necessary.
- Use **Building Blocks** computer activities beginning with Pizza Pizzazz 1 to develop and reinforce numeration and addition concepts.

Basic Understanding
- Repeat Engage activities in subsequent weeks to reinforce basic concepts.
- Use **Building Blocks** computer activities beginning with Pizza Pizzazz 3 to reinforce numeration and addition concepts.

Secure Understanding
- Use Challenge variations of this week's activities.
- Use **Building Blocks** computer activities beginning with Under the Hat.

Subtracting and Predicting • Lesson 5 Review **265**

Week 21: Adding and Comparing

Week at a Glance

This week, children begin **Number Worlds,** Week 21, and continue to explore addition in Picture Land.

Background

In Picture Land, numbers are represented as numerals. This allows children to forge links between the world of real quantities (e.g., objects) and the world of formal symbols such as the written numerals and addition, subtraction, and equal signs used when writing equations.

How Children Learn

As children begin this week's lesson, they should be able to solve and write subtraction problems involving both single- and double-digit numbers and should be able to solve missing-addend problems involving small numbers.

At the end of this week, children should be more comfortable adding small quantities and should be able to compare two sums to say which is larger.

Skills Focus
- Compare and order numbers
- Add whole numbers
- Use math symbols
- Understand equality

Teaching for Understanding

As children engage in these activities, they will become more comfortable with basic addition problems.

Observe closely while evaluating the Engage activities assigned for this week.
- Are children correctly adding whole numbers?
- Are children using formal notation in their equations?
- Can children correctly compare sums?

Math at Home
Give one copy of the Letter to Home, page A21, to each child. Complete the activity in class, and then encourage children to share it with their caregivers.

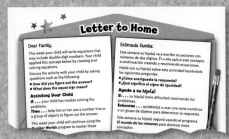

Letter to Home, Teacher Edition, p. A21

Math Vocabulary
compare To think about how things are alike and how they are different
add To put together
flip To turn over quickly
shuffle To mix together

English Learners
SPANISH COGNATES

English	Spanish
sign	signo
symbols	símbolos
activity	actividad

ALTERNATE VOCABULARY
stack A pile of flat things such as cards or books

Week 21 Planner: Adding and Comparing

PACING	LESSON	LEARNING GOALS	MATERIALS	TECHNOLOGY
DAY 1	Warm Up Teacher's Choice	Children reinforce counting and number sequence skills.	Materials will be selected from those used in previous weeks.	**Building Blocks** Double Compare 3
	Activity 38 **Picture Land:** Flip, Add, and Compare*	Children add small quantities and compare two quantities to say which is larger.	**Program Materials** • Number Cards (1–15), 2 sets per team • Number 0 Cards, 1 set • Symbol Cards: Plus Sign for each team **Additional Materials** Pencils	
DAY 2	Warm Up Teacher's Choice	Children reinforce counting and number sequence skills.	Materials will be selected from those used in previous weeks.	**Building Blocks** Double Compare 3
	Activity 39 **Picture Land:** Roll, Add, and Compare*	Children add two quantities together and compare quantities to say which is larger.	**Program Materials** Number 1–6 Cube, 1 per child **Additional Materials** • Box top with divider, 1 per group • Paper and pencils	
DAY 3	Warm Up Teacher's Choice	Children reinforce counting and number sequence skills.	Materials will be selected from those used in previous weeks.	**Building Blocks** Double Compare 3
	Activity 38 **Picture Land:** Flip, Add, and Compare*	Children add small quantities and compare two quantities to say which is larger.	Same as Lesson 1	
DAY 4	Warm Up Teacher's Choice	Children reinforce counting and number sequence skills.	Materials will be selected from those used in previous weeks.	**Building Blocks** Double Compare 3
	Activity 39 **Picture Land:** Roll, Add, and Compare*	Children add two quantities together and compare quantities to say which is larger.	**Program Materials** Number 1–6 Cube, 1 per child **Additional Materials** • Box top with divider, 1 per group • Paper and pencils	
DAY 5	Warm Up Teacher's Choice	Children reinforce counting and number sequence skills.	Materials will be selected from those used in previous weeks.	**Building Blocks** Review previous activities
	Review and Assess	Children review their favorite activities to improve understanding.	Materials will be selected from those used in previous weeks.	

* Includes Challenge Variations

Week 21 — Adding and Comparing
Lesson 1

Objective
Children practice counting and number-sequencing skills and add and compare quantities.

Program Materials
Warm Up
See materials for previous weeks.
Engage
- Number Cards (1–15), 2 sets for each pair of children
- Number 0 Cards, 1 set
- Symbol Cards: Plus Sign, 1 for each pair of children
- Picture Land Activity Sheet 13 for a Challenge
- Magnetic Number Line (optional)

Additional Materials
Engage
Pencils

Prepare Ahead
Add a Number 0 Card to each set of Number Cards.

Access Vocabulary
compare To think about how two things are alike and how they are different
flip To turn over quickly

Creating Context
In this lesson, children add to see which team has the highest number. Review comparative and superlative forms of the following words:

high—higher—highest
low—lower—lowest
big—bigger—biggest
small—smaller—smallest

1 Warm Up

Skill Building COMPUTING
Teacher's Choice
Before beginning the small-group activity **Flip, Add, and Compare,** choose one of the Warm-Up activities from previous weeks to use with the whole group. These activities include the following:
- Object Land: **Pointing and Winking**
- Picture Land: **Catch the Teacher**
- Line Land: **The Neighborhood**
- Sky Land: **Thermometer Counting**
- Object Land: **What Number Am I?**
- Circle Land: **Count Up**
- Circle Land: **Blastoff!**

Warm-Up Card 1

Purpose Children reinforce counting and number-sequence skills.

Monitoring Student Progress
If . . . children need help with basic skills,
Then . . . practice counting and number-sequence skills with them throughout the day.

2 Engage

Concept Building UNDERSTANDING
Flip, Add, and Compare
"Today we are going to practice addition and to see which team can get the highest number."

Follow the instructions on the Activity Card to play **Flip, Add, and Compare.** As children play, ask questions about what is happening in the activity.

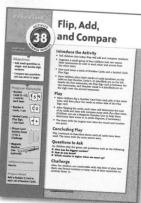

Activity Card 38

268 Number Worlds • Week 21

Purpose Children add small quantities to single- and double-digit numbers and compare two quantities to see which is larger.

Monitoring Student Progress

| If... children have trouble determining the correct sums, | Then... allow them to use the Magnetic Number Line or a desktop number line to help them. |

eMathTools Use the Number Line Tool to demonstrate and explore number sequence.

Building Blocks For further practice with addition and numeration, children should complete **Building Blocks** Double Compare 3.

3 Reflect — 10

Extended Response REASONING
Ask questions such as the following:

- **Who had the biggest number? How did you figure that out?** Answers will vary. Children may indicate that the biggest number is higher on the number line or higher when they count up.

- **How many more would your team (or the other team) have needed to have the biggest number? How did you figure that out?** Accept all reasonable answers.

Using Student Pages
Have each child complete *Workbook,* page 65. Did children circle the correct sums?

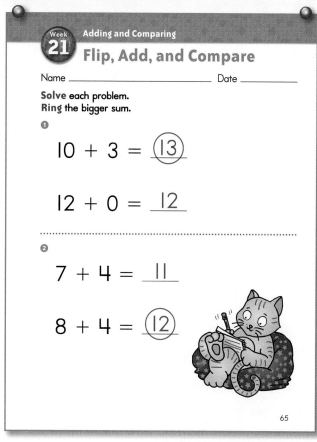

Workbook, p. 65

4 Assess

Informal Assessment
Use the Student Assessment Record, **Assessment,** page 100, to record informal observations.

COMPUTING	UNDERSTANDING
Teacher's Choice Did the child ❏ respond accurately? ❏ respond quickly? ❏ respond with confidence? ❏ self-correct?	**Flip, Add, and Compare** Did the child ❏ make important observations? ❏ extend or generalize learning? ❏ provide insightful answers? ❏ pose insightful questions?

Adding and Comparing • Lesson 1   269

Week 21 — Adding and Comparing
Lesson 2

Objective
Children reinforce counting and number-sequence skills as they add and compare quantities.

Program Materials
Warm Up
See materials from previous weeks.

Engage
Number 1–6 Cube, 1 per child
(Number 7–12 Cube for a Challenge)

Additional Materials
Engage
- Strong box top with divider for each group
- Paper and pencils for each group

Access Vocabulary
add To put together
roll To drop Number Cubes or Dot Cubes onto a game board

Creating Context
In the Reflect section of this activity, we ask children, "What is the biggest number your team could roll?" This grammatical structure may be unfamiliar to some English Learners. Explain to children that you are asking for the biggest possible number. The word *could* is the clue that tells what is being asked.

1 Warm Up

Skill Building — COMPUTING
Teacher's Choice

Before beginning the small-group activity **Roll, Add, and Compare**, choose one of the Warm-Up activities from previous weeks to use with the whole group. These activities include the following:
- Object Land: **Pointing and Winking**
- Picture Land: **Catch the Teacher**
- Line Land: **The Neighborhood**
- Sky Land: **Thermometer Counting**
- Object Land: **What Number Am I?**
- Circle Land: **Count Up**
- Circle Land: **Blastoff!**

Purpose Children reinforce counting and number-sequence skills.

Warm-Up Card 1

Monitoring Student Progress
If . . . children need help with basic skills,
Then . . . practice counting and number-sequence skills with them throughout the school day.

2 Engage

Concept Building — ENGAGING
Roll, Add, and Compare

"Today you will practice writing equations to see who has the biggest answer."

Follow the instructions on the Activity Card to play **Roll, Add, and Compare**. As children play, ask questions about what is happening in the activity.

Activity Card 39

270 Number Worlds • Week 21

Purpose Children reinforce addition and number-sequence skills as they write equations and compare sums.

Monitoring Student Progress

If . . . children forget to use formal notation when writing their equations,

Then . . . post a model equation in the classroom for children to use as a reference.

 Teacher's Note This is a fun activity to play with a small group or with the whole class.

 Building Blocks For further practice with addition and numeration, children should complete **Building Blocks** Double Compare 3.

3 Reflect — 10

Extended Response REASONING
Ask questions such as the following:
- **What numbers did your team roll? What do they equal? How did you figure that out?** Accept all reasonable answers.
- **What is the biggest number your team could roll?** 12 **How did you figure that out?** It is the sum of the biggest number on each of the two Number Cubes.

Using Student Pages
Have each child complete **Workbook,** page 66. Did children circle the correct sums?

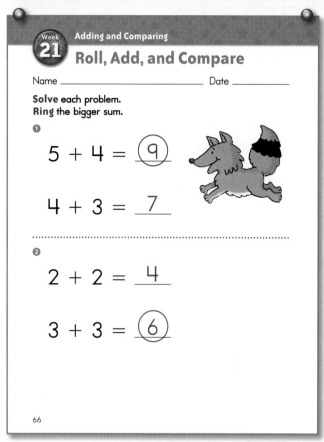

Workbook, p. 66

4 Assess

Informal Assessment
Use the Student Assessment Record, **Assessment,** page 100, to record informal observations.

COMPUTING	ENGAGING
Teacher's Choice Did the child	**Roll, Add, and Compare** Did the child
❏ respond accurately?	❏ pay attention to the contributions of others?
❏ respond quickly?	
❏ respond with confidence?	
❏ self-correct?	❏ contribute information and ideas?
	❏ improve on a strategy?
	❏ reflect on and check accuracy of work?

Adding and Comparing • Lesson 2   271

Week 21 — Adding and Comparing
Lesson 3

Objective
Children practice counting and number-sequencing skills and add and compare quantities.

Program Materials
Warm Up
See materials for previous weeks.

Engage
- Number Cards (1–15), 2 sets for each pair of children
- Number 0 Cards, 1 set
- Symbol Cards: Plus Sign, 1 for each pair of children
- Picture Land Activity Sheet 13 for a Challenge
- Magnetic Number Line (optional)

Additional Materials
Engage
Pencils

Prepare Ahead
Add a Number 0 Card to each set of Number Cards.

Access Vocabulary
shuffle To mix together
stack A pile of flat things like cards or books

Creating Context
Help English Learners successfully write number stories by reading aloud a few models first. To focus on the math concept without worrying about the language needed to create a story, provide a story frame in which students fill in key information.

1 Warm Up 5

Skill Building COMPUTING
Teacher's Choice
Before beginning the small-group activity **Flip, Add, and Compare,** choose one of the Warm-Up activities from previous weeks to use with the whole group. These activities include the following:
- Object Land: **Pointing and Winking**
- Picture Land: **Catch the Teacher**
- Line Land: **The Neighborhood**
- Sky Land: **Thermometer Counting**
- Object Land: **What Number Am I?**
- Circle Land: **Count Up**
- Circle Land: **Blastoff!**

Warm-Up Card 1

Purpose Children reinforce counting and number-sequence skills.

Monitoring Student Progress
| If . . . children need help with basic skills, | Then . . . practice counting and number-sequence skills with them throughout the day. |

2 Engage 30

Concept Building REASONING
Flip, Add, and Compare
"Today we are going to practice addition to see which team can get the highest number."

Follow the instructions on the Activity Card to play **Flip, Add, and Compare.** As children play, ask questions about what is happening in the activity.

Activity Card 38

272 Number Worlds • Week 21

Purpose Children add small quantities to single- and double-digit numbers and compare two quantities to see which is larger.

Monitoring Student Progress

| If . . . children are comfortable with this level of play, | Then . . . have them use formal notation to keep track of their equations as they play the Challenge variation of the game. |

Teacher's Note Encourage children to use this week's vocabulary words as they engage in the activities, discuss math concepts, and make predictions.

Building Blocks For further practice with addition and numeration, children should complete **Building Blocks** Double Compare 3.

4 Assess

Informal Assessment

Use the Student Assessment Record, **Assessment**, page 100, to record informal observations.

COMPUTING	REASONING
Teacher's Choice	**Flip, Add, and Compare**
Did the child	Did the child
❑ respond accurately?	❑ provide a clear explanation?
❑ respond quickly?	❑ communicate reasons and strategies?
❑ respond with confidence?	❑ choose appropriate strategies?
❑ self-correct?	❑ argue logically?

3 Reflect 10

Extended Response REASONING

Ask questions such as the following:

- **What would happen if you used Number Cards 0–10 instead of 0–15?** Accept all reasonable answers. Children may say that the sums of their equations would be smaller.
- **How can you check your answers?** Possible answers: have a teammate check it or check it yourself; use a number line; and so on.

Adding and Comparing • Lesson 3 **273**

Week 21 Adding and Comparing

Lesson 4

Objective
Children reinforce counting and number-sequence skills as they add and compare quantities.

Program Materials
Warm Up
See materials from previous weeks.

Engage
Number 1–6 Cube, 1 per student
(Number 7–12 Cube for a Challenge)

Additional Materials
Engage
- Strong box top with divider for each group
- Paper and pencils for each group

Access Vocabulary
sum Answer to an addition problem

Creating Context
Some English Learners may be confused by the homophones *sum* and *some*. Help distinguish the two words by explaining that when we talk about the sum of two numbers, we always use the article *the* before the word. *Sum* is a noun. The describing word *some* will not have an article.

1 Warm Up

Skill Building COMPUTING
Teacher's Choice
Before beginning the small-group activity **Roll, Add, and Compare**, choose one of the Warm-Up activities from previous weeks to use with the whole group. These activities include the following:
- Object Land: **Pointing and Winking**
- Picture Land: **Catch the Teacher**
- Line Land: **The Neighborhood**
- Sky Land: **Thermometer Counting**
- Object Land: **What Number Am I?**
- Circle Land: **Count Up**
- Circle Land: **Blastoff!**

Warm-Up Card 1

Purpose Children reinforce counting and number-sequence skills.

Monitoring Student Progress
If . . . children can work fluently with numbers 1 through 10,

Then . . . challenge them to work with bigger numbers.

2 Engage

Concept Building APPLYING
Roll, Add, and Compare
"Today you will practice writing equations to see who has the biggest answer."

Follow the instructions on the Activity Card to play **Roll, Add, and Compare**. As children play, ask questions about what is happening in the activity.

Activity Card 39

274 Number Worlds • Week 21

Purpose Children reinforce addition and number-sequence skills as they write equations and compare sums.

Monitoring Student Progress

| If . . . children are comfortable with this level of play, | Then . . . challenge them to play the game with a Number 7–12 Cube. |

Teacher's Note You might like to add a fifth child to each group as a designated team leader to oversee the rolling, adding, and comparing and to settle any disputes that may arise.

Building Blocks For further practice with addition and numeration, children should complete **Building Blocks** Double Compare 3.

3 Reflect   10

Extended Response   REASONING
Ask questions such as the following:
- **What strategy did you use to solve these number stories? Why did you choose that strategy?** Accept all reasonable answers.
- **What patterns do you see?** Accept all reasonable answers.

4 Assess

Informal Assessment
Use the Student Assessment Record, **Assessment,** page 100, to record informal observations.

COMPUTING	APPLYING
Teacher's Choice Did the child ❏ respond accurately? ❏ respond quickly? ❏ respond with confidence? ❏ self-correct?	**Roll, Add, and Compare** Did the child ❏ apply learning to new situations? ❏ contribute concepts? ❏ contribute answers? ❏ connect mathematics to the real world?

Adding and Comparing • Lesson 4   **275**

Week 21 — Adding and Comparing

Lesson 5
Review

Objective
Children review the material presented in Week 21 of **Number Worlds**.

Program Materials
Warm Up
See materials from previous weeks.

Engage
See materials from previous weeks.

Creating Context
Playing games is an excellent way to give English Learners practice listening to and speaking English. The natural repetition of procedural and counting language replaces tedious drills with authentic, active experience. Wait time is built into the process, and the low-stakes environment makes the use of games an enjoyable and efficient learning tool in math.

1 Warm Up

Skill Building — COMPUTING
Teacher's Choice
Before beginning the **Free-Choice** activity, choose one of the Warm-Up activities from previous weeks to use with the whole group. These activities include the following:

- Object Land: **Pointing and Winking**
- Picture Land: **Catch the Teacher**
- Line Land: **The Neighborhood**
- Sky Land: **Thermometer Counting**
- Object Land: **What Number Am I?**
- Circle Land: **Count Up**
- Circle Land: **Blastoff!**

Warm-Up Card 1

Purpose Children reinforce counting and number-sequence skills.

2 Engage

Concept Building — APPLYING
Free-Choice Activity

For the last day of the week, allow children to choose an activity from the previous weeks. Some activities they may choose include the following:

- Object Land: **Add Them Up**
- Picture Land: **Making and Writing Equations**
- Picture Land: **Adding Concentration**

Make a note of the activities children select. Do they prefer easy or challenging activities? Continue to provide Challenge opportunities for children who have mastered the basic activities.

Monitoring Student Progress

If... children	Then... choose an activity
would benefit from extra practice on specific skills,	for them.

Reflect

Extended Response REASONING

Ask questions such as the following:

- **What did you like about playing** Flip, Add, and Compare?
- **Was there anything about playing this game you didn't like?**
- **Did this game help you do something you couldn't do before? What did it help you do?**
- **What was easy when you were playing** Roll, Add, and Compare?
- **What was hard when you were playing** Roll, Add, and Compare?
- **What do you remember most about this game?**
- **What was your favorite part?**

Using Student Pages

Have each child complete *Workbook,* page 67. Is the child correctly applying the skills learned in Week 21?

Workbook, p. 67

Assess

A Gather Evidence

Formal Assessment

Have students complete the weekly test on *Assessment,* page 65. Record formal assessment scores on the Student Assessment Record, *Assessment,* page 100.

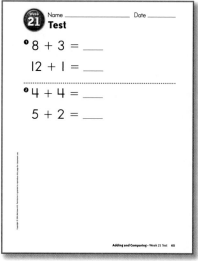

Assessment, p. 65

B Summarize Findings

Review the Student Assessment Records. Determine whether children have Minimal, Basic, or Secure understanding of the concepts presented in Week 21.

C Differentiate Instruction

Based on your observations, use these teaching strategies next week to follow up.

Minimal Understanding

- Repeat the Warm-Up and Engage activities in subsequent weeks to develop numeration concepts.
- Repeat the activities from this week. Observe children's understanding of these concepts, and reinforce as necessary.
- Use **Building Blocks** computer activities beginning with Number Compare 1 to develop and reinforce numeration concepts.

Basic Understanding

- Repeat Engage activities in subsequent weeks to reinforce basic concepts.
- Use **Building Blocks** computer activities beginning with Double Compare 3 to reinforce numeration and addition concepts.

Secure Understanding

- Use Challenge variations of this week's activities.
- Use **Building Blocks** computer activities beginning with Double Compare 4.

Adding and Comparing • Lesson 5 Review 277

Week 22: Subtracting to Zero

Week at a Glance

This week, children begin **Number Worlds**, Week 22, and continue to explore subtraction in Picture Land and Line Land.

Skills Focus
- Add whole numbers
- Subtract whole numbers
- Use math symbols
- Create and solve word problems

Background

In Line Land, a decrease in quantity is shown as backward movement along a number line. In Picture Land, numbers are represented as numerals. This allows children to forge links between the world of objects and the world of formal symbols, in which a decrease in quantity is represented by the minus sign.

How Children Learn

As children begin this week's lesson, they should be able to solve and to write basic subtraction problems.

At the end of this week, children should have gained experience writing and solving story problems and should understand that a number subtracted from itself equals zero.

Teaching for Understanding

As children engage in these activities, they will deepen their understanding of subtraction and will learn that any number subtracted from itself equals zero.

Observe closely while evaluating the Engage activities for this week.

- Can children successfully add and subtract whole numbers?
- Are children correctly using math symbols such as the plus sign and minus sign?

Math at Home

Give one copy of the Letter to Home, page A22, to each child. Complete the activity in class, and then encourage children to share it with their caregivers.

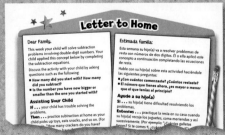

Letter to Home, Teacher Edition, p. A22

Math Vocabulary

zero None
equation Number story
subtract To take away
minus sign Symbol that means "take away"
pattern Something that repeats the same way each time

English Learners

SPANISH COGNATES

English	Spanish
equal	igual
magnetic	magnético
minus	menos
sign	signo
subtraction	sustracción
zero	cero

Week 22 Planner: Subtracting to Zero

PACING	LESSON	LEARNING GOALS	MATERIALS	TECHNOLOGY
DAY 1	Warm Up Teacher's Choice	Children reinforce counting and number-sequence skills.	Materials will be selected from those used in previous weeks.	Building Blocks Count Free Explore
	Activity 40 **Picture Land:** Subtracting to Zero	Children gain experience writing and solving subtraction word problems.	**Program Materials** Counters	
DAY 2	Warm Up Teacher's Choice	Children reinforce counting and number-sequence skills.	Materials will be selected from those used in previous weeks.	Building Blocks Function Machine 1
	Activity 41 **Line Land:** Which Way Should I Go?	Children subtract, read plus and minus signs, and decide which action to take.	**Program Materials** • Magnetic Number Line, 1 per child • Magnetic Chips • +\− Cube • Number 1–6 Cube	
DAY 3	Warm Up Teacher's Choice	Children reinforce counting and number-sequence skills.	Materials will be selected from those used in previous weeks.	Building Blocks Count Free Explore
	Activity 40 **Picture Land:** Subtracting to Zero	Children gain experience writing and solving subtraction word problems.	**Program Materials** Counters	
DAY 4	Warm Up Teacher's Choice	Children reinforce counting and number-sequence skills.	Materials will be selected from those used in previous weeks.	Building Blocks Function Machine 1
	Activity 41 **Line Land:** Which Way Should I Go?	Children subtract, read plus and minus signs, and decide which action to take.	**Program Materials** • Magnetic Number Line, 1 per child • Magnetic Chips • +\− Cube • Number 1–6 Cube	
DAY 5	Warm Up Teacher's Choice	Children reinforce counting and number-sequence skills.	Materials will be selected from those used in previous weeks.	Building Blocks Review previous activities
	Review and Assess	Children review their favorite activities to improve their understanding of more difficult concepts.	Materials will be selected from those used in previous weeks.	

Week 22 — Subtracting to Zero
Lesson 1

Objective
Children reinforce counting and number-sequence skills and gain experience writing and solving subtraction word problems.

Program Materials
Warm Up
See materials from previous weeks.

Engage
Counters

Prepare Ahead
Engage
Use index cards to create several subtraction problems with an answer of 0 that are suitable for your class (e.g., 9 − 9 = ___, 4 − 4 = ___, and so on). You may create problems with both vertical and horizontal formats.

Access Vocabulary
zero None

Creating Context
English Learners may have some difficulty creating story problems in English. Before you begin, help children brainstorm math-process terms they will need such as *minus, equals, take away,* and so on. Work as a group to create a story frame for English Learners who need some language support.

---

1. Warm Up — 5

Skill Building — COMPUTING
Teacher's Choice
Before beginning the whole group activity **Subtracting to Zero,** choose one of the Warm-Up activities from previous weeks to use with the whole group. These activities include the following:

- Object Land: **Pointing and Winking**
- Picture Land: **Catch the Teacher**
- Line Land: **The Neighborhood**
- Sky Land: **Thermometer Counting**
- Object Land: **What Number Am I?**
- Circle Land: **Count Up**
- Circle Land: **Blastoff!**

Warm-Up Card 1

Purpose Children reinforce counting and number-sequence skills.

Monitoring Student Progress
If . . . children need help with basic skills,
Then . . . practice counting and number-sequence skills with them throughout the day.

---

2. Engage — 30

Concept Building — UNDERSTANDING
Subtracting to Zero
"Today we are going to make up and solve subtraction stories."

Follow the instructions on the Activity Card to play **Subtracting to Zero.** As children play, ask questions about what is happening in the activity.

Activity Card 40

280 Number Worlds • Week 22

Purpose Children gain experience writing and solving subtraction word problems and realize that any number subtracted from itself will equal 0.

Monitoring Student Progress

If . . . children have trouble translating the formal subtraction problem into a story,

Then . . . remind them to use the Counters or to draw pictures.

Building Blocks For additional practice with subtraction, children should complete **Building Blocks** Count Free Explore.

3 Reflect 10

Extended Response REASONING
Ask questions such as the following:
- **How many did you start with? How many did you take away? How many are left? How did you figure that out?** Answers will vary; accept all reasonable explanations.
- **What was your favorite story? Why?** Accept all reasonable answers.
- **Will any number take away itself be 0?** yes **Why?** because there is nothing left over

Using Student Pages
Have each child complete **Workbook,** page 68. Did the child identify the missing number?

Week 22 Subtracting to Zero

Subtracting to Zero

Name _____ Date _____

Write the answer.

① $12 - 12 = \underline{0}$

② $7 - \underline{7} = 0$

③ $4 - 4 = \underline{0}$

④ $15 - \underline{15} = 0$

68

Workbook, p. 68

4 Assess

Informal Assessment
Use the Student Assessment Record, **Assessment,** page 100, to record informal observations.

COMPUTING	UNDERSTANDING
Teacher's Choice	**Subtracting to Zero**
Did the child	Did the child
❑ respond accurately?	❑ make important observations?
❑ respond quickly?	
❑ respond with confidence?	❑ extend or generalize learning?
❑ self-correct?	❑ provide insightful answers?
	❑ pose insightful questions?

Subtracting to Zero • Lesson 1 281

Week 22 — Subtracting to Zero
Lesson 2

Objective
Children reinforce counting and subtraction skills, read formally notated plus and minus signs, and determine which action to take.

Program Materials
Warm Up
See materials from previous weeks.

Engage
- Magnetic Number Line, 1 per child
- Magnetic Chips
- +\– Cube
- Number 1–6 Cube

Access Vocabulary
subtract To take away
minus sign Symbol that means "take away"

Creating Context
Help English Learners distinguish between the words *should* and *could* by explaining that *should* asks for advice, and *could* describes something they are capable of doing. Give children practice answering similar questions such as "Which _____ should I _____?" and "Which _____ could I _____?"

1. Warm Up

Skill Building COMPUTING
Teacher's Choice
Before beginning the small-group activity **Which Way Should I Go?** choose one of the Warm-Up activities from previous weeks to use with the whole group. These activities include the following:
- Object Land: **Pointing and Winking**
- Picture Land: **Catch the Teacher**
- Line Land: **The Neighborhood**
- Sky Land: **Thermometer Counting**
- Object Land: **What Number Am I?**
- Circle Land: **Count Up**
- Circle Land: **Blastoff!**

Warm-Up Card 1

Purpose Children reinforce counting and number-sequence skills.

> **Monitoring Student Progress**
> **If . . .** children need help with basic skills,
> **Then . . .** practice counting and number-sequence skills with them throughout the day.

2. Engage

Concept Building ENGAGING
Which Way Should I Go?
"Today we're going to use plus and minus signs to see who can get to zero first."

Follow the instructions on the Activity Card to play **Which Way Should I Go?** As children play, ask questions about what is happening in the activity.

Activity Card 41

Purpose Children subtract small quantities from single- and double-digit numbers, read formally notated plus and minus signs, and determine which action to take.

Monitoring Student Progress

| If... children have trouble determining whether they should add or subtract, | Then... post a plus sign and minus sign in the classroom and refer to them throughout the day. |

eMathTools Use the Number Line Tool to demonstrate and explore number sequence.

 Building Blocks For additional practice with subtraction, children should complete **Building Blocks** Function Machine 1.

3 Reflect 10

Extended Response REASONING

Ask questions such as the following:

- **Did you roll a plus sign or a minus sign?** Answers will vary. **Did you put Chips on the Number Line or take them off?** Answers will vary. **How did you know to do that?** The plus sign means "add something on;" the minus sign means "take something away."

- **Is the number you had after you added or subtracted bigger or smaller than the one you started with? How do you know?** Accept all reasonable answers. Children may say that because they added (or subtracted), their new number is bigger (or smaller) than the original number.

Using Student Pages

Have each child complete **Workbook,** page 69. Did the child ring the smaller number?

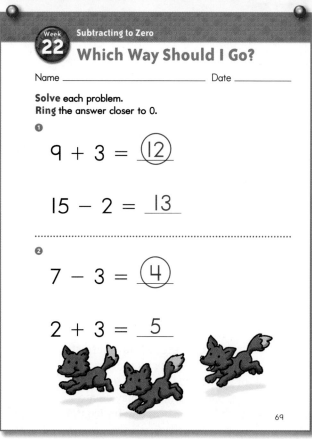

Workbook, p. 69

4 Assess

Informal Assessment

Use the Student Assessment Record, **Assessment,** page 100, to record informal observations.

COMPUTING	ENGAGING
Teacher's Choice Did the child	**Which Way Should I Go?** Did the child
❏ respond accurately?	❏ pay attention to the contributions of others?
❏ respond quickly?	
❏ respond with confidence?	
❏ self-correct?	❏ contribute information and ideas?
	❏ improve on a strategy?
	❏ reflect on and check accuracy of work?

Subtracting to Zero • Lesson 2 283

Week 22 — Subtracting to Zero

Objective
Children reinforce counting and number-sequence skills and gain experience writing and solving subtraction word problems.

Program Materials
Warm Up
See materials from previous weeks.

Engage
Counters

Prepare Ahead
Engage
Use index cards to create several subtraction problems with an answer of 0 that are suitable for your class (e.g., 9 − 9 = ___, 4 − 4 = ___, and so on). You may create problems with both vertical and horizontal formats.

Access Vocabulary
pattern Something that repeats the same way each time

Creating Context
Help English Learners expand their vocabulary by brainstorming different ways to say *zero*. Start a word wall with a large *0* at the top, and list the words used to describe this concept. Help early English Learners by asking questions that include physical cues, such as "Show me 0 fingers."

Math-a-saurus

1 Warm Up 5

Skill Building — COMPUTING
Teacher's Choice
Before beginning the whole-group activity **Subtracting to Zero,** choose one of the Warm-Up activities from previous weeks to use with the whole group. These activities include the following:

- Object Land: **Pointing and Winking**
- Picture Land: **Catch the Teacher**
- Line Land: **The Neighborhood**
- Sky Land: **Thermometer Counting**
- Object Land: **What Number Am I?**
- Circle Land: **Count Up**
- Circle Land: **Blastoff!**

Purpose Children reinforce counting and number-sequence skills.

Warm-Up Card 1

Monitoring Student Progress
If . . . children need help with basic skills,
Then . . . practice counting and number-sequence skills with them throughout the day.

2 Engage 30

Concept Building — REASONING
Subtracting to Zero
"Today we are going to make up and solve subtraction stories."

Follow the instructions on the Activity Card to play **Subtracting to Zero.** As children play, ask questions about what is happening in the activity.

Activity Card 40

284 Number Worlds • Week 22

Purpose Children gain experience writing and solving subtraction word problems and realize that any number subtracted from itself will equal 0.

Monitoring Student Progress

| **If** . . . children can fluently construct a story to describe the formal equation, | **Then** . . . challenge them to write their own equations and to make up a story to describe them. |

Teacher's Note Encourage children to use this week's vocabulary words as they engage in the activities, discuss math concepts, and make predictions.

 For additional practice with subtraction, children should complete **Building Blocks** Count Free Explore.

3 Reflect 10

Extended Response REASONING
Ask questions such as the following:

- **What patterns do you see?** Accept all reasonable answers. Children may observe that any number minus itself equals 0.
- **What is 3 − 0?** 3 **10 − 0?** 10 **Does a number minus 0 always equal itself?** yes **Why?** Accept all reasonable answers. 0 means there is nothing to take away.

4 Assess

Informal Assessment
Use the Student Assessment Record, **Assessment,** page 100, to record informal observations.

COMPUTING **Teacher's Choice** Did the child	REASONING **Subtracting to Zero** Did the child
❏ respond accurately? ❏ respond quickly? ❏ respond with confidence? ❏ self-correct?	❏ provide a clear explanation? ❏ communicate reasons and strategies? ❏ choose appropriate strategies? ❏ argue logically?

Subtracting to Zero • Lesson 3

Week 22 — Subtracting to Zero

Objective
Children reinforce counting and subtraction skills, read formally notated plus and minus signs, and determine which action to take.

Program Materials
Warm Up
See materials from previous weeks.

Engage
- Magnetic Number Line, 1 per child
- Magnetic Chips
- +\− Cube
- Number 1–6 Cube

Access Vocabulary
equation Number story

Creating Context
Symbols can be very helpful for English Learners. The cube in this activity has plus and minus symbols that tell players whether to move up or down the Magnetic Number Line. Ask children about other symbols that help them know which way to go, such as street signs that give directions, exit signs, and so on.

1 Warm Up

Skill Building — COMPUTING
Teacher's Choice

Before beginning the small-group activity **Which Way Should I Go?** choose one of the Warm-Up activities from previous weeks to use with the whole group. These activities include the following:

- Object Land: **Pointing and Winking**
- Picture Land: **Catch the Teacher**
- Line Land: **The Neighborhood**
- Sky Land: **Thermometer Counting**
- Object Land: **What Number Am I?**
- Circle Land: **Count Up**
- Circle Land: **Blastoff!**

Warm-Up Card 1

Purpose Children reinforce counting and number-sequence skills.

Monitoring Student Progress
If . . . children can work fluently with numbers 1 through 10,

Then . . . challenge them to work with bigger numbers.

2 Engage

Concept Building — APPLYING
Which Way Should I Go?

"Today we're going to use plus and minus signs to see who can get to zero first."

Follow the instructions on the Activity Card to play **Which Way Should I Go?** As children play, ask questions about what is happening in the activity.

Activity Card 41

Purpose Children subtract small quantities from single- and double-digit numbers, read formally notated plus and minus signs, and determine which action to take.

Monitoring Student Progress

| **If . . .** children are comfortable with this level of play, | **Then . . .** challenge them to write equations that correspond to their moves along the Magnetic Number Line. |

 For additional practice with subtraction, children should complete **Building Blocks** Function Machine 1.

3 Reflect 10

Extended Response REASONING
Ask questions such as the following:

- **If you roll a minus sign, do you want to roll a big number or a little number?** a big number
  Why? Possible answer: because it will help you get to 0 faster

- **If you roll a plus sign, do you want to roll a big number or a little number?** a little number
  Why? Possible answer: because a big number will take you farther away from 0

4 Assess

Informal Assessment
Use the Student Assessment Record, **Assessment**, page 100, to record informal observations.

COMPUTING	APPLYING
Teacher's Choice Did the child	**Which Way Should I Go?** Did the child
❏ respond accurately?	❏ apply learning to new situations?
❏ respond quickly?	❏ contribute concepts?
❏ respond with confidence?	❏ contribute answers?
❏ self-correct?	❏ connect mathematics to the real world?

Subtracting to Zero • Lesson 4

Week 22: Subtracting to Zero

Lesson 5

Review

Objective
Children review the material presented in Week 22 of *Number Worlds*.

Program Materials

Warm Up
See materials from previous weeks.

Engage
See materials from previous weeks.

Creating Context
English Learners often understand math concepts better when they can interact with other students who speak the same primary language. Allow children to work with partners or cross-age helpers who speak the same primary language so they can check understanding with one another in team situations.

1. Warm Up

Skill Building — COMPUTING

Teacher's Choice
Before beginning the **Free-Choice** activity, choose one of the Warm-Up activities from previous weeks to use with the whole group. These activities include the following:

- Object Land: **Pointing and Winking**
- Picture Land: **Catch the Teacher**
- Line Land: **The Neighborhood**
- Sky Land: **Thermometer Counting**
- Object Land: **What Number Am I?**
- Circle Land: **Count Up**
- Circle Land: **Blastoff!**

Warm-Up Card 1

Purpose Children reinforce counting and number-sequence skills.

2. Engage

Concept Building — APPLYING

Free-Choice Activity
For the last day of the week, allow children to choose an activity from the previous weeks. Some activities they may choose include the following:

- Object Land: **More Food Fun**
- Picture Land: **Making and Writing Equations**
- Line Land: **Tear It Up and Throw It Away**

Make a note of the activities children select. Do they prefer easy or challenging activities? Continue to provide Challenge opportunities for children who have mastered the basic activities.

Monitoring Student Progress

| If . . . children would benefit from extra practice on specific skills, | Then . . . choose an activity for them. |

288 Number Worlds • Week 22

Reflect

Extended Response — REASONING
Ask questions such as the following:
- **What did you like about playing** Subtracting to Zero?
- **Was there anything about playing this game you didn't like?**
- **Did this game help you do something you couldn't do before? What did it help it do?**
- **What was easy when you were playing** Which Way Should I Go?
- **What was hard when you were playing** Which Way Should I Go?
- **What do you remember most about this game?**
- **What was your favorite part?**

Using Student Pages
Have each child complete **Workbook,** page 70. Is the child correctly applying the skills learned in Week 22?

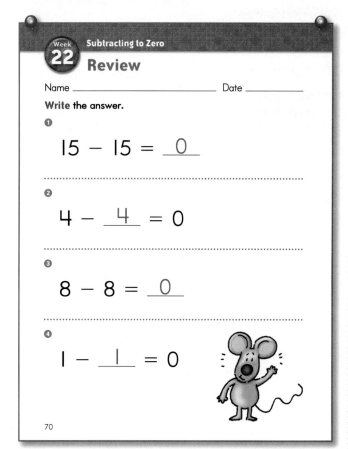

Workbook, p. 70

 ## Assess

A Gather Evidence
Formal Assessment
Have children complete the weekly test on *Assessment,* page 67. Record formal assessment scores on the Student Assessment Record, *Assessment,* page 100.

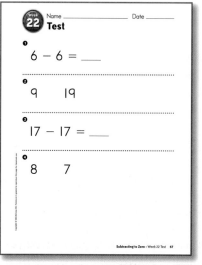

Assessment, p. 67

B Summarize Findings
Review the Student Assessment Records. Determine whether children have Minimal, Basic, or Secure understanding of the concepts presented in Week 22.

C Differentiate Instruction
Based on your observations, use these teaching strategies next week to follow up.

Minimal Understanding
- Repeat the Warm-Up and Engage activities in subsequent weeks to develop numeration concepts.
- Repeat the activities from this week. Observe children's understanding of these concepts and reinforce as necessary.
- Use **Building Blocks** computer activities beginning with Count Free Explore to develop and reinforce numeration concepts.

Basic Understanding
- Repeat Engage activities in subsequent weeks to reinforce basic concepts.
- Use **Building Blocks** computer activities beginning with Count Free Explore to reinforce subtraction concepts.

Secure Understanding
- Use Challenge variations of this week's activities.
- Use **Building Blocks** computer activities beginning with Function Machine 1.

Subtracting to Zero • Lesson 5 Review 289

Week 23: More Adding and Subtracting

Week at a Glance

This week, children begin **Number Worlds,** Week 23, and continue to explore Line Land and Picture Land.

Background

In Line Land, a decrease in quantity is shown as backward movement along a number line. In Picture Land, numbers are represented as numerals. This allows children to forge links between the world of objects and the world of formal symbols, in which a decrease in quantity is represented by the minus sign.

How Children Learn

As children begin this week's lesson, they should be able to solve and write basic addition and subtraction problems.

By the end of this week, children should have gained fluency adding and subtracting small quantities from single- and double-digit numbers and writing addition and subtraction equations.

Skills Focus
- Add and subtract whole numbers
- Write addition and subtraction equations

Teaching for Understanding

As children engage in these activities, they should deepen their understanding of addition, subtraction, and the formal notation used when writing equations.

Observe closely while evaluating the Engage activities for this week.

- Can children correctly add and subtract quantities from single- and double-digit numbers?
- Can children correctly write addition and subtraction equations?

Math at Home

Give one copy of the Letter to Home, page A23, to each child. Complete the activity in class, and then encourage children to share it with their caregivers.

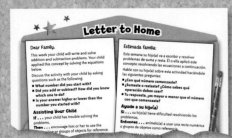

Letter to Home, Teacher Edition, p. A23

Math Vocabulary
facedown Cards on a table with the dots or numerals down (not showing)
stack Pile of flat things such as cards or books
shuffle To mix together
pawn Game piece

English Learners
SPANISH COGNATES

English	Spanish
addition	adición
equal	igual
line	línea
minus	menos
zero	cero

ALTERNATE VOCABULARY
quest A search or adventure

Week 23 Planner: More Adding and Subtracting

PACING	LESSON	LEARNING GOALS	MATERIALS	TECHNOLOGY
DAY 1	**Warm Up** Teacher's Choice	Children reinforce counting and number-sequence skills.	Materials will be selected from those used in previous weeks.	**Building Blocks** Count Free Explore
	Activity 42 **Line Land:** Counting Back*	Children subtract quantities from single- and double-digit numbers and write corresponding equations.	**Program Materials** • Magnetic Number Line, 1 per child • Magnetic Chips • Number Cards (1–10) **Additional Materials** Paper and pencils	
DAY 2	**Warm Up** Teacher's Choice	Children reinforce counting and number-sequence skills.	Materials will be selected from those used in previous weeks.	**Building Blocks** Function Machine 1
	Activity 32 **Picture Land:** Dragon Quest 2*	Children use addition and subtraction in successive operations.	**Program Materials** • Dragon Quest Game Board • Pawns, Spinner • Dragon Quest Cards +1 through +5, −1 and −2 • Picture Land Activity Sheet 10, 1 per child **Additional Materials** Paper and pencils	
DAY 3	**Warm Up** Teacher's Choice	Children reinforce counting and number-sequence skills.	Materials will be selected from those used in previous weeks.	**Building Blocks** Count Free Explore
	Activity 42 **Line Land:** Counting Back*	Children subtract quantities from single- and double-digit numbers.	**Program Materials** Same as Lesson 1	
DAY 4	**Warm Up** Teacher's Choice	Children reinforce counting and number-sequence skills.	Materials will be selected from those used in previous weeks.	**Building Blocks** Function Machine 1
	Activity 32 **Picture Land:** Dragon Quest 2*	Children use addition and subtraction in successive operations.	**Program Materials** Same as Lesson 2	
DAY 5	**Warm Up** Teacher's Choice	Children reinforce counting and number-sequence skills.	Materials will be selected from those used in previous weeks.	**Building Blocks** Review previous activities.
	Review and Assess	Children review their favorite activities to improve understanding.	Materials will be selected from those used in previous weeks.	

* Includes Challenge Variations

Week 23 — Lesson 1
More Adding and Subtracting

Objective
Children practice numeration skills and subtract quantities from single- and double-digit numbers.

Program Materials
Warm Up
See materials for previous weeks.

Engage
- Magnetic Number Line, 1 per child
- Magnetic Chips
- Number Cards (1–10)

Additional Materials
Engage
Paper and pencils

Access Vocabulary
shuffle To mix together
facedown Cards on a table with the dots or numerals down (not showing)

Creating Context
Reinforce the concept of predicting. Ask, "How many of you watch the weather report on television? The reporters predict what the weather will be like based on information they have. In this activity, we will predict where we will be on a number line at the end of each turn."

1 Warm Up — 5

Skill Building — COMPUTING
Teacher's Choice
Before beginning the small-group activity **Counting Back**, choose one of the Warm-Up activities from previous weeks to use with the whole group. These activities include the following:
- Object Land: **Pointing and Winking**
- Picture Land: **Catch the Teacher**
- Line Land: **The Neighborhood**
- Sky Land: **Thermometer Counting**
- Object Land: **What Number Am I?**
- Circle Land: **Count Up**
- Circle Land: **Blastoff!**

Purpose Children reinforce counting and number-sequence skills.

Monitoring Student Progress
If . . . children need help with basic skills,
Then . . . practice counting and number-sequence skills with them throughout the day.

Warm-Up Card 1

2 Engage — 30

Concept Building — UNDERSTANDING
Counting Back
"Today we are going to play a subtraction game."

Follow the instructions on the Activity Card to play **Counting Back**. As children play, ask questions about what is happening in the activity.

Activity Card 42

292 Number Worlds • Week 23

Purpose Children use formal notation to track backward progression on a number line.

Monitoring Student Progress

| **If . . .** children have trouble with the subtraction operation, | **Then . . .** have them play the game with Number Cards (1–5) and gradually work up to Number Cards (1–10). |

 Teacher's Note Encourage children to use this week's vocabulary words as they engage in the activities, discuss math concepts, and make predictions.

Building Blocks For additional practice with subtraction, children should complete *Building Blocks* Count Free Explore.

3 Reflect — 10

Extended Response REASONING
Ask questions such as the following:
- **On your first turn, how many numbers did you need to move?** Answers will vary.
- **Where were you after you did that?** Answers will vary. **How did you figure that out?** Accept all reasonable answers. Children may have counted backward from the biggest number on which a Magnetic Chip rests. They may also have counted off the number of chips indicated on the Number Card, starting with the chip that rests on the biggest number on the Magnetic Number Line.

Using Student Pages
Have each child complete *Workbook,* page 71. Did the child write the correct number story?

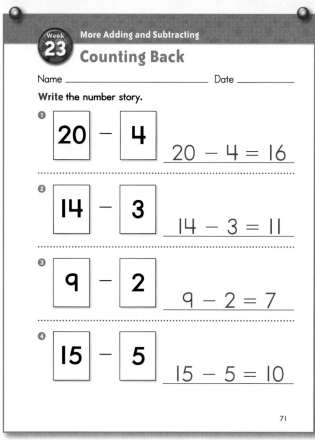

Workbook, p. 71

4 Assess

Informal Assessment
Use the Student Assessment Record, **Assessment,** page 100, to record informal observations.

COMPUTING	**UNDERSTANDING**
Teacher's Choice	**Counting Back**
Did the child	Did the child
❏ respond accurately?	❏ make important observations?
❏ respond quickly?	❏ extend or generalize learning?
❏ respond with confidence?	❏ provide insightful answers?
❏ self-correct?	❏ pose insightful questions?

More Adding and Subtracting • Lesson 1 293

Week 23 — More Adding and Subtracting

Lesson 2

Objective

Children practice numeration skills and use addition and subtraction in successive operations.

Program Materials

Warm Up
See materials from previous weeks.

Engage
- Dragon Quest Game Board
- Pawns
- Spinner
- Dragon Quest Cards +1 through +5, −1 and −2
- Picture Land Activity Sheet 10, 1 per child

Additional Materials

Engage
Paper and pencils

Prepare Ahead

Make sure the pile of Dragon Quest Cards contains only Addition Cards 1–5 and Subtraction Cards −1 and −2. Make sure the top six cards in the facedown pile are addition cards so that children will not have to subtract from 0 until they are ready to handle this challenge.

Access Vocabulary

quest A search or adventure
pawn Game piece

Creating Context

Be sensitive to the fact that some children may not have experience with board games. Discuss the routines used in board games such as determining who starts the game, taking turns, and knowing how the game is won.

1 Warm Up

Skill Building — COMPUTING

Teacher's Choice

Before beginning the small-group activity **Dragon Quest 2,** choose one of the Warm-Up activities from previous weeks to use with the whole group. These activities include the following:

- Object Land: **Pointing and Winking**
- Picture Land: **Catch the Teacher**
- Line Land: **The Neighborhood**
- Sky Land: **Thermometer Counting**
- Object Land: **What Number Am I?**
- Circle Land: **Count Up**
- Circle Land: **Blastoff!**

Warm-Up Card 1

Purpose Children reinforce counting and number-sequence skills.

Monitoring Student Progress

If . . . children need help with basic skills,

Then . . . practice counting and number-sequence skills with them throughout the day.

2 Engage

Concept Building — ENGAGING

Dragon Quest 2

"In today's game, you will collect buckets of water that you can use to put out a dragon's fire."

Follow the instructions on the Activity Card to play **Dragon Quest 2: The Dragon's Sister.** As children play, ask questions about what is happening in the activity.

Activity Card 32

294 Number Worlds • Week 23

Purpose Children use addition and subtraction in successive operations.

> **Monitoring Student Progress**
>
> **If . . .** children have trouble doing the subtraction problems,
>
> **Then . . .** allow them to use a number line for reference.

 For further practice with addition and subtraction, children should complete **Building Blocks** Function Machine 1.

3 Reflect — 10

Extended Response REASONING

Ask questions such as the following:

- **What symbol do we use to show we are adding (or subtracting) something?** plus sign (minus sign)
- **If you had 12 buckets of water, how many more would you need to put out the dragon's fire?** 3 **How did you figure that out?** Children may count on to determine the number of buckets still needed.

Using Student Pages

Have each child complete **Workbook,** page 72. Did the child write the correct answer?

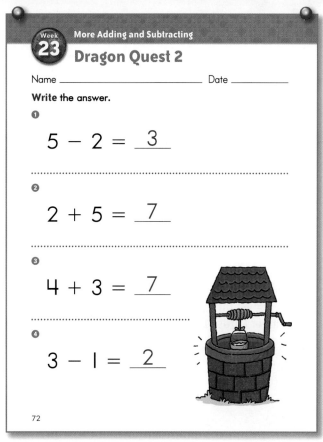

Workbook, p. 72

4 Assess

Informal Assessment

Use the Student Assessment Record, **Assessment,** page 100, to record informal observations.

COMPUTING	ENGAGING
Teacher's Choice	**Dragon Quest 2**
Did the child	Did the child
❏ respond accurately?	❏ pay attention to the contributions of others?
❏ respond quickly?	
❏ respond with confidence?	❏ contribute information and ideas?
❏ self-correct?	
	❏ improve on a strategy?
	❏ reflect on and check accuracy of work?

More Adding and Subtracting • Lesson 2 **295**

Week 23: More Adding and Subtracting

Lesson 3

Objective

Children practice numeration skills and subtract quantities from single- and double-digit numbers.

Program Materials

Warm Up
See materials for previous weeks.

Engage
- Magnetic Number Line, 1 per child
- Magnetic Chips
- Number Cards (1–10)

Additional Materials
Engage
Paper and pencils

Access Vocabulary

stack Pile of flat things such as cards or books

Creating Context

The variety of words used to describe subtraction (*take away, minus, less, subtracted from,* and so on) may be puzzling for English Learners. Help children brainstorm a list of words and phrases used to describe subtraction.

1 Warm Up — 5

Skill Building COMPUTING
Teacher's Choice

Before beginning the small-group activity **Counting Back**, choose one of the Warm-Up activities from previous weeks to use with the whole group. These activities include the following:

- Object Land: **Pointing and Winking**
- Picture Land: **Catch the Teacher**
- Line Land: **The Neighborhood**
- Sky Land: **Thermometer Counting**
- Object Land: **What Number Am I?**
- Circle Land: **Count Up**
- Circle Land: **Blastoff!**

Purpose Children reinforce counting and number-sequence skills.

Warm-Up Card 1

Monitoring Student Progress

If . . . children need help with basic skills,

Then . . . practice counting and number-sequence skills with them throughout the day.

2 Engage — 30

Concept Building REASONING
Counting Back

"Today we are going to play a subtraction game."

Follow the instructions on the Activity Card to play **Counting Back**. As children play, ask questions about what is happening in the activity.

Activity Card 42

296 Number Worlds • Week 23

Purpose Children use formal notation to track backward progression on a number line.

Monitoring Student Progress

| If ... children can fluently write and solve the subtraction problems, | Then ... challenge them to play the game with Number Cards (1–20). |

eMathTools Use the Number Line Tool to demonstrate and explore number sequence.

Building Blocks For additional practice with subtraction and addition, children should complete **Building Blocks** Count Free Explore.

3 Reflect 10

Extended Response REASONING

Ask questions such as the following:

- **What would happen if you played the game with bigger Number Cards? Would you get to 0 faster or slower? Why?** Faster. Bigger numbers would allow players to make bigger jumps along the number line.

- **What would happen if you played the game with addition cards as well as subtraction cards? Would you get to 0 faster or slower? Why?** Slower. Addition cards would move a player farther from 0 instead of closer.

4 Assess

Informal Assessment

Use the Student Assessment Record, **Assessment,** page 100, to record informal observations.

COMPUTING	REASONING
Teacher's Choice	**Counting Back**
Did the child	Did the child
❑ respond accurately?	❑ provide a clear explanation?
❑ respond quickly?	❑ communicate reasons and strategies?
❑ respond with confidence?	
❑ self-correct?	❑ choose appropriate strategies?
	❑ argue logically?

More Adding and Subtracting • Lesson 3 297

Week 23 — More Adding and Subtracting
Lesson 4

Objective
Children practice numeration skills and use addition and subtraction in successive operations.

Program Materials
Warm Up
See materials from previous weeks.
Engage
- Dragon Quest Game Board
- Pawns
- Spinner
- Dragon Quest Cards +1 through +5, −1 and −2
- Picture Land Activity Sheet 10, 1 per child

Additional Materials
Engage
Paper and pencils

Prepare Ahead
Make sure the pile of Dragon Quest Cards contains only Addition Cards 1–5 and Subtraction Cards −1 and −2. Make sure the top six cards in the facedown pile are addition cards.

Access Vocabulary
add To put together
subtract To take away

Creating Context
Some English Learners may know what the plus, minus, and equal signs mean but may not know how to articulate this understanding. Ask questions that let children show comprehension by pointing or by manipulating objects to demonstrate meaning. Model sentence frames such as "____ + ____ = ____" and "____ − ____ = ____" for children to use as they talk about equations.

1 Warm Up

Skill Building COMPUTING
Teacher's Choice
Before beginning the small-group activity **Dragon Quest 2,** choose one of the Warm-Up activities from previous weeks to use with the whole group. These activities include the following:
- Object Land: **Pointing and Winking**
- Picture Land: **Catch the Teacher**
- Line Land: **The Neighborhood**
- Sky Land: **Thermometer Counting**
- Object Land: **What Number Am I?**
- Circle Land: **Count Up**
- Circle Land: **Blastoff!**

Warm-Up Card 1

Purpose Children reinforce counting and number-sequence skills.

Monitoring Student Progress
If . . . children can work fluently with numbers 1–10,
Then . . . challenge them to work with bigger numbers.

2 Engage

Concept Building APPLYING
Dragon Quest 2: The Dragon's Sister
"In today's game, you will collect buckets of water you can use to put out a dragon's fire."

Follow the instructions on the Activity Card to play **Dragon Quest 2: The Dragon's Sister.** As children play, ask questions about what is happening in the activity.

Activity Card 32

Purpose Children use addition and subtraction in successive operations.

Monitoring Student Progress

If . . . children can fluently perform the addition and subtraction required for the activity,

Then . . . have them play one of the Challenge variations.

 Building Blocks For further practice with addition and subtraction, children should complete **Building Blocks** Function Machine 1.

3 Reflect

Extended Response REASONING

Ask questions such as the following:

- **How do you know if you've won the game?** Children have won when they get at least fifteen buckets of water and land in the dragon's lair. **How can you prove you're the winner?** They may check their answers by adding and subtracting their cards a second time or by reviewing the Activity Sheet.

- **Did anyone get a lot of big numbers? Did this help you get a lot of buckets, or did it slow you down? Why?** Encourage children to refer to their completed Activity Sheets to explain their answers.

4 Assess

Informal Assessment

Use the Student Assessment Record, **Assessment,** page 100, to record informal observations.

COMPUTING	APPLYING
Teacher's Choice	**Dragon Quest 2**
Did the child	Did the child
❏ respond accurately?	❏ apply learning to new situations?
❏ respond quickly?	❏ contribute concepts?
❏ respond with confidence?	❏ contribute answers?
❏ self-correct?	❏ connect mathematics to the real world?

More Adding and Subtracting • Lesson 4   299

Week 23: More Adding and Subtracting

Lesson 5 Review

Objective
Children review the material presented in Week 23 of **Number Worlds**.

Program Materials
Warm Up
See materials from previous weeks.

Engage
See materials from previous weeks.

Creating Context
Although it is important to ask beginning English Learners questions that have a lower language demand, this does not mean that questions should have a lower cognitive load. Ask English Learners to demonstrate their understanding with manipulatives, models, and illustrations. Check understanding by asking children to show, demonstrate, or point to the answer.

1 Warm Up

Skill Building — COMPUTING
Teacher's Choice
Before beginning the **Free-Choice** activity, choose one of the Warm-Up activities from previous weeks to use with the whole group. These activities include the following:
- Object Land: **Pointing and Winking**
- Picture Land: **Catch the Teacher**
- Line Land: **The Neighborhood**
- Sky Land: **Thermometer Counting**
- Object Land: **What Number Am I?**
- Circle Land: **Count Up**
- Circle Land: **Blastoff!**

Warm-Up Card 1

Purpose Children reinforce counting and number-sequence skills.

2 Engage

Concept Building — APPLYING
Free-Choice Activity
For the last day of the week, allow children to choose an activity from the previous weeks. Some activities they may choose include the following:
- Picture Land: **Dragon Quest 1**
- Object Land: **They're Gone**
- Picture Land: **Subtracting to Zero**

Make a note of the activities children select. Do they prefer easy or challenging activities? Continue to provide Challenge opportunities for children who have mastered the basic activities.

Monitoring Student Progress
| If . . . children would benefit from extra practice on specific skills, | Then . . . choose an activity for them. |

Reflect

Extended Response REASONING
Ask questions such as the following:
- What did you like about playing Counting Back?
- Was there anything about playing this game you didn't like?
- Did this game help you do something you couldn't do before? What did it help you do?
- What was easy when you were playing Dragon Quest 2?
- What was hard when you were playing Dragon Quest 2?
- What do you remember most about this game?
- What was your favorite part?

Using Student Pages
Have each child complete **Workbook,** page 73. Is the child correctly applying the skills learned in Week 23?

Week 23 More Adding and Subtracting
Review

Name _____ Date _____
Write the answer.

1. $13 - 6 = \underline{7}$
2. $8 - 4 = \underline{4}$
3. $12 - 3 = \underline{9}$
4. $10 - 4 = \underline{6}$

73

Workbook, p. 73

Assess

A Gather Evidence
Formal Assessment
Have children complete the weekly test on *Assessment,* page 69. Record formal assessment scores on the Student Assessment Record, *Assessment,* page 100.

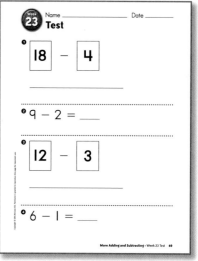

Assessment, p. 69

B Summarize Findings
Review the Student Assessment Records. Determine whether children have Minimal, Basic, or Secure understanding of the concepts presented in Week 23.

C Differentiate Instruction
Based on your observations, use these teaching strategies next week to follow up.

Minimal Understanding
- Repeat the Warm-Up and Engage activities in subsequent weeks to develop numeration concepts.
- Repeat the activities from this week. Observe children's understanding of these concepts and reinforce as necessary.
- Use **Building Blocks** computer activities beginning with Count Free Explore to develop and reinforce numeration concepts.

Basic Understanding
- Repeat Engage activities in subsequent weeks to reinforce basic concepts.
- Use **Building Blocks** computer activities beginning with Count Free Explore to reinforce numeration and addition concepts.

Secure Understanding
- Use Challenge variations of this week's activities.
- Use **Building Blocks** computer activities beginning with Function Machine 1.

More Adding and Subtracting • Lesson 5 Review 301

Week 24: Numbers to 100

Week at a Glance

This week, children begin **Number Worlds**, Week 24, and continue to explore Line Land and Sky Land.

Background

In Sky Land, numbers are represented as positions on a scale. When children are asked "How high are you now?" they might think of the height on the scale, the set size that corresponds to that position, or the number that shows the position on the scale.

How Children Learn

As children begin this week's lesson, they should be able to move up and down a number line to and from 20 and should be able to subtract small quantities from single- and double-digit numbers.

At the end of this week, children should be able to identify numerals to 100 and know the location of each number in the sequence.

Skills Focus

- Identify numerals 1–100
- Compare and order numbers
- Subtract whole numbers

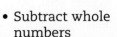

Teaching for Understanding

As children engage in these activities, they will explore the concept of magnitude and will begin to understand that rolling or choosing large numbers allows a player to move through the number sequence more quickly and in fewer steps.

Observe closely while evaluating the Engage activities assigned for this week.

- Can children correctly order numbers to 100?
- Can children accurately subtract whole numbers?

Math at Home

Give one copy of the Letter to Home, page A24, to each child. Complete the activity in class, and then encourage children to share it with their caregivers.

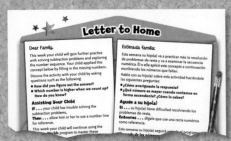

Letter to Home, Teacher Edition, p. A24

Math Vocabulary

position Where something is

equation Number story

pawn Game piece

most More than anyone else

English Learners

SPANISH COGNATES

English	Spanish
turn	turno
counters	contadores
count	contar
distance	distancia
line	línea

ALTERNATE VOCABULARY

elevator A small room that moves from one floor to another in a building

302 Number Worlds • Week 24

Week 24 Planner — Numbers to 100

PACING	LESSON	LEARNING GOALS	MATERIALS	TECHNOLOGY
DAY 1	Warm Up Teacher's Choice	Children reinforce counting and number-sequence skills.	Materials will be selected from those used in previous weeks.	**Building Blocks** Function Machine 1
	Activity 43 **Sky Land:** Elevator Game*	Children subtract small quantities from 10.	**Program Materials** • Elevator Game Board • Elevator Cards • Counters (10 of each color) • Multi-Land Activity Sheet 2	
DAY 2	Warm Up Teacher's Choice	Children reinforce counting and number-sequence skills.	Materials will be selected from those used in previous weeks.	**Building Blocks** Counting Activity 6
	Activity 44 **Line Land:** Number Line to 100	Children reinforce number sequence skills and identify numerals from 1 to 100.	**Program Materials** • Number Line Game Board to 100 • Pawns • Dot Cubes, or Number 7–12 Cubes **Additional Materials** Paper and pencils	
DAY 3	Warm Up Teacher's Choice	Children reinforce counting and number-sequence skills.	Materials will be selected from those used in previous weeks.	**Building Blocks** Function Machine 1
	Activity 43 **Sky Land:** Elevator Game*	Children subtract small quantities from 10.	**Program Materials** Same as Lesson 1	
DAY 4	Warm Up Teacher's Choice	Children reinforce counting and number-sequence skills.	Materials will be selected from those used in previous weeks.	**Building Blocks** Counting Activity 6
	Activity 44 **Line Land:** Number Line to 100	Children reinforce number sequence skills and identify numerals from 1 to 100.	**Program Materials** Same as Lesson 2	
DAY 5	Warm Up Teacher's Choice	Children reinforce counting and number-sequence skills.	Materials will be selected from those used in previous weeks.	**Building Blocks** Review previous activities
	Review and Assess	Children review their favorite activities to improve understanding.	Materials will be selected from those used in previous weeks.	

* Includes Challenge Variations

Week 24 — Numbers to 100
Lesson 1

Objective
Children reinforce counting and number-sequence skills and subtract small quantities from 10 on a vertical number line.

Program Materials
Warm Up
See materials from previous weeks.
Engage
- Elevator Game Board
- Elevator Cards
- Counters (10 of each color)
- Multi-Land Activity Sheet 2 (optional)

Access Vocabulary
elevator A small room that moves from one floor to another in a building
equation Number story

Creating Context
Demonstration and modeling are excellent comprehension strategies to use with English Learners. The Elevator Game Board provides a graphic representation of subtraction that allows children to be actively involved in learning the concept. Teaching the concept in a game format helps English Learners enjoy increased comprehension due to a lower stress level and purposeful repetition of the exercise.

---

1 Warm Up

Skill Building — COMPUTING
Teacher's Choice
Before beginning the small-group activity **Elevator Game**, choose one of the Warm-Up activities from previous weeks to use with the whole group. These activities include the following:
- Object Land: **Pointing and Winking**
- Picture Land: **Catch the Teacher**
- Line Land: **The Neighborhood**
- Sky Land: **Thermometer Counting**
- Object Land: **What Number Am I?**
- Circle Land: **Count Up**
- Circle Land: **Blastoff!**

Purpose Children reinforce counting and number-sequence skills.

Warm-Up Card 1

Monitoring Student Progress
If . . . children need help with basic skills,
Then . . . practice counting and number-sequence skills with them throughout the day.

2 Engage

Concept Building — UNDERSTANDING
Elevator Game
"Today we will use subtraction to get from the top of a building to the ground floor."

Follow the instructions on the Activity Card to play **Elevator Game.** As children play, ask questions about what is happening in the activity.

Purpose Children subtract small quantities from 10.

Activity Card 43

304 Number Worlds • Week 24

Monitoring Student Progress

If . . . children have trouble correctly subtracting quantities from 10,

Then . . . have them start on a lower number such as 5.

 For additional practice with subtraction, children should complete **Building Blocks** Function Machine 1.

3 Reflect — 10

Extended Response REASONING

Ask questions such as the following:

- **What number did you start on? Did you move up or down on your first turn? What number did you land on? How did you figure that out?** Accept all reasonable answers.
- **If you're on the seventh floor, how many do you need to take away to reach the ground floor? How did you figure that out?** Children should indicate that they need to subtract the number that will leave them with no Counters.

Using Student Pages

Have each child complete **Workbook,** page 74. Did the child write the correct answer?

Week 24 — Numbers to 100

Elevator Game

Name _____ Date _____

Write the answer.

1. $8 - 4 = \underline{4}$
2. $10 - 3 = \underline{7}$
3. $4 - 2 = \underline{2}$
4. $7 - 3 = \underline{4}$

74

Workbook, p. 74

4 Assess

Informal Assessment

Use the Student Assessment Record, **Assessment,** page 100, to record informal observations.

COMPUTING	UNDERSTANDING
Teacher's Choice	**Elevator Game**
Did the child	Did the child
❏ respond accurately?	❏ make important observations?
❏ respond quickly?	❏ extend or generalize learning?
❏ respond with confidence?	❏ provide insightful answers?
❏ self-correct?	❏ pose insightful questions?

Numbers to 100 • Lesson 1 305

Week 24 — Numbers to 100
Lesson 2

Objective
Children reinforce counting and number-sequence skills to 100.

Program Materials
Warm Up
See materials for previous weeks.

Engage
- Number Line Game Board to 100
- Pawns
- Dot Cubes or Number 7–12 cubes

Additional Materials
Engage
Paper and pencils

Access Vocabulary
pawn Game piece
most More than anyone else

Creating Context
Children will work with large and small numbers as they move along the number line. To help English Learners, use hand gestures that emphasize *big* and *small*. Use gestures while saying the instructions aloud so English Learners understand the directions and learn the terminology at the same time.

1 Warm Up 5

Skill Building — COMPUTING
Teacher's Choice

Before beginning the small-group activity **Number Line to 100,** choose one of the Warm-Up activities from previous weeks to use with the whole group. These activities include the following:
- Object Land: **Pointing and Winking**
- Picture Land: **Catch the Teacher**
- Line Land: **The Neighborhood**
- Sky Land: **Thermometer Counting**
- Object Land: **What Number Am I?**
- Circle Land: **Count Up**
- Circle Land: **Blastoff!**

Warm-Up Card 1

Purpose Children reinforce counting and number-sequence skills.

Monitoring Student Progress
If . . . children need help with basic skills,
Then . . . practice counting and number-sequence skills with them throughout the day.

2 Engage 30

Concept Building — ENGAGING
Number Line to 100

"Today we will play a game and try to get all the way to 100."

Follow the instructions on the Activity Card to play **Number Line to 100.** As children play, ask questions about what is happening in the activity.

Activity Card 44

Purpose Children reinforce number-sequence skills to 100, identify numerals from 1 to 100, and predict the results of adding whole numbers.

Monitoring Student Progress

| If... children have trouble remembering the number sequence, | Then... help them identify and discuss the patterns in the ones and tens places. |

eMathTools Use the Number Line Tools to demonstrate and explore number sequence.

Building Blocks For additional practice with number sequence, children should complete **Building Blocks** Off the Tree: Add Apples.

3 Reflect — 10

Extended Response REASONING
Toward the end of the game, ask questions such as the following:

- **What number are you on? What number is your classmate on? Which number is higher when we count up?** Answers will vary.
- **Who has gone the farthest? Who has gone the shortest distance? How did you figure that out?** Answers will vary. Children may indicate that the player with the biggest number has gone the farthest, and the player with the smallest number has gone the shortest distance along the number line.

Using Student Pages
Have each child complete **Workbook,** page 75. Did the child correctly identify the missing number?

Week 24 — Numbers to 100
Number Line to 100

Name _____ Date _____

Write the missing number.

① 43 44 _45_ 46 47

② 24 _25_ 26 27 28

③ 68 69 _70_ 71 72

④ 96 97 98 _99_ 100

Workbook, p. 75

4 Assess

Informal Assessment
Use the Student Assessment Record, **Assessment,** page 100, to record informal observations.

COMPUTING	ENGAGING
Teacher's Choice	**Number Line to 100**
Did the child	Did the child
❏ respond accurately?	❏ pay attention to the contributions of others?
❏ respond quickly?	
❏ respond with confidence?	❏ contribute information and ideas?
❏ self-correct?	
	❏ improve on a strategy?
	❏ reflect on and check accuracy of work?

Numbers to 100 • Lesson 2 307

Week 24 — Numbers to 100
Lesson 3

Objective
Children reinforce counting and number-sequence skills and subtract small quantities from 10 on a vertical number line.

Program Materials
Warm Up
See materials from previous weeks.
Engage
- Elevator Game Board
- Elevator Cards
- Counters (10 of each color)
- Multi-Land Activity Sheet 2 (optional)

Access Vocabulary
plus sign Symbol that means "add"
minus sign Symbol that means "take away"

Creating Context
In the Elevator Game the words *top* and *bottom* are used. Help children brainstorm phrases or expressions that include position words, and make a list to display on the classroom wall.

1 Warm Up

Skill Building — COMPUTING
Teacher's Choice
Before beginning the small-group activity **Elevator Game**, choose one of the Warm-Up activities from previous weeks to use with the whole group. These activities include the following:
- Object Land: **Pointing and Winking**
- Picture Land: **Catch the Teacher**
- Line Land: **The Neighborhood**
- Sky Land: **Thermometer Counting**
- Object Land: **What Number Am I?**
- Circle Land: **Count Up**
- Circle Land: **Blastoff!**

Purpose Children reinforce counting and number-sequence skills.

Warm-Up Card 1

Monitoring Student Progress
| If . . . children need help with basic skills, | Then . . . practice counting and number sequence-skills with them throughout the day. |

2 Engage

Concept Building — REASONING
Elevator Game
"Today we will use subtraction to get from the top of a building to the ground floor."

Follow the instructions on the Activity Card to play **Elevator Game**. As children play, ask questions about what is happening in the activity.

Purpose Children subtract small quantities from 10.

Activity Card 43

Monitoring Student Progress

If . . . children can fluently subtract quantities from 10,

Then . . . have them play the Challenge variation of the game by writing the equation that corresponds to each move.

 Teacher's Note Encourage children to use this week's vocabulary words as they engage in the activities, discuss math concepts, and make predictions.

Building Blocks For additional practice with subtraction, children should complete **Building Blocks** Function Machine 1.

3 Reflect 10

Extended Response REASONING
Ask questions such as the following:

- **Would you rather pick big numbers or little numbers? Why?** Accept all reasonable answers. Discuss with children that big numbers help them move down the elevator faster, but little numbers may be more useful when they are very close to the ground floor.
- **Can you think of other times at school or at home when you subtract?** Accept all reasonable answers.

4 Assess

Informal Assessment
Use the Student Assessment Record, **Assessment,** page 100, to record informal observations.

COMPUTING	REASONING
Teacher's Choice	**Elevator Game**
Did the child	Did the child
❑ respond accurately?	❑ provide a clear explanation?
❑ respond quickly?	❑ communicate reasons and strategies?
❑ respond with confidence?	❑ choose appropriate strategies?
❑ self-correct?	❑ argue logically?

Numbers to 100 • Lesson 3 309

Week 24 — Numbers to 100

Lesson 4

Objective
Children reinforce counting and number-sequence skills to 100.

Program Materials
Warm Up
See materials for previous weeks.

Engage
- Number Line Game Board to 100
- Pawns
- Dot Cubes, or Number 7–12 Cubes

Additional Materials
Engage
Paper and pencils

Access Vocabulary
position Where something is

Creating Context
Playing games is an excellent way to give English Learners practice listening to and speaking English. The natural repetition of procedural and counting language replaces tedious drills with authentic, active experience. Wait time is built into the process, and the low-stakes environment makes the use of games an enjoyable and efficient learning tool in math.

1 Warm Up

Skill Building — COMPUTING
Teacher's Choice

Before beginning the small-group activity **Number Line to 100,** choose one of the Warm-Up activities from previous weeks to use with the whole group. These activities include the following:

- Object Land: **Pointing and Winking**
- Picture Land: **Catch the Teacher**
- Line Land: **The Neighborhood**
- Sky Land: **Thermometer Counting**
- Object Land: **What Number Am I?**
- Circle Land: **Count Up**
- Circle Land: **Blastoff!**

Warm-Up Card 1

Purpose Children reinforce counting and number-sequence skills.

Monitoring Student Progress
If . . . children can work fluently with numbers 1 through 10,
Then . . . challenge them to work with bigger numbers.

2 Engage

Concept Building — APPLYING
Number Line to 100

"Today we will play a game and try to get all the way to 100."

Follow the instructions on the Activity Card to play **Number Line to 100.** As children play, ask questions about what is happening in the activity.

Activity Card 44

310 Number Worlds • Week 24

Purpose Children reinforce number-sequence skills to 100, identify numerals from 1 to 100, and predict the results of adding whole numbers.

Monitoring Student Progress

| If . . . children are comfortable with the number sequence, | Then . . . challenge them to write the addition equation that corresponds to each roll of the Dot Cube. |

 Building Blocks For additional practice with number sequence, children should complete *Building Blocks* Off the Tree: Add Apples.

3 Reflect 10

Extended Response REASONING

Ask questions such as the following:

- **What patterns do you see?** Accept all reasonable answers. Encourage children to discuss how the numbers 0 through 9 repeat along the number line.
- **Would you rather roll big numbers or little numbers? Why?** Accept all reasonable answers. Lead students to see that rolling large numbers helps them move along the number line faster.

4 Assess

Informal Assessment

Use the Student Assessment Record, **Assessment,** page 100, to record informal observations.

COMPUTING	APPLYING
Teacher's Choice	**Number Line to 100**
Did the child	Did the child
❏ respond accurately?	❏ apply learning to new situations?
❏ respond quickly?	❏ contribute concepts?
❏ respond with confidence?	❏ contribute answers?
❏ self-correct?	❏ connect mathematics to the real world?

Numbers to 100 • Lesson 4   311

Week 24 — Numbers to 100

Lesson 5: Review

Objective
Children review the concepts presented in Week 24 of **Number Worlds**.

Program Materials
Warm Up
See materials for previous weeks.

Engage
See materials for previous weeks.

Creating Context
Ask English Learners to show their understanding with manipulatives or models. Check understanding by asking children to demonstrate the answer.

✓ Cumulative Assessment
After Lesson 5 is complete, children should complete the Cumulative Assessment, **Assessment**, pages 94–95. Using the key on Assessment, page 93, identify incorrect responses. Reteach and review the activities to reinforce concept understanding.

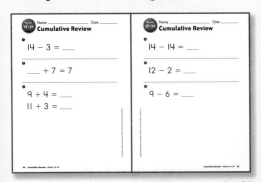

Assessment, pp. 94–95

---

1 Warm Up — 5

Skill Building COMPUTING
Teacher's Choice
Before beginning the **Free-Choice** activity, choose one of the Warm-Up activities from previous weeks to use with the whole group. These activities include the following:

- Object Land: **Pointing and Winking**
- Picture Land: **Catch the Teacher**
- Line Land: **The Neighborhood**
- Sky Land: **Thermometer Counting**
- Object Land: **What Number Am I?**
- Circle Land: **Count Up**
- Circle Land: **Blastoff!**

Warm-Up Card 1

Purpose Children reinforce counting and number-sequence skills.

---

2 Engage — 20

Concept Building APPLYING
Free-Choice Activity
For the last day of the week, allow children to choose an activity from the previous weeks. Some activities they may choose include the following:

- Picture Land: **Meet Your Neighbors**
- Picture Land: **Can You Guess My Number?**
- Picture Land: **Subtracting to Zero**

Make a note of the activities children select. Do they prefer easy or challenging activities? Continue to provide Challenge opportunities for children who have mastered the basic activities..

Monitoring Student Progress
If . . . children would benefit from extra practice on specific skills,

Then . . . choose an activity for them.

---

312 Number Worlds • Week 24

3. Reflect

Extended Response REASONING

Ask questions such as the following:

- **What did you like about playing** Elevator Game?
- **Was there anything about playing this game you didn't like?**
- **Did this game help you do something you couldn't do before? What did it help you do?**
- **What was easy when you were playing** Number Line to 100?
- **What was hard when you were playing** Number Line to 100?
- **What do you remember most about this game?**
- **What was your favorite part?**

Using Student Pages

Have each child complete **Workbook,** page 76. Is the child correctly applying the skills learned in Week 24?

Workbook, p. 76

4. Assess

A Gather Evidence

Formal Assessment

Have students complete the weekly test on **Assessment,** page 71. Record formal assessment scores on the Student Assessment Record, **Assessment,** page 100.

Assessment, p. 71

B Summarize Findings

Review the Student Assessment Records. Determine whether children have Minimal, Basic, or Secure understanding of the concepts presented in Week 24.

C Differentiate Instruction

Based on your observations, use these teaching strategies next week to follow up.

Minimal Understanding
- Repeat the Warm-Up and Engage activities in subsequent weeks to develop concepts of counting objects.
- Repeat the activities from this week. Observe children's understanding of these concepts and reinforce as necessary.
- Use **Building Blocks** computer activities beginning with Count Free Explore to develop and reinforce numeration concepts.

Basic Understanding
- Repeat Engage activities in subsequent weeks to reinforce basic concepts.
- Use **Building Blocks** computer activities beginning with Counting Activity 6 to reinforce numeration concepts.

Secure Understanding
- Use Challenge variations of this week's activities.
- Use **Building Blocks** computer activities beginning with Math-O-Scope.

Numbers to 100 • Lesson 5 Review 313

Week 25: More Numbers to 100

Week at a Glance

This week, children begin **Number Worlds**, Week 25, and continue to explore Line Land.

Background

In Line Land, numbers are represented as positions on a line. When a child is asked, "Where are you now?" he or she might think of a position on the number line, a set size that corresponds to that position, or a number that shows the position on the number line.

How Children Learn

At the beginning of this week, children should be able to identify numerals to 100 and know the location of each number in the sequence.

By the end of this week, children should have strengthened their understanding of the base-ten number system and should be able to identify and to write numerals to 100.

Skills Focus

- Count on
- Compare and order numbers
- Solve problems

Teaching for Understanding

As children engage in these activities, they will identify and sequence numbers to 100 while increasing their knowledge of the base-ten number system, or "number neighborhoods." This knowledge paves the way for more advanced work with addition and subtraction.

Observe closely while evaluating the Engage activities assigned for this week.

- Are children counting to 100?
- Are children correctly identifying relative positions of numbers?
- Can children correctly write numbers to 100?

Math at Home

Give one copy of the Letter to Home, page A25, to each child. Complete the activity in class, and then encourage children to share it with their caregivers.

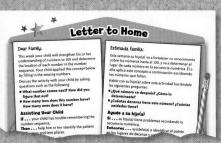

Letter to Home, Teacher Edition, p. A25

Math Vocabulary

before In front of; coming first

after Behind; coming later

near Close to

far Not close to; a long way away

English Learners

SPANISH COGNATES

English	Spanish
activity	activdad
compare	comparar
movement	movimiento
position	posición
progress	progreso

ALTERNATE VOCABULARY

deliver To take a message or package to someone

314 Number Worlds • Week 25

Week 25 Planner: More Numbers to 100

PACING	LESSON	LEARNING GOALS	MATERIALS	TECHNOLOGY
DAY 1	**Warm Up** Teacher's Choice	Children reinforce counting and number-sequence skills.	Materials will be selected from those used in previous weeks.	**Building Blocks** Before and After Math
	Activity 23 **Line Land:** Steve's New Bike*	Children identify and sequence numerals to 100 and relate number magnitude to distance traveled.	**Program Materials** • Neighborhood Number Line • Steve's Bike Pawn • Number Tiles (1–100) • Delivery Building Picture **Additional Materials** Small stickers	
DAY 2	**Warm Up** Teacher's Choice	Children reinforce counting and number-sequence skills.	Materials will be selected from those used in previous weeks.	**Building Blocks** Space Race: Number Choice
	Activity 45 **Line Land:** Writing Numbers to 100	Children write numbers 1–100 and know the positions of the ones and tens columns.	**Program Materials** Line Land Activity Sheet 4, 1 per child **Additional Materials** Pencils, 1 per child	**e MathTools** 100s Chart
DAY 3	**Warm Up** Teacher's Choice	Children reinforce counting and number-sequence skills.	Materials will be selected from those used in previous weeks.	**Building Blocks** Before and After Math
	Activity 23 **Line Land:** Steve's New Bike*	Children identify and sequence numerals to 100 and relate number magnitude to distance traveled.	**Program Materials** • Neighborhood Number Line • Steve's Bike Pawn • Number Tiles (1–100) • Delivery Building Picture **Additional Materials** Small stickers	**e MathTools** Number Line
DAY 4	**Warm Up** Teacher's Choice	Children reinforce counting and number-sequence skills.	Materials will be selected from those used in previous weeks.	**Building Blocks** Space Race: Number Choice
	Activity 45 **Line Land:** Writing Numbers to 100	Children write numbers 1–100 and know the positions of the ones and tens columns.	**Program Materials** Line Land Activity Sheet 4, 1 per child **Additional Materials** Pencils, 1 per child	
DAY 5	**Warm Up** Teacher's Choice	Children reinforce counting and number-sequence skills.	Materials will be selected from those used in previous weeks.	**Building Blocks** Review previous activities
	Review and Assess	Children review their favorite activities to improve understanding.	Materials will be selected from those used in previous weeks.	

*Includes Challenge Variations

Week 25 — More Numbers to 100

Lesson 1

Objective
Children identify and sequence numerals to 100 and relate number magnitude to distance traveled.

Program Materials
Warm Up
See materials from previous weeks.

Engage
- Neighborhood Number Line (all sections)
- Steve's Bike Pawn
- Number Tiles (1–100)
- Delivery Building Picture

Additional Materials

Engage
Small stickers

Access Vocabulary
near Close to
far Not close to; a long way away

Creating Context
Throughout this lesson, children will make comparisons such as describing Steve's longest and shortest deliveries. Review comparative and superlative forms of the following words:

long—longer—longest
short—shorter—shortest
high—higher—highest
near—nearer—nearest
far—farther—farthest

1 Warm Up — 5

Skill Building COMPUTING
Teacher's Choice
Before beginning the whole-class activity **Steve's New Bike,** choose one of the Warm-Up activities from previous weeks to use with the whole group. These activities include the following:
- Object Land: **Pointing and Winking**
- Picture Land: **Catch the Teacher**
- Line Land: **The Neighborhood**
- Sky Land: **Thermometer Counting**
- Object Land: **What Number Am I?**
- Circle Land: **Count Up**
- Circle Land: **Blastoff!**

Warm-Up Card 1

Purpose Children reinforce counting and number-sequence skills.

Monitoring Student Progress
If . . . children need help with basic skills,
Then . . . practice counting and number-sequence skills with them throughout the day.

2 Engage — 30

Concept Building UNDERSTANDING
Steve's New Bike

"Today you will help Steve deliver messages in the neighborhood."

Follow the instructions on the Activity Card to play **Steve's New Bike** with all sections of the Neighborhood Number Line. As children play, ask questions about what is happening in the activity.

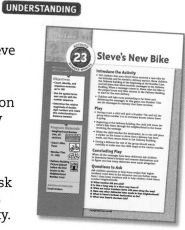

Activity Card 23

316 Number Worlds • Week 25

Purpose Children count, identify, and sequence numerals to 100; determine the relative magnitude of double-digit numbers; and relate this understanding to distance traveled.

Monitoring Student Progress

| If . . . a child delivers a message to the wrong address, | Then . . . ask the child to name the number he or she drew, and write it on the board. Ask the child what block of houses the number is in and how he or she knows. Then ask which house in that block the child will visit. If the child has trouble answering, call on other children to help. |

 For additional practice with the number sequence, children should complete **Building Blocks** Before and After Math.

3 Reflect 10

Extended Response REASONING
Ask questions such as the following:

- **Is the number 43 a long way or a short way from 0? 63?** (Repeat the question with several different numbers.) Accept all reasonable answers. Help children understand that higher numbers are farther from 0 than lower numbers.

- **What are some numbers Steve will pass along the way?** Answers will vary based on the number selected. Encourage children to discuss their answers in terms of number neighborhoods (the tens place).

Using Student Pages
Have each child complete **Workbook,** page 77. Did the child circle the correct answer?

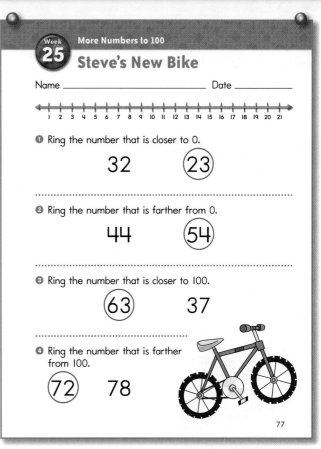

Workbook, p. 77

4 Assess

Informal Assessment
Use the Student Assessment Record, **Assessment,** page 100, to record informal observations.

COMPUTING	UNDERSTANDING
Teacher's Choice	**Steve's New Bike**
Did the child	Did the child
❏ respond accurately?	❏ make important observations?
❏ respond quickly?	
❏ respond with confidence?	❏ extend or generalize learning?
❏ self-correct?	❏ provide insightful answers?
	❏ pose insightful questions?

More Numbers to 100 • Lesson 1

Week 25 — More Numbers to 100

Lesson 2

Objective
Children work with numerals to 100 and know the positions of the ones and tens columns.

Program Materials
Warm Up
See materials from previous weeks.
Engage
Line Land Activity Sheet 4, 1 per child

Additional Materials
Engage
Pencils, 1 per child

Access Vocabulary
ones column Numeral on the far right of a number that tells how many ones are in the number
tens column Numeral to the left of the ones column that tells how many tens are in a number

Creating Context
English Learners often understand math concepts better when they can interact with other children who speak the same primary language. Allow children to work with partners or cross-age helpers who speak the same primary language so they can check understanding with one another in team situations.

1 Warm Up — 5

Skill Building COMPUTING
Teacher's Choice
Before beginning the small-group activity **Writing Numbers to 100,** choose one of the Warm-Up activities from previous weeks to use with the whole group. These activities include the following:

- Object Land: **Pointing and Winking**
- Picture Land: **Catch the Teacher**
- Line Land: **The Neighborhood**
- Sky Land: **Thermometer Counting**
- Object Land: **What Number Am I?**
- Circle Land: **Count Up**
- Circle Land: **Blastoff!**

Warm-Up Card 1

Purpose Children reinforce counting and number-sequence skills.

Monitoring Student Progress
If . . . children need help with basic skills,
Then . . . practice counting and number-sequence skills with them throughout the day.

2 Engage — 30

Concept Building ENGAGING
Writing Numbers to 100
"Today we are going to write numbers to 100."

Follow the instructions on the Activity Card to play **Writing Numbers to 100.** As children play, ask questions about what is happening in the activity.

Activity Card 45

318 Number Worlds • Week 25

Purpose Children know and write the number sequence 1–100 and know the positions of the ones and tens columns.

Monitoring Student Progress

If . . . children have trouble determining what neighborhood they are in (what the number in the tens place should be),

Then . . . remind them to look at the numbers that end in 0 on the number line. The first digit in those numbers tells children what neighborhood they are in.

MathTools Use the 100s Chart to demonstrate and explore place value.

Teacher's Note If children are not ready to write numbers to 100, begin by having them write numbers to 50.

Building Blocks For additional practice with numeration, children should complete **Building Blocks** Space Race: Number Choice.

Reflect   10

Extended Response   REASONING
Ask questions such as the following:

- **How did you know what number to write there?** Answers will vary. Possible answer: I counted up from 40: 41, 42, and so on.
- **How did you know what neighborhood you were in?** Possible answer: I was between 20 and 30 on the number line, so I was in the twenties. All of these numbers start with 2.

Using Student Pages
Have each child complete **Workbook,** page 78. Did the child write the correct numbers?

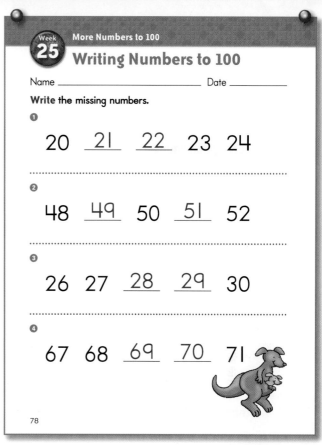

Week 25 More Numbers to 100
Writing Numbers to 100
Name _____ Date _____
Write the missing numbers.

1. 20 _21_ _22_ 23 24
2. 48 _49_ 50 _51_ 52
3. 26 27 _28_ _29_ 30
4. 67 68 _69_ _70_ 71

78

Workbook, p. 78

Assess

Informal Assessment
Use the Student Assessment Record, **Assessment,** page 100, to record informal observations.

COMPUTING	ENGAGING
Teacher's Choice Did the child ❑ respond accurately? ❑ respond quickly? ❑ respond with confidence? ❑ self-correct?	**Writing Numbers to 100** Did the child ❑ pay attention to the contributions of others? ❑ contribute information and ideas? ❑ improve on a strategy? ❑ reflect on and check accuracy of work?

More Numbers to 100 • Lesson 2

Week 25 — More Numbers to 100
Lesson 3

Objective
Children identify and sequence numerals to 100 and relate number magnitude to distance traveled.

Program Materials
Warm Up
See materials from previous weeks.

Engage
- Neighborhood Number Line (all sections)
- Steve's Bike Pawn
- Number Tiles (1–100)
- Delivery Building Picture

Additional Materials
Engage
small stickers

Access Vocabulary
pawn Game piece
deliver To take a message or package to someone

Creating Context
Using gestures is an excellent way to enhance comprehension for English Learners. In **Steve's New Bike,** children are asked questions about distance and comparison. Use hand gestures to help English Learners understand terms such as *nearest* and *farthest*.

1 Warm Up — 5

Skill Building COMPUTING
Teacher's Choice
Before beginning the whole-class activity **Steve's New Bike,** choose one of the Warm-Up activities from previous weeks to use with the whole group. These activities include the following:
- Object Land: **Pointing and Winking**
- Picture Land: **Catch the Teacher**
- Line Land: **The Neighborhood**
- Sky Land: **Thermometer Counting**
- Object Land: **What Number Am I?**
- Circle Land: **Count Up**
- Circle Land: **Blastoff!**

Warm-Up Card 1

Purpose Children reinforce counting and number-sequence skills.

Monitoring Student Progress
If . . . children need help with basic skills,
Then . . . practice counting and number-sequence skills with them throughout the day.

2 Engage — 30

Concept Building REASONING
Steve's New Bike
"Today you will help Steve deliver messages in the neighborhood."

Follow the instructions on the Activity Card to play **Steve's New Bike.** As children play, ask questions about what is happening in the activity.

Activity Card 23

320 Number Worlds • Week 25

Purpose Children count, identify, and sequence numerals to 100; determine the relative magnitude of double-digit numbers; and relate this understanding to distance traveled.

Monitoring Student Progress

| If . . . children have mastered the game, | Then . . . have them play the Challenge variation of **Steve's New Bike.** Encourage children to discuss alternative ways to figure out how much money Steve earned, such as counting by twos, creating a multiple-addend addition problem, and so on. |

 MathTools Use the Number Line Tool to demonstrate and explore number sequence.

Teacher's Note Children may enjoy playing this in small, independent groups when they are familiar with the procedure. Have them figure out Steve's longest and shortest deliveries, how many deliveries he made altogether, and how much money Steve earned.

Building Blocks For additional practice with the number sequence, children should complete **Building Blocks** Before and After Math.

 ## Reflect   10

Extended Response   REASONING

Ask questions such as the following:

- **What was Steve's longest ride so far? How do you know?** Answers will vary. Help children understand that larger numbers indicate a longer distance traveled along the number line.
- **What was Steve's shortest ride? How did you figure that out?** Answers will vary. Help children understand that smaller numbers indicate a shorter distance traveled along the number line.

4 Assess

Informal Assessment

Use the Student Assessment Record, **Assessment,** page 100, to record informal observations.

COMPUTING	REASONING
Teacher's Choice	**Steve's New Bike**
Did the child	Did the child
❏ respond accurately?	❏ provide a clear explanation?
❏ respond quickly?	❏ communicate reasons and strategies?
❏ respond with confidence?	
❏ self-correct?	❏ choose appropriate strategies?
	❏ argue logically?

More Numbers to 100 • Lesson 3   **321**

Week 25 — More Numbers to 100
Lesson 4

Objective
Children work with numerals to 100 and know the positions of the ones and tens columns.

Program Materials
Warm Up
See materials from previous weeks.

Engage
Line Land Activity Sheet 4, 1 per child

Additional Materials
Engage
Pencils, 1 per child

Access Vocabulary
before In front of; coming first
after Behind; coming later

Creating Context
English prepositions present a challenge to English Learners, but demonstrating their meaning can help make them understandable. Have several children line up and point out who is in line before and after each child. Then go to a number line, and point to a number that comes before or after another number.

1 Warm Up   5

Skill Building   COMPUTING
Teacher's Choice
Before beginning the small-group activity **Writing Numbers to 100,** choose one of the Warm-Up activities from previous weeks to use with the whole group. These activities include the following:

- Object Land: **Pointing and Winking**
- Picture Land: **Catch the Teacher**
- Line Land: **The Neighborhood**
- Sky Land: **Thermometer Counting**
- Object Land: **What Number Am I?**
- Circle Land: **Count Up**
- Circle Land: **Blastoff!**

Warm-Up Card 1

Purpose Children reinforce counting and number-sequence skills.

Monitoring Student Progress
If . . . children can work fluently with numbers 1–10,
Then . . . challenge them to work with bigger numbers.

2 Engage   30

Concept Building   APPLYING
Writing Numbers to 100
"Today we are going to write numbers to 100."

Follow the instructions on the Activity Card to play **Writing Numbers to 100.** As children play, ask questions about what is happening in the activity.

Activity Card 45

322 Number Worlds • Week 25

Purpose Children know and write the number sequence 1–100 and know the positions of the ones and tens columns.

Monitoring Student Progress

| If . . . children can fluently write numbers to 100, | Then . . . invite them to work with a partner to create and solve number riddles. For example, "I have 5 tens and 4 ones. What number am I?" |

Teacher's Note Encourage children to use this week's vocabulary words as they engage in the activities, discuss math concepts, and make predictions.

Building Blocks For additional practice with numeration, children should complete **Building Blocks** Space Race: Number Choice.

3 Reflect 10

Extended Response REASONING

Ask questions such as the following:

- **How many tens are in this number? How did you figure that out?** Answers will vary. Children should indicate that the left digit shows how many tens are in the number.
- **How many ones are in this number? How did you figure that out?** Answers will vary. Children should indicate that the right digit shows how many ones are in the number.
- **Which numbers were the hardest to write?** Accept all reasonable answers.
- **Which numbers were the easiest to write? Why?** Accept all reasonable answers.

4 Assess

Informal Assessment

Use the Student Assessment Record, **Assessment**, page 100, to record informal observations.

COMPUTING	APPLYING
Teacher's Choice Did the child ❏ respond accurately? ❏ respond quickly? ❏ respond with confidence? ❏ self-correct?	**Writing Numbers to 100** Did the child ❏ apply learning to new situations? ❏ contribute concepts? ❏ contribute answers? ❏ connect mathematics to the real world?

Week 25 — More Numbers to 100

Lesson 5

Review

Objective
Children review the concepts presented in Week 25 of **Number Worlds**.

Program Materials

Warm Up
See materials from previous weeks.

Engage
See materials from previous weeks.

Creating Context
In addition to assessing and observing children using math concepts, complete an observation of children's language proficiency with academic English for mathematics. If your designated English Language Proficiency Test has an observation checklist, use it to determine which English forms, functions, and vocabulary are still needed for academic language proficiency.

1. Warm Up

Skill Building — COMPUTING
Teacher's Choice

Before beginning the **Free-Choice** activity, choose one of the Warm-Up activities from previous weeks to use with the whole group. These activities include the following:

- Object Land: **Pointing and Winking**
- Picture Land: **Catch the Teacher**
- Line Land: **The Neighborhood**
- Sky Land: **Thermometer Counting**
- Object Land: **What Number Am I?**
- Circle Land: **Count Up**
- Circle Land: **Blastoff!**

Warm-Up Card 1

Purpose Children reinforce counting and number-sequence skills.

2. Engage

Concept Building — APPLYING
Free-Choice Activity

For the last day of the week, allow children to choose an activity from the previous weeks. Some activities they may choose include the following:

- Line Land: **Numbering the Neighborhood**
- Picture Land: **Building a Neighborhood**
- Line Land: **Number Line to 100**

Make a note of the activities children select. Do they prefer easy or challenging activities? Continue to provide Challenge opportunities for children who have mastered the basic activities.

Monitoring Student Progress

If . . . children	Then . . . choose an activity
would benefit from extra practice on specific skills,	for them.

324 Number Worlds • Week 25

3 Reflect

Extended Response REASONING
Ask questions such as the following:
- What did you like about playing Steve's New Bike?
- Was there anything about playing this game you didn't like?
- Did this game help you do something you couldn't do before? What did it help you do?
- What was easy when you were playing Writing Numbers to 100?
- What was hard when you were playing Writing Numbers to 100?
- What do you remember most about this game?
- What was your favorite part?

Using Student Pages
Have each child complete **Workbook,** page 79. Is the child correctly applying the skills learned in Week 25?

Workbook, p. 79

4 Assess

A Gather Evidence

Formal Assessment
Have children complete the weekly test on **Assessment,** page 73. Record formal assessment scores on the Student Assessment Record, **Assessment,** page 100.

Assessment, p. 73

B Summarize Findings
Review the Student Assessment Records. Determine whether students have Minimal, Basic, or Secure understanding of the concepts presented in Week 25.

C Differentiate Instruction
Based on your observations, use these teaching strategies next week to follow up.

Minimal Understanding
- Repeat the Warm-Up and Engage activities in subsequent weeks to develop numeration concepts.
- Repeat the activities from this week. Observe children's understanding of these concepts, and reinforce as necessary.
- Use **Building Blocks** computer activities beginning with Counting Activity 3 to develop and reinforce numeration concepts.

Basic Understanding
- Repeat Engage activities in subsequent weeks to reinforce basic concepts.
- Use **Building Blocks** computer activities beginning with Before and After Math to reinforce numeration concepts.

Secure Understanding
- Use Challenge variations of this week's activities.
- Use **Building Blocks** computer activities beginning with Math-O-Scope.

More Numbers to 100 • Lesson 5 Review   **325**

Week 26: Addition Stories

Week at a Glance

This week, children begin **Number Worlds,** Week 26, and continue to explore addition in Picture Land.

Background

In Picture Land, numbers are represented as numerals. This allows children to forge links between the world of real quantities (e.g., objects) and the world of formal symbols such as written numerals and addition, subtraction, and equal signs used when writing equations.

How Children Learn

As children begin this week's lesson, they should be able to write and solve addition problems involving both single- and double-digit numbers.

At the end of this week, children should be able to add two quantities up to 20 and to solve addition story problems involving two or more addends.

Skills Focus
- Add whole numbers
- Use math symbols
- Understand equality
- Solve problems

Teaching for Understanding

As children engage in these activities, they will begin to understand how addition applies to real-world problems and everyday situations.

Observe closely while evaluating the Engage activities assigned for this week.

- Are children correctly adding whole numbers?
- Are children correctly using math symbols in addition problems?
- Can children apply addition concepts to real-world situations?

Math at Home

Give one copy of the Letter to Home, page A26, to each child. Complete the activity in class, and then encourage children to share it with their caregivers.

Letter to Home, Teacher Edition, p. A26

Math Vocabulary

word problem Problem that has a story as well as numbers

equation Number story

discard Set cards aside where they won't be used anymore

facedown Cards on a table with the dots or numerals down (not showing)

English Learners

SPANISH COGNATES

English	Spanish
addition	adición
count	contar
equal	igual
number	número
signs	signos

ALTERNATE VOCABULARY

altogether In all; total

326 Number Worlds • Week 26

Week 26 Planner — Addition Stories

PACING	LESSON	LEARNING GOALS	MATERIALS	TECHNOLOGY
DAY 1	Warm Up Teacher's Choice	Children reinforce counting and number-sequence skills.	Materials will be selected from those used in previous weeks.	Building Blocks Barkley's Bones 2
	Activity 46 **Picture Land:** How Many Balloons?*	Children interpret and solve addition word problems and write the corresponding equations.	**Program Materials** Counters **Additional Materials** • Chart paper and marker • Paper and pencils, for a Challenge	
DAY 2	Warm Up Teacher's Choice	Children reinforce counting and number-sequence skills.	Materials will be selected from those used in previous weeks.	Building Blocks Pizza Pizzazz 3
	Activity 47 **Picture Land:** And the Answer Is . . .	Children add two quantities up to 20.	**Program Materials** • Dot Set Cards (3–10), 4 sets, shuffled • Number Cards (6–20), 2 sets, shuffled	
DAY 3	Warm Up Teacher's Choice	Children reinforce counting and number-sequence skills.	Materials will be selected from those used in previous weeks.	Building Blocks Barkley's Bones 2
	Activity 46 **Picture Land:** How Many Balloons?*	Children interpret and solve addition word problems and write the corresponding equations.	**Program Materials** Counters **Additional Materials** • Chart paper and marker • Paper and pencils, for a Challenge	
DAY 4	Warm Up Teacher's Choice	Children reinforce counting and number-sequence skills.	Materials will be selected from those used in previous weeks.	Building Blocks Pizza Pizzazz 3
	Activity 47 **Picture Land:** And the Answer Is . . .	Children gain additional practice adding two quantities up to 20.	**Program Materials** • Dot Set Cards (3–10), 4 sets, shuffled • Number Cards (6–20), 2 sets, shuffled	
DAY 5	Warm Up Teacher's Choice	Children reinforce counting and number-sequence skills.	Materials will be selected from those used in previous weeks.	Building Blocks Review previous activities
	Review and Assess	Children review their favorite activities to improve their understanding of more difficult concepts.	Materials will be selected from those used in previous weeks.	

* Includes Challenge Variations

Week 26 Lesson 1: Addition Stories

Objective
Children reinforce number-sequence skills and solve addition word problems.

Program Materials
Warm Up
See materials for previous weeks.

Engage
Counters

Additional Materials
Engage
- Chart paper and marker
- Paper and pencils, for a Challenge

Access Vocabulary
word problem Problem that has a story as well as numbers
altogether In all; total

Creating Context
To prepare English Learners for this activity, organize children into small groups, and have them talk about things they do at home that require addition. Have children work together to create a group story, complete with illustrations to show clearly that addition was involved. Have children use manipulatives or props to act out their story.

1 Warm Up — 5

Skill Building COMPUTING
Teacher's Choice
Before beginning the whole-class activity **How Many Balloons?** choose one of the Warm-Up activities from previous weeks to use with the whole group. These activities include the following:
- Object Land: **Pointing and Winking**
- Picture Land: **Catch the Teacher**
- Line Land: **The Neighborhood**
- Sky Land: **Thermometer Counting**
- Object Land: **What Number Am I?**
- Circle Land: **Count Up**
- Circle Land: **Blastoff!**

Warm-Up Card 1

Purpose Children reinforce counting and number-sequence skills.

Monitoring Student Progress
If . . . children need help with basic skills,
Then . . . use the **eMathTools** Set Tool to demonstrate and explore counting.

2 Engage — 30

Concept Building UNDERSTANDING
How Many Balloons?
"Today we are going to solve addition stories."

Follow the instructions on the Activity Card to play **How Many Balloons?** As children play, ask questions about what is happening in the activity.

Activity Card 46

328 Number Worlds • Week 26

Purpose Children interpret and solve addition word problems and write the corresponding equations.

Monitoring Student Progress

If . . . children have trouble identifying which numbers should be addends in the story problem,

Then . . . discuss the meaning of *altogether* with them. Explain that the phrase *"How many altogether?"* is a clue that they should add the numbers they know from the story problem to find the answer.

 Teacher's Note If your children have trouble writing and solving the story problem equations, provide extra modeling as needed. Encourage children to use manipulatives or to draw pictures to show what is happening in the story problem.

 Building Blocks For further practice with addition, children should complete **Building Blocks** Barkley's Bones 2.

3 Reflect 10

Extended Response REASONING
Ask questions such as the following:
- **In the last word problem, what did we want to find out? What facts did we know?** Answers will vary for each word problem.
- **Did we add or take away to find the answer?** add **Why?** Discuss with children that they were trying to find out how many total items there are, and point out "clue words" in the story problem such as *in all* and *altogether*.

Using Student Pages
Have each child complete **Workbook,** page 80. Did the child correctly write and solve the equation?

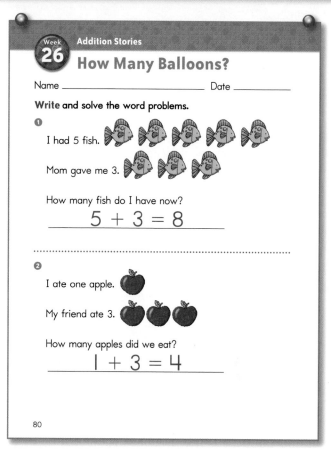

Workbook, p. 80

4 Assess

Informal Assessment
Use the Student Assessment Record, **Assessment,** page 100, to record informal observations.

COMPUTING	UNDERSTANDING
Teacher's Choice	**How Many Balloons?**
Did the child	Did the child
❏ respond accurately?	❏ make important observations?
❏ respond quickly?	❏ extend or generalize learning?
❏ respond with confidence?	❏ provide insightful answers?
❏ self-correct?	❏ pose insightful questions?

Addition Stories • Lesson 1   **329**

Week 26 Addition Stories
Lesson 2

Objective
Children practice counting and number-sequence skills and add two quantities up to 20.

Program Materials
Warm Up
See materials from previous weeks.
Engage
- Dot Set Cards (3–10), 4 sets, shuffled
- Number Cards (6–20), 2 sets, shuffled

Access Vocabulary
equation Number story
facedown Cards on a table with the dots or numerals down (not showing)

Creating Context
When using the game cards, explain the concept of *faceup* and *facedown*. These idioms may not be familiar to children. The main, or most meaningful, side of the card is the face. A card that is faceup communicates specific information, while a facedown card gives little information.

1. Warm Up

Skill Building — COMPUTING
Teacher's Choice
Before beginning the small-group activity **And the Answer Is . . .**, choose one of the Warm-Up activities from previous weeks to use with the whole group. These activities include the following:
- Object Land: **Pointing and Winking**
- Picture Land: **Catch the Teacher**
- Line Land: **The Neighborhood**
- Sky Land: **Thermometer Counting**
- Object Land: **What Number Am I?**
- Circle Land: **Count Up**
- Circle Land: **Blastoff!**

Purpose Children reinforce counting and number-sequence skills.

Warm-Up Card 1

Monitoring Student Progress
If . . . children need help with basic skills, **Then . . .** practice counting and number-sequence skills with them throughout the day.

2. Engage

Concept Building — ENGAGING
And the Answer Is . . .
"Today we are going to see who can be the first player to run out of cards."

Follow the instructions on the Activity Card to play **And the Answer Is . . . .** As children play, ask questions about what is happening in the activity.

Purpose Children add two quantities up to 20.

Activity Card 47

Monitoring Student Progress

If . . . children have trouble identifying the correct addend cards,

Then . . . play the game with cards up to 10 instead of 20.

 Teacher's Note This small-group activity might work best with you or a classroom helper leading the first few rounds. A designated dealer should be able to take over that role after children are familiar with the game.

Building Blocks For further practice with addition, children should complete **Building Blocks** Pizza Pizzazz 3.

3 Reflect 10

Extended Response REASONING
Ask questions such as the following:
- **On your last turn, what number were you trying to get?** Answers will vary.
- **Did your two Dot Cards add up to that number? How did you figure that out?** Accept all reasonable answers.

Using Student Pages
Have each child complete **Workbook,** page 81. Say, "Draw a line from each group of Dot Set Cards to show how much these cards equal altogether." Did the child match the Dot Set Cards to the correct Number Card?

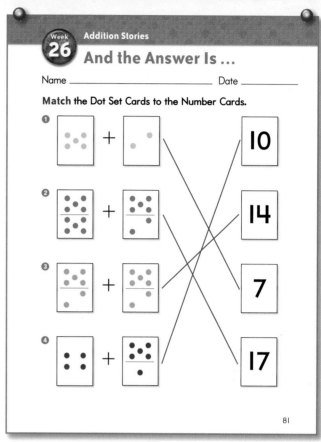

Workbook, p. 81

4 Assess

Informal Assessment
Use the Student Assessment Record, **Assessment,** page 100, to record informal observations.

COMPUTING	ENGAGING
Teacher's Choice	**And the Answer Is . . .**
Did the child	Did the child
❏ respond accurately?	❏ pay attention to the contributions of others?
❏ respond quickly?	
❏ respond with confidence?	❏ contribute information and ideas?
❏ self-correct?	
	❏ improve on a strategy?
	❏ reflect on and check accuracy of work?

Addition Stories • Lesson 2   **331**

Week 26 Addition Stories
Lesson 3

Objective
Children continue to practice number-sequence skills and solve addition word problems.

Program Materials
Warm Up
See materials for previous weeks.

Engage
Counters

Additional Materials
Engage
- Chart paper and marker
- Paper and pencils, for a Challenge

Access Vocabulary
plus sign Symbol that means "add"
equal sign Symbol that means "having the same amount"

Creating Context
Primary language can be a helpful tool for ensuring comprehension of math concepts. Pair English Learners of beginning English proficiency with children of the same primary-language background who have greater English proficiency. Have them work through the activity using their primary language to support their understanding of the word problems.

1 Warm Up

Skill Building COMPUTING
Teacher's Choice
Before beginning the whole-class activity **How Many Balloons?** choose one of the Warm-Up activities from previous weeks to use with the whole group. These activities include the following:
- Object Land: **Pointing and Winking**
- Picture Land: **Catch the Teacher**
- Line Land: **The Neighborhood**
- Sky Land: **Thermometer Counting**
- Object Land: **What Number Am I?**
- Circle Land: **Count Up**
- Circle Land: **Blastoff!**

Warm-Up Card 1

Purpose Children reinforce counting and number-sequence skills.

Monitoring Student Progress
If . . . children need help with basic skills,
Then . . . practice counting and number-sequence skills with them throughout the day.

2 Engage

Concept Building REASONING
How Many Balloons?
"Today we are going to solve addition stories."

Follow the instructions on the Activity Card to play **How Many Balloons?** As children play, ask questions about what is happening in the activity.

Purpose Children interpret and solve addition word problems and write the corresponding equations.

Activity Card 46

332 Number Worlds • Week 26

Monitoring Student Progress

| If . . . children are comfortable with this level of play, | Then . . . have them create and solve their own story problems by playing the Challenge variation of the game. |

Teacher's Note When a problem has more than two addends, show the children how to solve successive addition problems one step at a time. Remind them that it is faster to start with the larger addend and count on the smaller addend.

 For further practice with addition, children should complete **Building Blocks** Barkley's Bones 2.

3 Reflect 10

Extended Response REASONING

Ask questions such as the following:

- **How can you check your answer?** Accept all reasonable answers. Children may suggest using manipulatives, trading problems with a classmate, and so on.
- **What does this remind you of?** Accept all reasonable answers. Discuss with children how the story problems show real-life situations in which they use addition.

4 Assess

Informal Assessment

Use the Student Assessment Record, **Assessment**, page 100, to record informal observations.

COMPUTING	REASONING
Teacher's Choice	**How Many Balloons?**
Did the child	Did the child
❏ respond accurately?	❏ provide a clear explanation?
❏ respond quickly?	❏ communicate reasons and strategies?
❏ respond with confidence?	❏ choose appropriate strategies?
❏ self-correct?	❏ argue logically?

Addition Stories • Lesson 3 **333**

Week 26 Addition Stories
Lesson 4

Objective
Children continue to practice counting and number-sequence skills and add two quantities up to 20.

Program Materials
Warm Up
See materials from previous weeks.
Engage
- Dot Set Cards (3–10), 4 sets, shuffled
- Number Cards (6–20), 2 sets, shuffled

Access Vocabulary
discard Set cards aside where they won't be used anymore
shuffle To mix together

Creating Context
Discuss the concept of *equality* with children. Organize groups of classroom objects (pencils, books, and so on) into sets. Make some of the sets equal and some unequal. Discuss with children what makes two sets equal. *(The sets have the same number of objects in them.)* Can children think of times when it is important for things to be equal?

1 Warm Up 5

Skill Building COMPUTING
Teacher's Choice
Before beginning the small-group activity **And the Answer Is . . .** , choose one of the Warm-Up activities from previous weeks to use with the whole group. These activities include the following:
- Object Land: **Pointing and Winking**
- Picture Land: **Catch the Teacher**
- Line Land: **The Neighborhood**
- Sky Land: **Thermometer Counting**
- Object Land: **What Number Am I?**
- Circle Land: **Count Up**
- Circle Land: **Blastoff!**

Purpose Children reinforce counting and number-sequence skills.

Warm-Up Card 1

Monitoring Student Progress
If . . . children can work fluently with numbers 1–10, **Then . . .** challenge them to work with bigger numbers.

2 Engage 30

Concept Building APPLYING
And the Answer Is . . .
"Today we are going to see who can be the first player to run out of cards."

Follow the instructions on the Activity Card to play **And the Answer Is . . . .** As children play, ask questions about what is happening in the activity.

Purpose Children add two quantities up to 20.

Activity Card 47

Monitoring Student Progress

If . . . children are comfortable with this level of play,

Then . . . challenge them to play some rounds with three addend Dot Set Cards in their hands.

Teacher's Note Encourage children to use this week's vocabulary words as they engage in the activities, discuss math concepts, and make predictions.

 For further practice with addition, children should complete **Building Blocks** Pizza Pizzazz 3.

3 Reflect 10

Extended Response REASONING

Ask questions such as the following:

- **What strategy did you use to solve this problem? Why did you choose that strategy?** Accept all reasonable answers.
- **How can you check the answer?** Accept all reasonable answers.

4 Assess

Informal Assessment

Use the Student Assessment Record, **Assessment,** page 100, to record informal observations.

COMPUTING	APPLYING
Teacher's Choice	**And the Answer Is . . .**
Did the child	Did the child
❏ respond accurately?	❏ apply learning to new situations?
❏ respond quickly?	❏ contribute concepts?
❏ respond with confidence?	❏ contribute answers?
❏ self-correct?	❏ connect mathematics to the real world?

Addition Stories • Lesson 4   **335**

Week 26 — Addition Stories

Lesson 5
Review

Objective
Children review the material presented in Week 26 of *Number Worlds*.

Program Materials
Warm Up
See materials from previous weeks.

Engage
See materials from previous weeks.

Creating Context
In addition to assessing and observing children using math concepts, complete an observation of children's language proficiency with academic English for mathematics. If your designated English Language Proficiency Test has an observation checklist, use it to determine which English forms, functions, and vocabulary are still needed for academic language proficiency.

1 Warm Up — 5

Skill Building — COMPUTING
Teacher's Choice
Before beginning the **Free-Choice** activity, choose one of the Warm-Up activities from previous weeks to use with the whole group. These activities include the following:

- Object Land: **Pointing and Winking**
- Picture Land: **Catch the Teacher**
- Line Land: **The Neighborhood**
- Sky Land: **Thermometer Counting**
- Object Land: **What Number Am I?**
- Circle Land: **Count Up**
- Circle Land: **Blastoff!**

Warm-Up Card 1

Purpose Children reinforce counting and number-sequence skills.

2 Engage — 20

Concept Building — APPLYING
Free-Choice Activity
For the last day of the week, allow children to choose an activity from the previous weeks. Some activities they may choose include the following:

- Picture Land: **Counter Dropping**
- Picture Land: **Adding Concentration**
- Picture Land: **Flip, Add, and Compare**

Make a note of the activities children select. Do they prefer easy or challenging activities? Continue to provide Challenge opportunities for children who have mastered the basic activities.

Monitoring Student Progress
| If . . . children would benefit from extra practice on specific skills, | Then . . . choose an activity for them. |

Reflect

Extended Response REASONING
Ask questions such as the following:
- **What did you like about playing** How Many Balloons?
- **Was there anything about playing this game you didn't like?**
- **Did this game help you do something you couldn't do before? What did it help you do?**
- **What was easy when you were playing** And the Answer Is . . . ?
- **What was hard when you were playing** And the Answer Is . . . ?
- **What do you remember most about this game?**
- **What was your favorite part?**

Using Student Pages
Have each child complete **Workbook,** page 82. Is the child correctly applying the skills learned in Week 26?

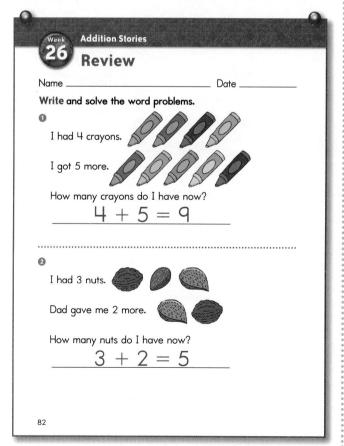

Workbook, p. 82

Assess

A Gather Evidence
Formal Assessment
Have children complete the weekly test on **Assessment,** page 75. Record formal assessment scores on the Student Assessment Record, **Assessment,** page 100.

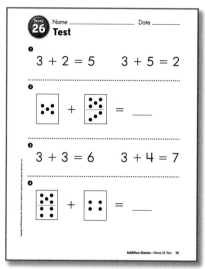

Assessment, p. 75

B Summarize Findings
Review the Student Assessment Records. Determine whether children have Minimal, Basic, or Secure understanding of the concepts presented in Week 26.

C Differentiate Instruction
Based on your observations, use these teaching strategies next week to follow up.

Minimal Understanding
- Repeat the Warm-Up and Engage activities in subsequent weeks to develop numeration concepts.
- Repeat the activities from this week. Observe children's understanding of these concepts, and reinforce as necessary.
- Use **Building Blocks** computer activities beginning with Count Free Explore to develop and reinforce numeration concepts.

Basic Understanding
- Repeat Engage activities in subsequent weeks to reinforce basic concepts.
- Use **Building Blocks** computer activities beginning with Counting Activity 6 to reinforce numeration and addition concepts.

Secure Understanding
- Use Challenge variations of this week's activities.
- Use **Building Blocks** computer activities beginning with Math-O-Scope.

Addition Stories • Lesson 5 Review   337

Week 27 Tens and Ones

Week at a Glance

This week, children begin **Number Worlds,** Week 27, and continue to explore Line Land and Picture Land.

Background

In Line Land, numbers are represented as positions on a line. When a child is asked, "Where are you now?" he or she might think of the position on the number line, the set size corresponding to that position, or the number that shows the position on the number line.

How Children Learn

At the beginning of this week, children should be able to add two quantities—up to 20—and to solve addition story problems involving two or more addends.

By the end of the week, children should be able to discuss the position of a number in the number sequence in terms of tens and ones.

Skills Focus

- Compare and order numbers
- Add whole numbers
- Count and identify numbers to 100
- Skip count by tens

Teaching for Understanding

As children engage in these activities, they should gain experience articulating number positions in terms of tens and ones. This allows children to build on their knowledge of number neighborhoods by using formal terminology.

Observe closely while evaluating the Engage activities assigned for this week.

- Are children correctly identifying numbers to 100?
- Can children add successive quantities to single- and double-digit numbers?

Math at Home

Give one copy of the Letter to Home, page A27, to each child Complete the activity in class, and then encourage children to share it with their caregivers.

Letter to Home, Teacher Edition, p. A27

Math Vocabulary

ones column Numeral on the far right of a number that tells how many ones are in the number

tens column Numeral to the left of the ones column that tells how many tens are in a number

pawn Game piece

deliver To take a message or package to someone

English Learners

SPANISH COGNATES

English	Spanish
addition	adición
count	contar
2-digit number	2-dígitos número

ALTERNATE VOCABULARY

quest A search or adventure

338 Number Worlds • Week 27

Week 27 Planner: Tens and Ones

PACING	LESSON	LEARNING GOALS	MATERIALS	TECHNOLOGY
DAY 1	Warm Up Teacher's Choice	Children reinforce counting and number-sequence skills.	Materials will be selected from those used in previous weeks.	Building Blocks School Supply Shop
	Activity 48 **Line Land:** Rosemary's Super Shoes	Children coordinate tens and ones to arrive at a one- or two-digit number.	**Program Materials** • Neighborhood Number Line • Rosemary's Shoes Pawn • Number Tiles (1–100) • Delivery Building Picture **Additional Materials** Small stickers	
DAY 2	Warm Up Teacher's Choice	Children reinforce counting and number-sequence skills.	Materials will be selected from those used in previous weeks.	Building Blocks Function Machine 1
	Activity 49 **Picture Land:** Dragon Quest 3	Children add successive quantities to single- and double-digit numbers.	**Program Materials** • Dragon Quest Game Board • Pawns, 1 per child • Spinner (1–4) • Dragon Quest Cards (+1 through +9 and −1 through −4) • Picture Land Activity Sheet 10 (optional) **Additional Materials** Paper and pencil	
DAY 3	Warm Up Teacher's Choice	Children reinforce counting and number-sequence skills.	Materials will be selected from those used in previous weeks.	Building Blocks School Supply Shop
	Activity 48 **Line Land:** Rosemary's Super Shoes	Children coordinate tens and ones to arrive at a one- or two-digit number.	**Program Materials** Same as Lesson 1	
DAY 4	Warm Up Teacher's Choice	Children reinforce counting and number-sequence skills.	Materials will be selected from those used in previous weeks.	Building Blocks Function Machine 1
	Activity 49 **Picture Land:** Dragon Quest 3	Children add quantities to single- and double-digit numbers.	**Program Materials** Same as Lesson 2	
DAY 5	Warm Up Teacher's Choice	Children reinforce number-sequence skills.	Materials will be selected from those used in previous weeks.	Building Blocks Review previous activities
	Review and Assess	Children review their favorite activities to improve understanding.	Materials will be selected from those used in previous weeks.	

Week 27 — Tens and Ones
Lesson 1

Objective
Children practice counting and number-sequence skills and coordinate tens and ones to arrive at a one- or two-digit number.

Program Materials
Warm Up
See materials for previous weeks.
Engage
- Neighborhood Number Line
- Rosemary's Shoes Pawn
- Number Tiles (1–100)
- Delivery Building Picture

Additional Materials
Engage
Small stickers

Access Vocabulary
ones column Numeral on the far right of a number that tells how many ones are in the number
tens column Numeral to the left of the ones column that tells how many tens are in a number

Creating Context
Children will make comparisons as they describe Rosemary's deliveries. Review comparative and superlative forms of the following words:

long—longer—longest
short—shorter—shortest
high—higher—highest
near—nearer—nearest
far—farther—farthest

1 Warm Up

Skill Building — COMPUTING
Teacher's Choice
Before beginning the whole-class activity **Rosemary's Super Shoes,** choose one of the Warm-Up activities from previous weeks to use with the whole group. These activities include the following:

- Object Land: **Pointing and Winking**
- Picture Land: **Catch the Teacher**
- Line Land: **The Neighborhood**
- Sky Land: **Thermometer Counting**
- Object Land: **What Number Am I?**
- Circle Land: **Count Up**
- Circle Land: **Blastoff!**

Warm-Up Card 1

Purpose Children reinforce counting and number-sequence skills.

Monitoring Student Progress
If . . . children need help with basic skills,
Then . . . practice counting and number-sequence skills with them throughout the day.

2 Engage

Concept Building — UNDERSTANDING
Rosemary's Super Shoes
"Today we are going to help Rosemary deliver packages in the neighborhood."

Follow the instructions on the Activity Card to play **Rosemary's Super Shoes.** As children play, ask questions about what is happening in the activity.

Activity Card 48

Purpose Children coordinate tens and ones to arrive at a single- or double-digit number.

Monitoring Student Progress

| If . . . children have trouble determining how many blocks they should jump over, | Then . . . remind them that the left digit tells how many blocks, or how many tens, are in each number. |

 Teacher's Note Model the process of determining and describing the delivery location as often as needed. Your children may enjoy playing this in independent small groups after they are familiar with the procedure. Have the small groups figure out and report to the whole group Rosemary's longest and shortest deliveries and how many deliveries were made altogether.

 Building Blocks For further practice with addition, children should complete **Building Blocks** School Supply Shop.

3 Reflect — 10

Extended Response REASONING

Ask questions such as the following:

- **What number did you pick? Is that a long way or a short way from 0?** Answers will vary.
- **What blocks did Rosemary pass along the way?** Answers will vary. Encourage children to discuss blocks in terms of the "tens column" or the "tens place."

Using Student Pages

Have each child complete **Workbook,** page 83. Did the child correctly identify the tens and ones in each number?

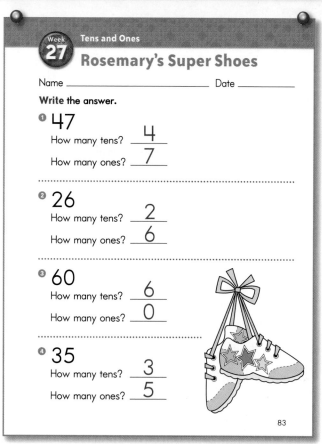

Workbook, p. 83

4 Assess

Informal Assessment

Use the Student Assessment Record, **Assessment,** page 100, to record informal observations.

COMPUTING	UNDERSTANDING
Teacher's Choice Did the child ❏ respond accurately? ❏ respond quickly? ❏ respond with confidence? ❏ self-correct?	**Rosemary's Super Shoes** Did the child ❏ make important observations? ❏ extend or generalize learning? ❏ provide insightful answers? ❏ pose insightful questions?

Tens and Ones • Lesson 1 341

Week 27 — Tens and Ones
Lesson 2

Objective
Children practice number-sequence skills and add successive quantities to single- and double-digit numbers.

Program Materials
Warm Up
See materials from previous weeks.

Engage
- Dragon Quest Game Board
- Pawns, 1 per child
- Spinner (1–4)
- Dragon Quest Cards (+1 through +9 and −1 through −4)
- Picture Land Activity Sheet 10 (optional)

Additional Materials
Engage
Paper and pencil

Access Vocabulary
quest A search or adventure
shuffle To mix together

Creating Context
The variety of words used to describe subtraction (such as *take away*, *minus*, *less*, and *subtracted from*) may be puzzling for English Learners. Help children brainstorm a list of words and phrases used to describe subtraction.

---

1. Warm Up

Skill Building — COMPUTING
Teacher's Choice
Before beginning the small-group activity **Dragon Quest 3**, choose one of the Warm-Up activities from previous weeks to use with the whole group. These activities include the following:
- Object Land: **Pointing and Winking**
- Picture Land: **Catch the Teacher**
- Line Land: **The Neighborhood**
- Sky Land: **Thermometer Counting**
- Object Land: **What Number Am I?**
- Circle Land: **Count Up**
- Circle Land: **Blastoff!**

Purpose Children reinforce counting and number-sequence skills.

Warm-Up Card 1

Monitoring Student Progress
If . . . children need help with basic skills,
Then . . . practice counting and number-sequence skills with them throughout the day.

---

2. Engage

Concept Building — ENGAGING
Dragon Quest 3
"Today we are going to collect 20 buckets of water to put out the dragon's fire."

Follow the instructions on the Activity Card to play **Dragon Quest 3**. As children play, ask questions about what is happening in the activity.

Activity Card 49

342   Number Worlds • Week 27

Purpose Children add successive quantities to single- and double-digit numbers.

Monitoring Student Progress

If . . . children have trouble doing the subtraction problems, | **Then . . .** allow them to use a number line for reference.

 Building Blocks For further practice with addition, children should complete **Building Blocks** Function Machine 1.

3 Reflect 10

Extended Response REASONING
Ask questions such as the following:
- If you have 12 buckets and you pick a +5 card, how many buckets will you have altogether? 17 How many buckets will you still need to put out the dragon's fire? 3
- **How did you figure that out?** Possible answers include adding or counting on.

Using Student Pages
Have each child complete **Workbook,** page 84. Did the child write the correct answer?

Week 27 Tens and Ones

Dragon Quest 3

Name _____ Date _____

Write the answer.

① $7 + 9 = \underline{16}$

② $14 - 4 = \underline{10}$

③ $12 + 8 = \underline{20}$

④ $13 + 3 = \underline{16}$

84

Workbook, p. 84

4 Assess

Informal Assessment
Use the Student Assessment Record, **Assessment,** page 100, to record informal observations.

COMPUTING	ENGAGING
Teacher's Choice Did the child	**Dragon Quest 3** Did the child
❏ respond accurately? ❏ respond quickly? ❏ respond with confidence? ❏ self-correct?	❏ pay attention to the contributions of others? ❏ contribute information and ideas? ❏ improve on a strategy? ❏ reflect on and check accuracy of work?

Tens and Ones • Lesson 2 343

Week 27 — Tens and Ones
Lesson 3

Objective
Children practice counting and number-sequence skills and coordinate tens and ones to arrive at a one- or two-digit number.

Program Materials
Warm Up
See materials for previous weeks.

Engage
- Neighborhood Number Line
- Rosemary's Shoes Pawn
- Number Tiles (1–100)
- Delivery Building Picture

Additional Materials
Engage
Small stickers

Access Vocabulary
pawn Game piece
deliver To take a message or package to someone

Creating Context
To reinforce the concepts of *near* and *far*, have children work in pairs to generate a list of places that are near to or far from the school. Make a chart with columns labeled *Near* and *Far* to record the answers, and have children illustrate or add a symbol for each location.

1 Warm Up • 5

Skill Building — COMPUTING
Teacher's Choice
Before beginning the whole-class activity **Rosemary's Super Shoes,** choose one of the Warm-Up activities from previous weeks to use with the whole group. These activities include the following:

- Object Land: **Pointing and Winking**
- Picture Land: **Catch the Teacher**
- Line Land: **The Neighborhood**
- Sky Land: **Thermometer Counting**
- Object Land: **What Number Am I?**
- Circle Land: **Count Up**
- Circle Land: **Blastoff!**

Warm-Up Card 1

Purpose Children reinforce counting and number-sequence skills.

Monitoring Student Progress
If . . . children need help with basic skills,
Then . . . practice counting and number-sequence skills with them throughout the day.

2 Engage • 30

Concept Building — REASONING
Rosemary's Super Shoes
"Today we are going to help Rosemary deliver packages in the neighborhood."

Follow the instructions on the Activity Card to play **Rosemary's Super Shoes.** As children play, ask questions about what is happening in the activity.

Activity Card 48

344 Number Worlds • Week 27

Purpose Children coordinate tens and ones to arrive at a single- or double-digit number.

Monitoring Student Progress

| If . . . children can fluently identify the correct number in terms of tens and ones, | Then . . . challenge them to draw a Number Tile and to use tens and ones to describe the number to a teammate who will move the pawn. (For example, "This number has 3 tens and 4 ones.") |

 MathTools Use the 100 Table to demonstrate and explore place value.

 Encourage children to use this week's vocabulary words as they engage in the activities, discuss math concepts, and make predictions.

 For further practice with addition and place value, children should complete **Building Blocks** School Supply Shop.

Reflect   10

Extended Response   REASONING

Ask questions such as the following:

- **What number did you pick? Have any other deliveries been made in that neighborhood?** Answers will vary.

- **What is Rosemary's farthest delivery so far?** Answers will vary. Encourage children to phrase their answers in terms of distance from 0 on the number line or a distance that is close to 100.

Assess

Informal Assessment

Use the Student Assessment Record, **Assessment,** page 100, to record informal observations.

COMPUTING	REASONING
Teacher's Choice Did the child ❏ respond accurately? ❏ respond quickly? ❏ respond with confidence? ❏ self-correct?	**Rosemary's Super Shoes** Did the child ❏ provide a clear explanation? ❏ communicate reasons and strategies? ❏ choose appropriate strategies? ❏ argue logically?

Tens and Ones • Lesson 3

Week 27 — Tens and Ones

Objective
Children practice number-sequence skills and add successive quantities to single- and double-digit numbers.

Program Materials

Warm Up
See materials from previous weeks.

Engage
- Dragon Quest Game Board
- Pawns, 1 per child
- Spinner (1–4)
- Dragon Quest Cards (+1 through +9 and −1 through −4)
- Picture Land Activity Sheet 10 (optional)

Additional Materials
Engage
Paper and pencil

Access Vocabulary
add Put together
subtract Take away

Creating Context
Work with children to brainstorm the many ways we describe the process of adding objects or numbers. For example, we might say, "1 plus 2," "1 and 2," "1 and 1 more," and so on. Make a word bank, and display it in the classroom so children can add other ways to talk about addition.

1. Warm Up

Skill Building — COMPUTING
Teacher's Choice

Before beginning the small-group activity **Dragon Quest 3**, choose one of the Warm-Up activities from previous weeks to use with the whole group. These activities include the following:

- Object Land: **Pointing and Winking**
- Picture Land: **Catch the Teacher**
- Line Land: **The Neighborhood**
- Sky Land: **Thermometer Counting**
- Object Land: **What Number Am I?**
- Circle Land: **Count Up**
- Circle Land: **Blastoff!**

Warm-Up Card 1

Purpose Children reinforce counting and number-sequence skills.

Monitoring Student Progress
If . . . children can work fluently with numbers 1–10,
Then . . . challenge them to work with bigger numbers.

2. Engage

Concept Building — APPLYING
Dragon Quest 3

"Today we are going to collect 20 buckets of water to put out the dragon's fire."

Follow the instructions on the Activity Card to play **Dragon Quest 3**. As children play, ask questions about what is happening in the activity.

Activity Card 49

Purpose Children add successive quantities to single- and double-digit numbers.

Monitoring Student Progress

| If . . . children can fluently write the equations for each operation performed on the game board, | Then . . . provide practice with mental addition and subtraction by having children eliminate the written record and arrange the acquired Dragon Quest Cards in an elimination format such as −4 cancels +4, and −2 and −1 cancel +3. |

 Building Blocks For further practice with addition and subtraction, children should complete **Building Blocks** Function Machine 1.

3 Reflect

Extended Response REASONING

Ask questions such as the following:

- **What symbol do we use to show we are adding something?** plus sign **Subtracting?** minus sign
- **What patterns do you see?** Accept all reasonable answers.

4 Assess

Informal Assessment

Use the Student Assessment Record, **Assessment**, page 100, to record informal observations.

COMPUTING	APPLYING
Teacher's Choice	**Dragon Quest 3**
Did the child	Did the child
❏ respond accurately?	❏ apply learning to new situations?
❏ respond quickly?	❏ contribute concepts?
❏ respond with confidence?	❏ contribute answers?
❏ self-correct?	❏ connect mathematics to the real world?

Tens and Ones • Lesson 4

Week 27 — Tens and Ones

Lesson 5
Review

Objective
Children review the material presented in Week 27 of *Number Worlds*.

Additional Materials
Warm Up
See materials from previous weeks.

Engage
See materials from previous weeks.

Creating Context
Discuss with children different meanings of the word *like*. Make a two-column chart on the board. Label one column *I Like* and the other column *Because*. Have children give reasons why they like something as you complete the chart on the board. This will help them describe why they like certain *Number Worlds* activities.

1 Warm Up

Skill Building — COMPUTING
Teacher's Choice
Before beginning the **Free-Choice** activity, choose one of the Warm-Up activities from previous weeks to use with the whole group. These activities include the following:

- Object Land: **Pointing and Winking**
- Picture Land: **Catch the Teacher**
- Line Land: **The Neighborhood**
- Sky Land: **Thermometer Counting**
- Object Land: **What Number Am I?**
- Circle Land: **Count Up**
- Circle Land: **Blastoff!**

Warm-Up Card 1

Purpose Children reinforce counting and number-sequence skills.

2 Engage

Concept Building — APPLYING
Free-Choice Activity
For the last day of the week, allow children to choose an activity from the previous weeks. Some activities they may choose include the following:

- Picture Land: **Dragon Quest 2**
- Line Land: **Number Line to 100**
- Line Land: **Writing Numbers to 100**

Make a note of the activities children select. Do they prefer easy or challenging activities? Continue to provide Challenge opportunities for children who have mastered the basic activities

Monitoring Student Progress
| If . . . children would benefit from extra practice on specific skills, | Then . . . choose an activity for them. |

348 Number Worlds • Week 27

3 Reflect

Extended Response REASONING
Ask questions such as the following:
- What did you like about playing Rosemary's Super Shoes?
- Was there anything about playing this game you didn't like?
- Did this game help you do something you couldn't do before? What did it help you do?
- What was easy when you were playing Dragon Quest 3?
- What was hard when you were playing Dragon Quest 3?
- What do you remember most about this game?
- What was your favorite part?

Using Student Pages
Have each child complete **Workbook,** page 85. Did the child correctly apply the concepts learned in Week 27?

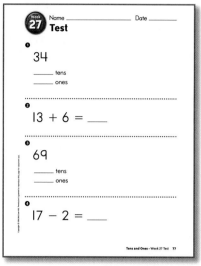

Workbook, p. 85

4 Assess

A Gather Evidence
Formal Assessment
Have students complete the weekly test on **Assessment,** page 77. Record formal assessment scores on the Student Assessment Record, **Assessment,** page 100.

Assessment, p. 77

B Summarize Findings
Review the Student Assessment Records. Determine whether students have Minimal, Basic, or Secure understanding of the concepts presented in Week 27.

C Differentiate Instruction
Based on your observations, use these teaching strategies next week to follow up.

Minimal Understanding
- Repeat the Warm-Up and Engage activities to develop numeration concepts.
- Repeat the activities from this week. Observe children's understanding of these concepts, and reinforce as necessary.
- Use **Building Blocks** computer activities beginning with Count Free Explore to develop and reinforce numeration concepts.

Basic Understanding
- Repeat Engage activities in subsequent weeks to reinforce basic concepts.
- Use **Building Blocks** computer activities beginning with School Supply Shop to reinforce numeration concepts.

Secure Understanding
- Use **Building Blocks** computer activities beginning with Function Machine 2.

Tens and Ones • Lesson 5 Review 349

Week 28: Adding and Subtracting Length

Week at a Glance

This week, children begin **Number Worlds**, Week 28, and continue to explore Line Land and Sky Land.

Background

In Sky Land, numbers are represented as positions on a scale. When children are asked, "How high are you now?" they might think of height on a scale, the set size that corresponds to that position, or the number that shows the position on the scale.

How Children Learn

At the beginning of this week, children should be able to add and subtract whole numbers up to 20.

By the end of the week, children should deepen their understanding of addition and subtraction by using subtraction to compare two numbers and by exploring the commutative property of addition with multiple addends.

Skills Focus
- Order and compare numbers
- Subtract whole numbers
- Add whole numbers
- Write addition equations

Teaching for Understanding

As children engage in these activities, they should come to understand that subtraction can be used as a tool to compare the magnitudes of two numbers.

Observe closely while evaluating the Engage activities assigned for this week.
- Are children correctly comparing measurements?
- Can children correctly add and subtract whole numbers?

Math at Home

Give one copy of the Letter to Home, page A28, to each child. Complete the activity in class, and then encourage children to share it with their caregivers.

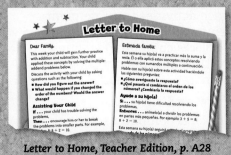

Letter to Home, Teacher Edition, p. A28

Math Vocabulary

measure To find out how long or how tall something is

solve To figure out; to find the answer

stack Pile of flat things such as cards or books

record To write down

English Learners

SPANISH COGNATES

English	Spanish
compare	comparar
count	contar
differences	diferencias
minus	menos

ALTERNATE VOCABULARY

compare To think about how things are the same and different.

Week 28 Planner: Adding and Subtracting Length

PACING	LESSON	LEARNING GOALS	MATERIALS	TECHNOLOGY
DAY 1	**Warm Up** Teacher's Choice	Children reinforce counting and number-sequence skills.	Materials will be selected from those used in previous weeks.	**Building Blocks** Gus' Garage
DAY 1	**Activity 50** Sky Land: How Much More?	Children use subtraction to determine how much longer one item is than another.	**Program Materials** Learning Links **Additional Materials** • Items to measure • Paper and pencils for each group	
DAY 2	**Warm Up** Teacher's Choice	Children reinforce counting and number-sequence skills.	Materials will be selected from those used in previous weeks.	**Building Blocks** Barkley's Bones 3
DAY 2	**Activity 51** Line Land: Let's Add Some More	Children use formal notation to indicate forward progress on a number line.	**Program Materials** • Plus Cards, 2 sets • Magnetic Number Line, 1 per child • Magnetic Chips **Additional Materials** Paper and pencils	
DAY 3	**Warm Up** Teacher's Choice	Children reinforce counting and number-sequence skills.	Materials will be selected from those used in previous weeks.	**Building Blocks** Gus' Garage
DAY 3	**Activity 50** Sky Land: How Much More?	Children use subtraction to determine how much longer one item is than another.	**Program Materials** Learning Links **Additional Materials** • Items to measure • Paper and pencils for each group	
DAY 4	**Warm Up** Teacher's Choice	Children reinforce counting and number-sequence skills.	Materials will be selected from those used in previous weeks.	**Building Blocks** Barkley's Bones 3
DAY 4	**Activity 51** Line Land: Let's Add Some More	Children use formal notation to indicate forward progress on a number line.	**Program Materials** • Plus Cards, 2 sets • Magnetic Number Line, 1 per child • Magnetic Chips **Additional Materials** Paper and pencils	
DAY 5	**Warm Up** Teacher's Choice	Children reinforce counting and number-sequence skills.	Materials will be selected from those used in previous weeks.	**Building Blocks** Review previous activities
DAY 5	**Review and Assess**	Children review favorite activities to improve understanding.	Materials will be selected from those used in previous weeks.	

Week 28 — Adding and Subtracting Length

Lesson 1

Objective
Children practice numeration skills and use subtraction to determine how much longer one item is than another.

Program Materials
Warm Up
See materials from previous weeks.

Engage
Learning Links

Additional Materials
Engage
- Items to measure
- Paper and pencils for each group

Access Vocabulary
measure To find out how long or how tall something is
compare To think about how things are the same and how they are different

Creating Context
Children will compare the lengths and heights of different objects. Review comparative and superlative forms of the following words:

long—longer—longest
short—shorter—shortest
high—higher—highest

1 Warm Up — 5

Skill Building — COMPUTING
Teacher's Choice

Before beginning the small-group activity **How Much More?** choose one of the Warm-Up activities from previous weeks to use with the whole group. These activities include the following:

- Object Land: **Pointing and Winking**
- Picture Land: **Catch the Teacher**
- Line Land: **The Neighborhood**
- Sky Land: **Thermometer Counting**
- Object Land: **What Number Am I?**
- Circle Land: **Count Up**
- Circle Land: **Blastoff!**

Purpose Children reinforce counting and number-sequence skills.

Monitoring Student Progress
If . . . children need help with basic skills,
Then . . . practice counting and number-sequence skills with them throughout the day.

Warm-Up Card 1

2 Engage — 30

Concept Building — UNDERSTANDING
How Much More?

"Today we are going to measure some things in the classroom."

Follow the instructions on the Activity Card to play **How Much More?** As children play, ask questions about what is happening in the activity.

Activity Card 50

352 Number Worlds • Week 28

Purpose Children use subtraction to determine how much longer one item is than another.

Monitoring Student Progress

If . . . children have trouble writing subtraction equations,

Then . . . remind them that they will always subtract the smaller number from the bigger number to find the difference in length or height.

Teacher's Note If some children in your class are sensitive about their height, demonstrate the activity by measuring classroom items of different heights, such as a door and a chair.

 Building Blocks For further practice with addition, children should complete **Building Blocks** Gus' Garage.

3 Reflect 10

Extended Response REASONING

Ask questions such as the following:
- **How long was the first item?** Answers will vary.
- **How long was the second item?** Answers will vary.
- **How much longer was one item than the other item?** Answers will vary. **How did you figure that out?** Children should indicate that they subtracted the smaller number from the bigger number.

Using Student Pages

Have each child complete **Workbook**, page 86. Did the child correctly compare the two heights?

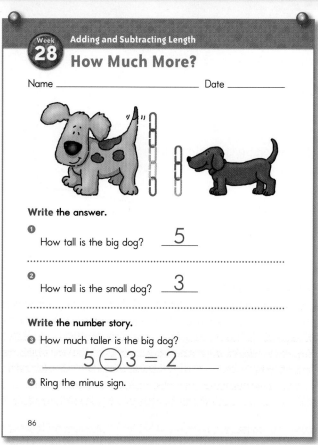

Workbook, p. 86

4 Assess

Informal Assessment

Use the Student Assessment Record, **Assessment**, page 100, to record informal observations.

COMPUTING	UNDERSTANDING
Teacher's Choice	**How Much More?**
Did the child	Did the child
❏ respond accurately?	❏ make important observations?
❏ respond quickly?	❏ extend or generalize learning?
❏ respond with confidence?	❏ provide insightful answers?
❏ self-correct?	❏ pose insightful questions?

Adding and Subtracting Length • Lesson 1   **353**

Week 28 — Adding and Subtracting Length

Lesson 2

Objective
Children practice numeration skills and use formal notation to indicate forward progress on a number line.

Program Materials
Warm Up
See materials from previous weeks.

Engage
- Plus Cards, 2 sets
- Magnetic Number Line, 1 per child
- Magnetic Chips

Additional Materials
Engage
Paper and pencils

Access Vocabulary
add To put together
sum Answer to an addition problem

Creating Context
Organize children into small groups, and have them talk about things they do at home that rely on addition. Have children work together to create a group story, complete with illustrations, to clearly show that addition was involved. Have children use manipulatives or props to act out their stories.

1 Warm Up

Skill Building — COMPUTING
Teacher's Choice
Before beginning the small-group activity **Let's Add Some More,** choose one of the Warm-Up activities from previous weeks to use with the whole group. These activities include the following:
- Object Land: **Pointing and Winking**
- Picture Land: **Catch the Teacher**
- Line Land: **The Neighborhood**
- Sky Land: **Thermometer Counting**
- Object Land: **What Number Am I?**
- Circle Land: **Count Up**
- Circle Land: **Blastoff!**

Purpose Children reinforce counting and number-sequence skills.

Warm-Up Card 1

Monitoring Student Progress
If . . . children need help with basic skills,
Then . . . practice counting and number-sequence skills with them throughout the day.

2 Engage

Concept Building — ENGAGING
Let's Add Some More
"Today we are going to see who can write the best addition number stories."

Follow the instructions on the Activity Card to play **Let's Add Some More.** As children play, ask questions about what is happening in the activity.

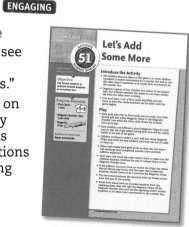

Activity Card 51

Purpose Children use formal notation to indicate forward progress on a number line.

Monitoring Student Progress

If... children have trouble retracing the opposing team's steps,	Then... invite them to divide the multiple-addend number story into smaller parts. For example, children can write 2 + 3 + 4 + 5 + 6 as 2 + 3 = 5; 5 + 4 = 9; 9 + 5 = 14, and so on.

 MathTools Use the Number Line Tool to demonstrate and explore number sequence.

Building Blocks For further practice with addition, children should complete **Building Blocks** Barkley's Bones 3.

3 Reflect

Extended Response REASONING

Ask questions such as the following:
- **What number did you pick last?** Answers will vary.
- **How many spaces did you move?** Answers will vary.
- **Where were you when you did that? How did you figure it out?** Answers will vary. Children may indicate that they added or counted on.
- **Did you show that move in the number story you wrote? Show me.** Answers will vary.

Using Student Pages

Have each child complete **Workbook,** page 87. Did the children write the correct answer?

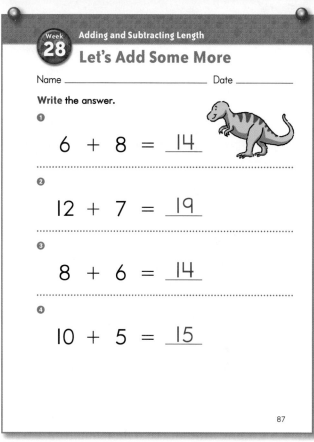

Workbook, p. 87

4 Assess

Informal Assessment

Use the Student Assessment Record, **Assessment,** page 100, to record informal observations.

COMPUTING	ENGAGING
Teacher's Choice	Let's Add Some More
Did the child	Did the child
❏ respond accurately?	❏ pay attention to the contributions of others?
❏ respond quickly?	
❏ respond with confidence?	❏ contribute information and ideas?
❏ self-correct?	
	❏ improve on a strategy?
	❏ reflect on and check accuracy of work?

Adding and Subtracting Length • Lesson 2 355

Week 28: Adding and Subtracting Length

Lesson 3

Objective
Children practice numeration skills and use subtraction to determine how much longer one item is than another.

Program Materials
Warm Up
See materials from previous weeks.

Engage
Learning Links

Additional Materials
Engage
- Items to measure
- Paper and pencils for each group

Access Vocabulary
minus sign Symbol that means "take away"
solve To figure out; to find the answer

Creating Context
Help English Learners successfully write subtraction number stories by first reading a few models aloud. To focus on the math concept without worrying too much about the language needed, provide a story frame in which students fill in key information: ____ − ____ = ____.

1. Warm Up 5

Skill Building — COMPUTING
Teacher's Choice
Before beginning the small-group activity **How Much More?** choose one of the Warm-Up activities from previous weeks to use with the whole group. These activities include the following:
- Object Land: **Pointing and Winking**
- Picture Land: **Catch the Teacher**
- Line Land: **The Neighborhood**
- Sky Land: **Thermometer Counting**
- Object Land: **What Number Am I?**
- Circle Land: **Count Up**
- Circle Land: **Blastoff!**

Purpose Children reinforce counting and number-sequence skills.

Warm-Up Card 1

Monitoring Student Progress
If . . . children need help with basic skills,
Then . . . practice counting and number-sequence skills with them throughout the day.

2. Engage 30

Concept Building — REASONING
How Much More?
"Today we are going to measure some things in the classroom."

Follow the instructions on the Activity Card to play **How Much More?** As children play, ask questions about what is happening in the activity.

Activity Card 50

356 Number Worlds • Week 28

Purpose Children use subtraction to determine how much longer one item is than another item.

Monitoring Student Progress

If . . . children are comfortable with this level of play,

Then . . . challenge them to measure longer or taller objects.

 Teacher's Note Encourage children to use this week's vocabulary words as they engage in the activities, discuss math concepts, and make predictions.

Building Blocks For further practice with addition, children should complete **Building Blocks** Gus' Garage.

3 Reflect — 10

Extended Response REASONING

Ask questions such as the following:

- **How can you check your answer?** Accept all reasonable answers. Children may suggest adding or counting up from the smaller number.
- **Could you have done this another way?** Accept all reasonable answers.

4 Assess

Informal Assessment

Use the Student Assessment Record, **Assessment,** page 100, to record informal observations.

COMPUTING	REASONING
Teacher's Choice	**How Much More?**
Did the child	Did the child
❏ respond accurately?	❏ provide a clear explanation?
❏ respond quickly?	❏ communicate reasons and strategies?
❏ respond with confidence?	❏ choose appropriate strategies?
❏ self-correct?	❏ argue logically?

Adding and Subtracting Length • Lesson 3   357

Week 28 — Adding and Subtracting Length

Lesson 4

Objective
Children practice numeration skills and use formal notation to indicate forward progress on a number line.

Program Materials
Warm Up
See materials from previous weeks.

Engage
- Plus Cards, 2 sets
- Magnetic Number Line, 1 per child
- Magnetic Chips

Additional Materials
Engage
Paper and pencils

Access Vocabulary
stack Pile of flat things such as cards or books
record To write down

Creating Context
Some English Learners may be confused by the homophones *sum* and *some*. Help distinguish the two words by explaining that when we talk about the sum of two numbers, we always use the article *the* before the word. *Sum* is a noun. The describing word *some* will not have an article.

1. Warm Up

Skill Building — COMPUTING
Teacher's Choice
Before beginning the small-group activity **Let's Add Some More,** choose one of the Warm-Up activities from previous weeks to use with the whole group. These activities include the following:
- Object Land: **Pointing and Winking**
- Picture Land: **Catch the Teacher**
- Line Land: **The Neighborhood**
- Sky Land: **Thermometer Counting**
- Object Land: **What Number Am I?**
- Circle Land: **Count Up**
- Circle Land: **Blastoff!**

Warm-Up Card 1

Purpose Children reinforce counting and number-sequence skills.

Monitoring Student Progress
If...	Then...
children can work fluently with numbers 1–10,	challenge them to work with bigger numbers.

2. Engage

Concept Building — APPLYING
Let's Add Some More
"Today we are going to see who can write the best addition number stories."

Follow the instructions on the Activity Card to play **Let's Add Some More.** As children play, ask questions about what is happening in the activity.

Activity Card 51

358 Number Worlds • Week 28

Purpose Children use formal notation to indicate forward progress on a number line.

Monitoring Student Progress

| **If . . .** children are comfortable with this level of play, | **Then . . .** challenge them to see how many combinations of the addends 2, 3, 4, 5, and 6 they can create that equal 20. What does this tell them about the order of addends in an addition problem? |

 For further practice with addition, children should complete **Building Blocks** Barkley's Bones 3.

3 Reflect 10

Extended Response REASONING

Ask questions such as the following:

- **What patterns do you see?** Accept all reasonable answers.
- **What would happen if you added a +1 Card to your stack?** The sum would be 21 instead of 20. Encourage children to experiment with the order of the addends.

4 Assess

Informal Assessment

Use the Student Assessment Record, **Assessment**, page 100, to record informal observations.

COMPUTING	APPLYING
Teacher's Choice	**Let's Add Some More**
Did the child	Did the child
❏ respond accurately?	❏ apply learning to new situations?
❏ respond quickly?	❏ contribute concepts?
❏ respond with confidence?	❏ contribute answers?
❏ self-correct?	❏ connect mathematics to the real world?

Adding and Subtracting Length • Lesson 4

Week 28: Adding and Subtracting Length

Lesson 5
Review

Objective
Children review the material presented in Week 28 of **Number Worlds**.

Program Materials
Warm Up
See materials from previous weeks.

Engage
See materials from previous weeks.

Creating Context
In addition to assessing and observing children using math concepts, complete an observation of children's language proficiency with academic English for mathematics. If your designated English Language Proficiency Test has an observation checklist, use it to determine which English forms, functions, and vocabulary are still needed for academic language proficiency.

1 Warm Up 5

Skill Building COMPUTING
Teacher's Choice
Before beginning the **Free-Choice** activity, choose one of the Warm-Up activities from previous weeks to use with the whole group. These activities include the following:

- Object Land: **Pointing and Winking**
- Picture Land: **Catch the Teacher**
- Line Land: **The Neighborhood**
- Sky Land: **Thermometer Counting**
- Object Land: **What Number Am I?**
- Circle Land: **Count Up**
- Circle Land: **Blastoff!**

Warm-Up Card 1

Purpose Children reinforce counting and number-sequence skills.

2 Engage 20

Concept Building APPLYING
Free-Choice Activity
For the last day of the week, allow children to choose an activity from previous weeks. Some activities they may choose include the following:

- Sky Land: **Linking and Thinking**
- Line Land: **Counting Back**
- Object Land: **Three-Area Counter Drop**

Make a note of the activities children select. Do they prefer easy or challenging activities? Continue to provide Challenge opportunities for children who have mastered the basic activities.

Monitoring Student Progress
| If . . . children would benefit from extra practice on specific skills, | Then . . . choose an activity for them. |

3 Reflect — 10

Extended Response — REASONING
Ask questions such as the following:
- **What did you like about playing** How Much More?
- **Was there anything about playing this game you didn't like?**
- **Did this game help you do something you couldn't do before? What did it help you do?**
- **What was easy when you were playing** Let's Add Some More?
- **What was hard when you were playing** Let's Add Some More?
- **What do you remember most about this game?**
- **What was your favorite part?**

Using Student Pages
Have each child complete **Workbook,** page 88. Is the child correctly applying the skills learned in Week 28?

Week 28 — Adding and Subtracting Length
Review

Name _____ Date _____
Write the answer.

1. 14 + 6 = **20**
2. 9 − 7 = **2**
3. 4 + 12 = **16**
4. 15 − 4 = **11**

88

Workbook, p. 88

4 Assess — 10

A Gather Evidence
Formal Assessment
Have children complete the weekly test on **Assessment,** page 79. Record formal assessment scores on the Student Assessment Record, **Assessment,** page 100.

Week 28 Test

Name _____ Date _____

1. 3 + 2 = 5   3 − 2 = 1
2. 2 + 3 + 4 = ___
3. 2 − 1 = 1   1 + 1 = 2
4. 4 + 2 + 3 = ___

Assessment, p. 79

B Summarize Findings
Review the Student Assessment Records. Determine whether students have Minimal, Basic, or Secure understanding of the concepts presented in Week 28.

C Differentiate Instruction
Based on your observations, use these teaching strategies next week to follow up.

Minimal Understanding
- Repeat the Warm-Up and Engage activities in subsequent weeks to develop concepts of counting objects.
- Repeat the activities from this week. Observe children's understanding of these concepts and reinforce as necessary.
- Use **Building Blocks** computer activities beginning with Count Free Explore to develop and reinforce numeration concepts.

Basic Understanding
- Repeat Engage activities in subsequent weeks to reinforce basic concepts.
- Use **Building Blocks** computer activities beginning with Gus' Garage to reinforce measurement concepts.

Secure Understanding
- Use Challenge variations of this week's activities.
- Use **Building Blocks** computer activities beginning with Multidigit Word Problems with Tools.

Adding and Subtracting Length • Lesson 5 Review

Week 29: Addition and Subtraction Stories

Week at a Glance

This week, children begin **Number Worlds,** Week 29, and continue to explore Picture Land.

Background

In Picture Land, numbers are represented as numerals. This allows children to forge links between the world of real quantities (e.g., objects) and the world of formal symbols (such as numerals and addition, subtraction, and equal signs) used when writing equations.

How Children Learn

As children begin this week's lesson, they should be comfortable writing and solving simple equations and solving addition problems with multiple addends.

By the end of the week, children will have gained additional practice with multiple-addend problems and should be able to solve two-step word problems.

Skills Focus
- Add whole numbers
- Subtract whole numbers
- Solve word problems
- Write addition and subtraction equations

Teaching for Understanding

As children engage in these activities, they should begin to see how math can be used in their everyday lives.

Observe closely while evaluating the Engage activities assigned for this week.
- Can children add whole numbers?
- Can children subtract whole numbers?
- Can children use critical thinking skills to write and solve word problems?

Math at Home

Give one copy of the Letter to Home, page A29, to each child. Complete the activity in class, and then encourage children to share it with their caregivers.

Letter to Home, Teacher Edition, p. A29

Math Vocabulary

word problem Problem that has a story instead of just numbers

equation Number story

left over Not used

fact Something you know is true

English Learners
SPANISH COGNATES

English	Spanish
addition	adición
cent	centavo
combine	combinar
equal	igual
minus	menos

ALTERNATE VOCABULARY

enough The amount needed

362　Number Worlds • Week 29

Week 29 Planner — Addition and Subtraction Stories

PACING	LESSON	LEARNING GOALS	MATERIALS	TECHNOLOGY
DAY 1	Warm Up — Teacher's Choice	Children reinforce counting and number-sequence skills.	Materials will be selected from those used in previous weeks.	**Building Blocks** Multidigit Word Problems with Tools
	Activity 52 — Picture Land: How Many Are Left?	Children use addition and subtraction to solve two-step word problems.	**Program Materials** Counters **Additional Materials** Paper and pencils for each child	
DAY 2	Warm Up — Teacher's Choice	Children reinforce counting and number-sequence skills.	Materials will be selected from those used in previous weeks.	**Building Blocks** Pizza Pizzazz Free Explore
	Activity 53 — Picture Land: Going Shopping*	Children solve successive addition and subtraction problems involving single-digit addends.	**Program Materials** • Multi-Land Activity Sheet 1, 1 per child • Counters (optional) **Additional Materials** Paper and pencils for each child	
DAY 3	Warm Up — Teacher's Choice	Children reinforce counting and number-sequence skills.	Materials will be selected from those used in previous weeks.	**Building Blocks** Multidigit Word Problems with Tools
	Activity 52 — Picture Land: How Many Are Left?	Children use addition and subtraction to solve two-step word problems.	**Program Materials** Counters **Additional Materials** Paper and pencils for each child	
DAY 4	Warm Up — Teacher's Choice	Children reinforce counting and number-sequence skills.	Materials will be selected from those used in previous weeks.	**Building Blocks** Pizza Pizzazz Free Explore
	Activity 53 — Picture Land: Going Shopping*	Children solve successive addition and subtraction problems involving single-digit addends.	**Program Materials** • Multi-Land Activity Sheet 1, 1 per child • Counters (optional) **Additional Materials** Paper and pencils for each child	
DAY 5	Warm Up — Teacher's Choice	Children reinforce counting and number-sequence skills.	Materials will be selected from those used in previous weeks.	**Building Blocks** Review previous activities
	Review and Assess	Children review their favorite activities to improve their understanding of more difficult concepts.	Materials will be selected from those used in previous weeks.	

*Includes Challenge Variations

Week 29 — Lesson 1: Addition and Subtraction Stories

Objective
Children practice numeration skills and solve two-step word problems.

Program Materials
Warm Up
See materials for previous weeks.

Engage
Counters

Additional Materials
Engage
Paper and pencils for each child

Access Vocabulary
word problem Problem that has a story instead of just numbers
equation Number story

Creating Context
Organize children into small groups, and have them talk about things they do at home that require addition. Have children work together to create a group story, complete with illustrations, to clearly show that addition was involved. Have children use manipulatives or props to act out their stories.

1 Warm Up (5)

Skill Building COMPUTING
Teacher's Choice
Before beginning the whole-class activity **How Many Are Left?** choose one of the Warm-Up activities from previous weeks to use with the whole group. These activities include the following:

- Object Land: **Pointing and Winking**
- Picture Land: **Catch the Teacher**
- Line Land: **The Neighborhood**
- Sky Land: **Thermometer Counting**
- Object Land: **What Number Am I?**
- Circle Land: **Count Up**
- Circle Land: **Blastoff!**

Warm-Up Card 1

Purpose Children reinforce counting and number-sequence skills.

Monitoring Student Progress
If . . . children need help with basic skills,
Then . . . practice counting and number-sequence skills with them throughout the day.

2 Engage (30)

Concept Building UNDERSTANDING
How Many Are Left?
"Today we are going to solve word problems."

Follow the instructions on the Activity Card to play **How Many Are Left?** As children play, ask questions about what is happening in the activity.

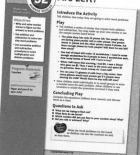
Activity Card 52

364 Number Worlds • Week 29

Purpose Children use addition and subtraction to solve two-step word problems. They use successive addition and subtraction procedures to solve multiplication and division problems.

Monitoring Student Progress

If . . . children have trouble solving the word problems and writing the corresponding equations,

Then . . . have them use Counters to model the problems. Guide children through the problem-solving process by discussing (1) what they want to find out, (2) what facts they know, and (3) what strategy they will use to find the answer.

eMathTools Use the Set Tool to model the word problems.

 Teacher's Note Model the problem-solving process as often as needed to help student understanding.

Building Blocks For further practice with word problems, children should complete **Building Blocks** Multidigit Word Problems with Tools.

3 Reflect 10

Extended Response REASONING
Ask questions such as the following:
- **What were we trying to find out?** Answers will vary. Discuss with children how to find the "question" part of the word problem.
- **What facts did we know?** Answers will vary.
- **What strategy did you use to solve this problem?** Accept all reasonable answers.

Using Student Pages
Have each child complete **Workbook,** page 89. Did the child write the correct equation?

 Week 29 Addition and Subtraction Stories
How Many Are Left?

Name _____ Date _____

Write the number stories.

① I had 14 flowers. I gave 5 to my friend. How many are left?

$14 - 5 = 9$

② Then, I gave 3 flowers to my brother. How many are left?

$9 - 3 = 6$

③ Then, I gave 2 flowers to my teacher. How many are left?

$6 - 2 = 4$

89

Workbook, p. 89

4 Assess

Informal Assessment
Use the Student Assessment Record, **Assessment,** page 100, to record informal observations.

COMPUTING	UNDERSTANDING
Teacher's Choice	**How Many Are Left?**
Did the child	Did the child
❏ respond accurately?	❏ make important observations?
❏ respond quickly?	
❏ respond with confidence?	❏ extend or generalize learning?
❏ self-correct?	❏ provide insightful answers?
	❏ pose insightful questions?

Addition and Subtraction Stories • Lesson 1

Week 29

Addition and Subtraction Stories

Lesson 2

Objective
Children practice numeration skills and solve successive addition and subtraction problems.

Program Materials
Warm Up
See materials from previous weeks.

Engage
- Multi-Land Activity Sheet 1, 1 per child
- Counters (optional)

Additional Materials
Paper and pencil for each child

Access Vocabulary
pretend To make believe

Creating Context
Many children have experience shopping with adults, but English Learners may not know the vocabulary for the products. Using pictures or props, review the names of the items on the Activity Sheet. For beginning English Learners, provide a sentence frame such as "I want _____." For more proficient English Learners, model the conditional tense, "I would pick _____ and _____."

1 Warm Up — 5

Skill Building COMPUTING
Teacher's Choice
Before beginning the whole-class activity **Going Shopping,** choose one of the Warm-Up activities from previous weeks to use with the whole group. These activities include the following:

- Object Land: **Pointing and Winking**
- Picture Land: **Catch the Teacher**
- Line Land: **The Neighborhood**
- Sky Land: **Thermometer Counting**
- Object Land: **What Number Am I?**
- Circle Land: **Count Up**
- Circle Land: **Blastoff!**

Purpose Children reinforce counting and number-sequence skills.

Warm-Up Card 1

Monitoring Student Progress
If... children need help with basic skills, **Then...** practice counting and number-sequence skills with them throughout the day.

2 Engage — 30

Concept Building ENGAGING
Going Shopping
"Today we are going on a pretend shopping trip."

Follow the instructions on the Activity Card to play **Going Shopping.** As children play, ask questions about what is happening in the activity.

Activity Card 53

Purpose Children solve successive addition and subtraction problems involving single-digit addends.

Monitoring Student Progress

If . . . children have trouble keeping track of how much money they have spent,

Then . . . invite them to model how much they spend on each purchase with Counters or to keep a running tally on a piece of paper.

Teacher's Note If you have created a poster-sized version of the Shopping Guide for the **Shopping Trip** activity, it will be useful for this activity as well. It can provide a focal point for group discussion and can therefore enhance the quality of children's math talk as they discuss possible combinations of purchases.

 For further practice with addition, children should complete **Building Blocks** Pizza Pizzazz Free Explore.

3 Reflect 10

Extended Response REASONING

Ask questions such as the following:

- **How much money did you have to start with?** 10 cents
- **What items did you buy?** Answers will vary.
- **Did you have enough money to buy all those items?** Answers will vary. **How did you figure that out?** Children may add the cost of their items and subtract the total from 10, or they may start at 10 and subtract or count back the cost of each individual item.

Using Student Pages

Have each child complete **Workbook,** page 90. Did the child write the correct answer?

Week 29 Addition and Subtraction Stories

Going Shopping

Name _____ Date _____

Write the answer.

① $5 + 3 + 2 = \underline{10}$

② $4 + 7 + 4 = \underline{15}$

③ $3 + 9 + 5 = \underline{17}$

④ $3 + 6 + 3 = \underline{12}$

90

Workbook, p. 90

4 Assess

Informal Assessment

Use the Student Assessment Record, **Assessment,** page 100, to record informal observations.

COMPUTING	ENGAGING
Teacher's Choice	**Shopping Trip**
Did the child	Did the child
❏ respond accurately?	❏ pay attention to the contributions of others?
❏ respond quickly?	
❏ respond with confidence?	❏ contribute information and ideas?
❏ self-correct?	
	❏ improve on a strategy?
	❏ reflect on and check accuracy of work?

Addition and Subtraction Stories • Lesson 2 367

Week 29 — Addition and Subtraction Stories
Lesson 3

Objective
Children practice numeration skills and solve two-step word problems.

Program Materials
Warm Up
See materials from previous weeks.

Engage
Counters

Additional Materials
Engage
Paper and pencils for each child

Access Vocabulary
fact Something you know is true
altogether In all; total

Creating Context
English Learners often understand math concepts better when they can interact with other children who speak the same primary language. Allow children to work with partners or cross-age helpers who speak the same primary language so that they can check understanding with one another in team situations.

1 Warm Up

Skill Building — COMPUTING
Teacher's Choice
Before beginning the whole-class activity **How Many Are Left?** choose one of the Warm-Up activities from previous weeks to use with the whole group. These activities include the following:
- Object Land: **Pointing and Winking**
- Picture Land: **Catch the Teacher**
- Line Land: **The Neighborhood**
- Sky Land: **Thermometer Counting**
- Object Land: **What Number Am I?**
- Circle Land: **Count Up**
- Circle Land: **Blastoff!**

Warm-Up Card 1

Purpose Children reinforce counting and number-sequence skills.

Monitoring Student Progress
If . . . children need help with basic skills,
Then . . . practice counting and number-sequence skills with them throughout the day.

2 Engage

Concept Building — REASONING
How Many Are Left?
"Today we are going to solve word problems."

Follow the instructions on the Activity Card to play **How Many Are Left?** As children play, ask questions about what is happening in the activity.

Activity Card 52

368 Number Worlds • Week 29

Purpose Children use addition and subtraction to solve two-step word problems. They use successive addition and subtraction procedures to solve multiplication and division problems.

Monitoring Student Progress

| If... children can fluently solve your word problems and write the corresponding equations, | Then... challenge them to work with a partner to create and solve their own word problems. |

 Teacher's Note Encourage children to use this week's vocabulary words as they engage in the activities, discuss math concepts, and make predictions.

Building Blocks For further practice with word problems, children should complete *Building Blocks* Multidigit Word Problems with Tools.

3 Reflect 10

Extended Response REASONING

Ask questions such as the following:

- **Which number did you put first in your number story?** Answers will vary based on the story problem involved. **Why?** Accept all reasonable answers. For example, in a "how many are left" problem, children may indicate that they put the biggest number first because that is how many they started with.
- **Did you add or subtract? Why?** Accept all reasonable answers.

4 Assess

Informal Assessment

Use the Student Assessment Record, **Assessment,** page 100, to record informal observations.

COMPUTING	REASONING
Teacher's Choice	**How Many Are Left?**
Did the child	Did the child
❏ respond accurately?	❏ provide a clear explanation?
❏ respond quickly?	❏ communicate reasons and strategies?
❏ respond with confidence?	❏ choose appropriate strategies?
❏ self-correct?	❏ argue logically?

Addition and Subtraction Stories • Lesson 3 369

Week 29 — Addition and Subtraction Stories

Objective
Children practice numeration skills and solve successive addition and subtraction problems.

Program Materials
Warm Up
See materials from previous weeks.

Engage
- Multi-Land Activity Sheet 1, 1 per child
- Counters (optional)

Additional Materials
Paper and pencils for each child

Access Vocabulary
enough The amount needed
left over Not used

Creating Context
One of the Reflect questions for this lesson is "What would happen if you had 20 cents to spend instead of 10 cents?" This grammatical construction will be new for some English Learners, but it is important for them to learn as they consider alternatives and practice critical thinking. Explain that "what if" questions let us talk about possibilities.

1 Warm Up — 5

Skill Building — COMPUTING
Teacher's Choice
Before beginning the whole-group activity **Going Shopping**, choose one of the Warm-Up activities from previous weeks to use with the whole group. These activities include the following:
- Object Land: **Pointing and Winking**
- Picture Land: **Catch the Teacher**
- Line Land: **The Neighborhood**
- Sky Land: **Thermometer Counting**
- Object Land: **What Number Am I?**
- Circle Land: **Count Up**
- Circle Land: **Blastoff!**

Warm-Up Card 1

Purpose Children reinforce counting and number-sequence skills.

Monitoring Student Progress
If . . . children can work fluently with numbers 1–10,
Then . . . challenge them to work with bigger numbers.

2 Engage — 30

Concept Building — APPLYING
Going Shopping
"Today we are going on a pretend shopping trip."

Follow the instructions on the Activity Card to play **Going Shopping**. As children play, ask questions about what is happening in the activity.

Activity Card 53

370 Number Worlds • Week 29

Purpose Children solve successive addition and subtraction problems involving single-digit addends.

Monitoring Student Progress

If . . . children are comfortable with this level of play,

Then . . . challenge them to write a multiple-addend equation that describes their purchases.

 Building Blocks For further practice with addition, children should complete **Building Blocks** Pizza Pizzazz Free Explore.

3 Reflect 10

Extended Response REASONING

Ask questions such as the following:

- **How much did your items cost altogether?** Answers will vary. **How did you figure that out?** Children should add the cost of each item.
- **What would happen if you had 20 cents to spend instead of 10 cents?** Children should indicate that they could buy more items or that they would have more money left over if they bought the same items.

4 Assess

Informal Assessment

Use the Student Assessment Record, **Assessment,** page 100, to record informal observations.

COMPUTING	APPLYING
Teacher's Choice	**Going Shopping**
Did the child	Did the child
❏ respond accurately?	❏ apply learning to new situations?
❏ respond quickly?	❏ contribute concepts?
❏ respond with confidence?	❏ contribute answers?
❏ self-correct?	❏ connect mathematics to the real world?

Addition and Subtraction Stories • Lesson 4   **371**

Week 29: Addition and Subtraction Stories

Lesson 5 Review

Objective
Children review the material presented in Week 29 of **Number Worlds**.

Program Materials

Warm Up
See materials from previous weeks.

Engage
See materials from previous weeks.

Creating Context
Although it is important to ask beginning English Learners questions that have a lower language demand, this does not mean that questions should have a lower cognitive load. Ask English Learners to demonstrate their understanding with manipulatives, models, and illustrations. Check understanding by asking children to show, to demonstrate, or to point to the answer.

1 Warm Up

Skill Building — COMPUTING

Teacher's Choice
Before beginning the **Free-Choice** activity, choose one of the Warm-Up activities from previous weeks to use with the whole group. These activities include the following:

- Object Land: **Pointing and Winking**
- Picture Land: **Catch the Teacher**
- Line Land: **The Neighborhood**
- Sky Land: **Thermometer Counting**
- Object Land: **What Number Am I?**
- Circle Land: **Count Up**
- Circle Land: **Blastoff!**

Warm-Up Card 1

Purpose Children reinforce counting and number-sequence skills.

2 Engage

Concept Building — APPLYING

Free-Choice Activity
For the last day of the week, allow children to choose an activity from the previous weeks. Some activities they may choose include the following:

- Object Land: **Shopping Trip**
- Picture Land: **How Many Balloons?**
- Line Land: **Let's Add Some More**

Make a note of the activities children select. Do they prefer easy or challenging activities? Continue to provide Challenge opportunities for children who have mastered the basic activities.

Monitoring Student Progress

If . . . children	Then . . . choose an activity
would benefit from extra practice on specific skills,	for them.

372 Number Worlds • Week 29

3 Reflect

Extended Response REASONING

Ask questions such as the following:

- What did you like about playing How Many Are Left?
- Was there anything about playing this game you didn't like?
- Did this game help you do something you couldn't do before? What did it help you do?
- What was easy when you were playing Going Shopping?
- What was hard when you were playing Going Shopping?
- What do you remember most about this game?
- What was your favorite part?

Using Student Pages

Have each child complete **Workbook,** page 91. Is the child correctly applying the skills learned in Week 29?

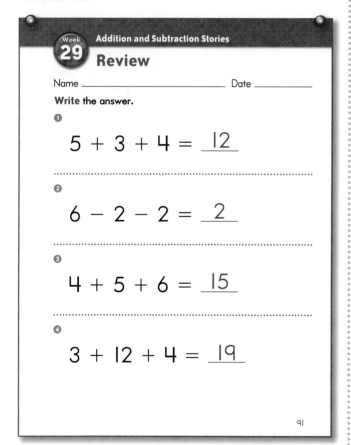

Workbook, p. 91

4 Assess

A Gather Evidence

Formal Assessment

Have students complete the weekly test on **Assessment,** page 81. Record formal assessment scores on the Student Assessment Record, **Assessment,** page 100.

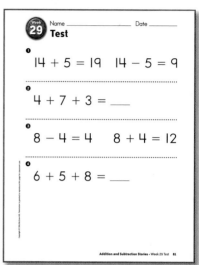

Assessment, p. 81

B Summarize Findings

Review the Student Assessment Records. Determine whether students have Minimal, Basic, or Secure understanding of the concepts presented in Week 29.

C Differentiate Instruction

Based on your observations, use these teaching strategies next week to follow up.

Minimal Understanding
- Repeat the Warm-Up and Engage activities in subsequent weeks to develop numeration concepts.
- Repeat the activities from this week. Observe children's understanding of these concepts, and reinforce as necessary.
- Use **Building Blocks** computer activities beginning with Count Free Explore to develop and reinforce numeration concepts.

Basic Understanding
- Repeat Engage activities in subsequent weeks to reinforce basic concepts.
- Use **Building Blocks** computer activities beginning with Pizza Pizzazz Free Explore to reinforce numeration and addition concepts.

Secure Understanding
- Use **Building Blocks** computer activities beginning with Multidigit Word Problems with Tools.

Addition and Subtraction Stories • Lesson 5 Review   373

Week 30: Making a Map

Week at a Glance

This week, children begin **Number Worlds**, Week 30, and continue to explore Line Land and Picture Land.

Background

In Line Land, numbers are represented as positions on a line. When a child is asked "Where are you now?" he or she might think of the position on the number line, the set size that corresponds to that position, or the number that shows the position on the number line.

How Children Learn

As children begin this week's activities, they should be able to solve multiple-addend problems and two-step word problems.

At the end of this week, children should be able to apply numeration concepts to real-world situations and should be able to solve word problems with multiple addends.

Skills Focus
- Compare and order numbers
- Add whole numbers
- Solve word problems
- Use spatial terms to describe a location
- Create a map

Teaching for Understanding

As children engage in these activities, they should begin to see how math can be used in their everyday lives.

Observe closely while evaluating the Engage activities assigned for this week.

- Can children add whole numbers?
- Can children compare and order numbers?
- Can children use critical-thinking skills to write and solve word problems?

Math at Home
Give one copy of the Letter to Home, page A30, to each child. Complete the activity in class, and then encourage children to share it with their caregivers.

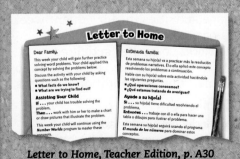

Letter to Home, Teacher Edition, p. A30

Math Vocabulary
direction Which way you move (right or left, up or down)

enough The amount that is needed

extra More than what is needed

far Not close to; a long way away

English Learners
SPANISH COGNATES

English	Spanish
distance	distancia
equal	igual
map	mapa
signs	signos

ALTERNATE VOCABULARY

before In front of; coming first

after Behind; coming later

Week 30 Planner: Making a Map

PACING	LESSON	LEARNING GOALS	MATERIALS	TECHNOLOGY
DAY 1	Warm Up Teacher's Choice	Children reinforce counting and number-sequence skills.	Materials will be selected from those used in previous weeks.	Building Blocks Barkley's Bones 3
DAY 1	Activity 54 **Picture Land:** Feed the Pets	Children solve successive addition problems with single-digit addends.	**Prepare Ahead** • Paper and pencils • Feeding Instructions Note	
DAY 2	Warm Up Teacher's Choice	Children reinforce counting and number-sequence skills.	Materials will be selected from those used in previous weeks.	Building Blocks Rocket Blast 1
DAY 2	Activity 55 **Line Land:** Hallway Map	Children create a map with numbered locations and use spatial terms to describe a location.	**Additional Materials** • Paper and pencils • Volunteers to follow the maps	
DAY 3	Warm Up Teacher's Choice	Children reinforce counting and number-sequence skills.	Materials will be selected from those used in previous weeks.	Building Blocks Barkley's Bones 3
DAY 3	Activity 54 **Picture Land:** Feed the Pets	Children solve successive addition problems with single-digit addends.	**Prepare Ahead** • Paper and pencils • Feeding Instructions Note	
DAY 4	Warm Up Teacher's Choice	Children reinforce counting and number-sequence skills.	Materials will be selected from those used in previous weeks.	Building Blocks Rocket Blast 1
DAY 4	Activity 55 **Line Land:** Hallway Map	Children create a map with numbered locations and use spatial terms to describe a location.	**Additional Materials** • Paper and pencils • Volunteers to follow the maps	
DAY 5	Warm Up Teacher's Choice	Children reinforce counting and number-sequence skills.	Materials will be selected from those used in previous weeks.	Building Blocks Review previous activities
DAY 5	Review and Assess	Children review their favorite activities to improve their understanding of more difficult concepts.	Materials will be selected from those used in previous weeks.	

Week 30
Making a Map
Lesson 1

Objective
Children practice numeration skills and solve successive addition problems.

Program Materials
Warm Up
See materials for previous weeks.

Prepare Ahead
Engage
- Paper and pencils
- Feeding Instructions Note

Access Vocabulary
enough The amount that is needed
extra More than what is needed

Creating Context
Organize children into small groups, and have them talk about things they do at home that require addition. Have children work together to create a group story, complete with illustrations to clearly show that addition was involved. Have children use manipulatives or props to act out their story.

1 Warm Up

Skill Building — COMPUTING
Teacher's Choice
Before beginning the whole-class activity **Feed the Pets,** choose one of the Warm-Up activities from previous weeks to use with the whole group. These activities include the following:
- Object Land: **Pointing and Winking**
- Picture Land: **Catch the Teacher**
- Line Land: **The Neighborhood**
- Sky Land: **Thermometer Counting**
- Object Land: **What Number Am I?**
- Circle Land: **Count Up**
- Circle Land: **Blastoff!**

Purpose Children reinforce counting and number-sequence skills.

Warm-Up Card 1

Monitoring Student Progress
If . . . children need help with basic skills,
Then . . . practice counting and number-sequence skills with them throughout the day.

2 Engage

Concept Building — UNDERSTANDING
Feed the Pets
"Today we are going to figure out if there is enough food to feed a neighbor's pets."

Follow the instructions on the Activity Card to play **Feed the Pets.** As children play, ask questions about what is happening in the activity.

Activity Card 54

Purpose Children solve successive addition problems with single-digit addends.

Monitoring Student Progress

| **If...** children have trouble remembering how much food each pet should receive, | **Then...** make a chart on the board showing a big dog, a little dog, three cats, and how many scoops of food each animal should receive. |

 For further practice with addition, children should complete **Building Blocks** Barkley's Bones 3.

3 Reflect — 10

Extended Response REASONING

Ask questions such as the following:

- **What were we trying to find out?** whether there is enough food to feed all the pets for 2 days
- **What facts did we know?** how much food there is and how much each pet eats in a day
- **What strategy did you use to solve this problem?** Accept all reasonable answers. Children may suggest adding the food each pet will eat and comparing that to the total amount of food.

Using Student Pages

Have each child complete **Workbook,** page 92. Did the child write the correct equations?

Week 30 Making a Map
Feed the Pets

Name _____ Date _____

Write and solve the number stories.

1. My dog eats 2 cups of food each day. How much does she eat in 2 days?

   $2 + 2 = 4$

2. My cat eats 1 cup of food each day. How much does he eat in 2 days?

   $1 + 1 = 2$

3. How much do they eat altogether?

   $4 + 2 = 6$

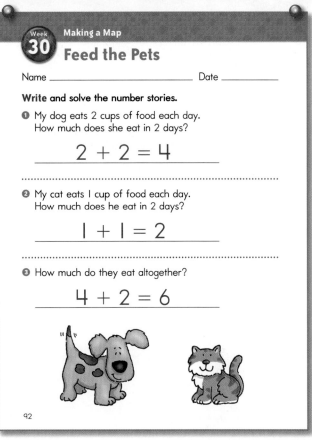

92

Workbook, p. 92

4 Assess

Informal Assessment

Use the Student Assessment Record, **Assessment,** page 100, to record informal observations.

COMPUTING	UNDERSTANDING
Teacher's Choice	**Feed the Pets**
Did the child	Did the child
❏ respond accurately?	❏ make important observations?
❏ respond quickly?	❏ extend or generalize learning?
❏ respond with confidence?	❏ provide insightful answers?
❏ self-correct?	❏ pose insightful questions?

Making a Map • Lesson 1 **377**

Week 30 — Making a Map
Lesson 2

Objective
Children practice counting and number-sequence skills and create a map with numbered locations.

Program Materials
Warm Up
See materials from previous weeks.

Additional Materials
Engage
- Paper and pencils
- Volunteers to follow the maps

Access Vocabulary
before In front of; coming first
after Behind; coming later

Creating Context
Prepositions are not in the beginning level of language proficiency. Help English Learners practice prepositions such as *before*, *after*, and *between* when they line up for recess. Have children identify who is before them in line, who is after them, and who is between two other children.

1 Warm Up 5

Skill Building COMPUTING
Teacher's Choice
Before beginning the small-group activity **Hallway Map,** choose one of the Warm-Up activities from previous weeks to use with the whole group. These activities include the following:
- Object Land: **Pointing and Winking**
- Picture Land: **Catch the Teacher**
- Line Land: **The Neighborhood**
- Sky Land: **Thermometer Counting**
- Object Land: **What Number Am I?**
- Circle Land: **Count Up**
- Circle Land: **Blastoff!**

Warm-Up Card 1

Purpose Children reinforce counting and number-sequence skills.

Monitoring Student Progress
| If . . . children need help with basic skills, | Then . . . practice counting and number-sequence skills with them throughout the day. |

2 Engage 30

Concept Building ENGAGING
Hallway Map
"Today we are going to make a map of the hallway outside our classroom."

Follow the instructions on the Activity Card to play **Hallway Map.** As children play, ask questions about what is happening in the activity.

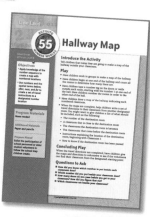
Activity Card 55

Purpose Children create a map with numbered locations and use spatial terms to describe a location.

Monitoring Student Progress

If . . . children have trouble drawing a map of the hallway,

Then . . . give children a model to work with. On the board or on handouts, draw a large, horizontal rectangle to represent the hallway and smaller vertical rectangles to represent the doors.

 MathTools Use the Number Line Tool to demonstrate and explore number sequence.

Teacher's Note You may choose to have children number only one side of the hallway, or you may have them number both sides by going down one side and up the other side.

Building Blocks For further practice with addition, children should complete **Building Blocks** Rocket Blast 1.

3 Reflect 10

Extended Response REASONING

Ask questions such as the following:

- **How did you know which number to put beside each classroom door?** Possible answer: I started with 1 at the first door, put 2 beside the next door, and counted up the rest of the hallway.
- **Which number did you put beside your classroom door?** Answers will vary.

Using Student Pages

Have each child complete **Workbook,** page 93. Did the child identify the correct room numbers?

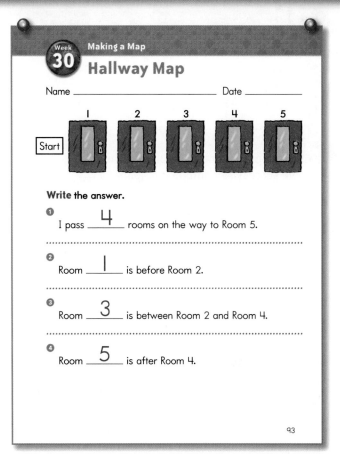

Workbook, p. 93

4 Assess

Informal Assessment

Use the Student Assessment Record, **Assessment,** page 100, to record informal observations.

COMPUTING	ENGAGING
Teacher's Choice	**Hallway Map**
Did the child	Did the child
❏ respond accurately?	❏ pay attention to the contributions of others?
❏ respond quickly?	
❏ respond with confidence?	❏ contribute information and ideas?
❏ self-correct?	
	❏ improve on a strategy?
	❏ reflect on and check accuracy of work?

Making a Map • Lesson 2 379

Week 30 — Making a Map

Lesson 3

Objective
Children practice numeration skills and solve successive addition problems.

Program Materials
Warm Up
See materials for previous weeks.

Prepare Ahead
Engage
- Paper and pencils
- Feeding Instructions Note

Access Vocabulary
left over Not used
too few Less than what is needed

Creating Context
Some English Learners may be confused by the homophones *sum* and *some*. Help distinguish the two words by explaining that when we talk about the sum of two numbers, we always use the article *the* before the word. *Sum* is a noun. The describing word *some* will not have an article.

1 Warm Up 5

Skill Building COMPUTING
Teacher's Choice
Before beginning the whole-class activity **Feed the Pets,** choose one of the Warm-Up activities from previous weeks to use with the whole group. These activities include the following:
- Object Land: **Pointing and Winking**
- Picture Land: **Catch the Teacher**
- Line Land: **The Neighborhood**
- Sky Land: **Thermometer Counting**
- Object Land: **What Number Am I?**
- Circle Land: **Count Up**
- Circle Land: **Blastoff!**

Warm-Up Card 1

Purpose Children reinforce counting and number-sequence skills.

> **Monitoring Student Progress**
>
> **If . . .** children need help with basic skills,
>
> **Then . . .** practice counting and number-sequence skills with them throughout the day.

2 Engage 30

Concept Building REASONING
Feed the Pets

"Today we are going to figure out if there is enough food to feed a neighbor's pets."

Follow the instructions on the Activity Card to play **Feed the Pets.** As children play, ask questions about what is happening in the activity.

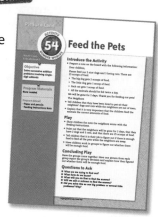

Activity Card 54

380 Number Worlds • Week 30

Purpose Children solve successive addition problems with single-digit addends.

Monitoring Student Progress

If... children are comfortable with this level of play,

Then... invite them to work with a partner to make up their own "feed the pet" word problems.

 Encourage children to use this week's vocabulary words as they engage in the activities, discuss math concepts, and make predictions.

 For further practice with addition, children should complete **Building Blocks** Barkley's Bones 3.

3 Reflect 10

Extended Response REASONING

Ask questions such as the following:

- **Did we add or subtract to find the answer?** Children may add how much food each pet will eat and compare that number to 25 or may suggest starting at 25 and subtract the amount of food each pet will eat.

- **Did you solve this as one big problem or several little problems? Why?** Accept all reasonable answers.

4 Assess

Informal Assessment

Use the Student Assessment Record, **Assessment,** page 100, to record informal observations.

COMPUTING	REASONING
Teacher's Choice	**Feed the Pets**
Did the child	Did the child
❑ respond accurately?	❑ provide a clear explanation?
❑ respond quickly?	❑ communicate reasons and strategies?
❑ respond with confidence?	
❑ self-correct?	❑ choose appropriate strategies?
	❑ argue logically?

Making a Map • Lesson 3   **381**

Week 30 — Making a Map
Lesson 4

Objective
Children practice counting and number-sequence skills and create a map with numbered locations.

Program Materials
Warm Up
See materials from previous weeks.

Additional Materials
Engage
- Paper and pencils
- Volunteers to follow the maps

Access Vocabulary
direction Which way you move (right or left, up or down)
far Not close to; a long way away

Creating Context
Using gestures is an excellent way to enhance comprehension for English Learners. This activity provides the opportunity to talk about distance and comparison. Use hand gestures to help English Learners understand terms such as *nearest* and *farthest*.

1 Warm Up — 5

Skill Building — COMPUTING
Teacher's Choice
Before beginning the small-group activity **Hallway Map**, choose one of the Warm-Up activities from previous weeks to use with the whole group. These activities include the following:
- Object Land: **Pointing and Winking**
- Picture Land: **Catch the Teacher**
- Line Land: **The Neighborhood**
- Sky Land: **Thermometer Counting**
- Object Land: **What Number Am I?**
- Circle Land: **Count Up**
- Circle Land: **Blastoff!**

Warm-Up Card 1

Purpose Children reinforce counting and number-sequence skills.

Monitoring Student Progress
If . . . children can work fluently with numbers 1–10,
Then . . . challenge them to work with bigger numbers.

2 Engage — 30

Concept Building — APPLYING
Hallway Map
"Today we are going to make a map of the hallway outside our classroom."

Follow the instructions on the Activity Card to play **Hallway Map**. As children play, ask questions about what is happening in the activity.

Activity Card 55

382 Number Worlds • Week 30

Purpose Children create maps with numbered locations and use spatial terms to describe a location.

Monitoring Student Progress

If . . . children can create an accurate hallway map and directions,

Then . . . challenge them to map other things, such as the route they take to school or the areas of the classroom. Have children write directions to one of the areas on the map (without identifying the area) and then trade maps with a classmate to see if that child can use the directions to locate the correct area.

 For further practice with numeration, children should complete **Building Blocks** Rocket Blast 1.

3 Reflect   10

Extended Response   REASONING

Ask questions such as the following:

- **How many doors do you pass before you get to your classroom?** Answers will vary. **How did you figure that out?** Accept all reasonable answers. Children may count the doors or may say a number 1 less than the number with which they labeled their classroom door.
- **What patterns do you see?** Accept all reasonable answers.

 Assess

Informal Assessment

Use the Student Assessment Record, **Assessment,** page 100, to record informal observations.

COMPUTING	APPLYING
Teacher's Choice	**Hallway Map**
Did the child	Did the child
❏ respond accurately?	❏ apply learning to new situations?
❏ respond quickly?	❏ contribute concepts?
❏ respond with confidence?	❏ contribute answers?
❏ self-correct?	❏ connect mathematics to the real world?

Making a Map • Lesson 4   **383**

Week 30 — Making a Map

Lesson 5 Review

Objective
Children review the material presented in Week 30 of **Number Worlds**.

Program Materials
See materials from previous weeks.

Creating Context
To help English Learners at early proficiency levels, remember to use questions during assessment that are not heavily dependent on language. Assessment activities that tell children to "Show me how many" or "Point to the answer" are good ways to include early English Learners.

✓ Cumulative Assessment
After Lesson 5 is complete, children should complete the Cumulative Assessment, **Assessment,** pages 97–98. Using the key on **Assessment,** page 96, identify incorrect responses. Reteach and review the Warm-Up and Engage activities to reinforce concept understanding.

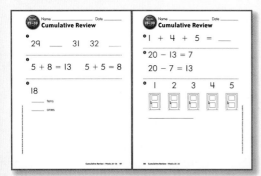

Assessment, pp. 97–98

1 Warm Up — 5 min

Skill Building — COMPUTING
Teacher's Choice
Before beginning the **Free-Choice** activity, choose one of the Warm-Up activities from previous weeks to use with the whole group. These activities include the following:

- Object Land: **Pointing and Winking**
- Picture Land: **Catch the Teacher**
- Line Land: **The Neighborhood**
- Sky Land: **Thermometer Counting**
- Object Land: **What Number Am I?**
- Circle Land: **Count Up**
- Circle Land: **Blastoff!**

Warm-Up Card 1

Purpose Children reinforce counting and number-sequence skills.

2 Engage — 20 min

Concept Building — APPLYING
Free-Choice Activity
For the last day of the week, allow children to choose an activity from previous weeks. Some activities they may choose include the following:

- Line Land: **Numbering the Neighborhood**
- Picture Land: **How Many Balloons?**
- Picture Land: **How Many Are Left?**

Make a note of the activities children select. Do they prefer easy or challenging activities? Continue to provide Challenge opportunities for children who have mastered the basic activities.

Monitoring Student Progress

If . . . children would benefit from extra practice on specific skills,

Then . . . choose an activity for them.

384 Number Worlds • Week 30

Reflect 10

Extended Response REASONING
Ask questions such as the following:
- What did you like about playing Feed the Pets?
- Was there anything about playing this game you didn't like?
- Did this game help you do something you couldn't do before? What did it help you do?
- What was easy when you were playing Hallway Map?
- What was hard when you were playing Hallway Map?
- What do you remember most about this game?
- What was your favorite part?

Using Student Pages
Have each child complete **Workbook,** page 94. Is the child correctly applying the skills learned in Week 30?

Workbook, p. 94

 Assess 10

A Gather Evidence
Formal Assessment
Have children complete the weekly test on **Assessment,** page 83. Record formal assessment scores on the Student Assessment Record, **Assessment,** page 100.

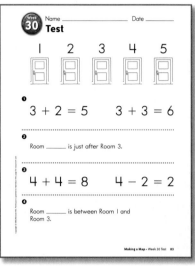

Assessment, p. 83

B Summarize Findings
Review the Student Assessment Records. Determine whether childen have Minimal, Basic, or Secure understanding of the concepts presented in Week 30.

C Differentiate Instruction
Based on your observations, use these teaching strategies next week to follow up.

Minimal Understanding
- Repeat the Warm-Up and Engage activities in subsequent weeks to develop numeration concepts.
- Repeat the activities from this week. Observe children's understanding of these concepts and reinforce as necessary.
- Use **Building Blocks** computer activities beginning with Count Free Explore to develop and reinforce numeration concepts.

Basic Understanding
- Repeat Engage activities in subsesquent weeks to reinforce basic concepts.
- Use **Building Blocks** computer activities beginning with Rocket Blast 1 to reinforce numeration concepts.

Secure Understanding
- Use **Building Blocks** computer activities beginning with Barkley's Bones 4.

Making a Map • Lesson 5 Review 385

Appendix

Appendix A
Letters to Home Week 1–5 .... **A1**
Letters to Home Week 6–10 .... **A6**
Letters to Home Week 11–15 .... **A11**
Letters to Home Week 16–20 .... **A16**
Letters to Home Week 21–25 .... **A21**
Letters to Home Week 26–30 .... **A26**

Appendix B
Object Land Activity Sheet 1:	Count and Compare	**B1**
Object Land Activity Sheet 2:	Party!	**B2**
Object Land Activity Sheet 3:	Subtracting Counters	**B3**
Picture Land Activity Sheet 1:	Dragon Quest 1 Record Form	**B4**
Picture Land Activity Sheet 2:	Building Numbers	**B5**
Picture Land Activity Sheet 3:	Five	**B6**
Picture Land Activity Sheet 4:	Six	**B7**
Picture Land Activity Sheet 5:	Seven	**B8**
Picture Land Activity Sheet 6:	Eight	**B9**
Picture Land Activity Sheet 7:	Nine	**B10**
Picture Land Activity Sheet 8:	Ten	**B11**
Picture Land Activity Sheet 9:	Counter Dropping Record Form	**B12**
Picture Land Activity Sheet 10:	Dragon Quest 2 and 3 Record Form	**B13**
Picture Land Activity Sheet 11:	Fish Pond	**B14**
Picture Land Activity Sheet 12:	Fish Pond Record Form	**B15**
Picture Land Activity Sheet 13:	Two-Team Record Form	**B16**
Line Land Activity Sheet 1:	Game Story for Addition	**B17**
Line Land Activity Sheet 2:	Number Lines to 20	**B18**
Line Land Activity Sheet 3:	Game Story for Subtraction	**B19**
Line Land Activity Sheet 4:	Multiples of 10 Number Line	**B20**
Sky Land Activity Sheet 1:	Graphing and Counting	**B21**
Multi-Land Activity Sheet 1:	Shopping Trip	**B22**
Multi-Land Activity Sheet 2:	Multiple-Players Record Form	**B23**
Multi-Land Activity Sheet 3:	Shape Blocks	**B24**

Appendix C
About Math Intervention **C1**
Building Number Sense with *Number Worlds* **C2**
Math Proficiencies **C7**
Content Strands of Mathematics **C9**
Technology Overview **C12**
Building Blocks **C13**
eMathTools **C15**
English Learners **C16**
Learning Trajectories **C17**
Trajectory Progress Chart **C31**
Glossary **C33**
Scope and Sequence **C37**

Letter to Home

Dear Family,

This week, your child will see numbers represented as groups of objects—the way children naturally learn about numbers. Below, your child applied this concept by counting the items in each row and writing that number in the correct space.

Discuss the activity with your child by asking questions such as the following:

- **How many bananas are in the second row?**
- **Are there more apples or more bananas?**
- **How do you know?**

Assisting Your Child

If . . . your child has difficulty counting correctly or writing the correct number,
Then . . . practice counting together at home. You can count plates on a table, steps on a staircase, and so on.

This week, your child will continue using the *Number Worlds* program to master these concepts.

Estimada familia:

Esta semana, su hijo(a) verá números representados como grupos de objetos, o sea, la forma en que los niños aprenden naturalmente los números. En el área de abajo, su hijo(a) aplicó este concepto contando los objetos de cada fila y escribiendo ese número en el lugar correcto.

Hable con su hijo(a) sobre esta actividad haciéndole las siguientes preguntas:

- **¿Cuántos plátanos hay en la segunda fila?**
- **¿Hay más manzanas o más plátanos?**
- **¿Cómo lo sabes?**

Ayude a su hijo(a)

Si . . . su hijo(a) tiene problemas contando correctamente o escribiendo el número correcto,
Entonces . . . practiquen contando en la casa. Pueden contar los platos sobre la mesa, los escalones de una escalera y así sucesivamente.

Esta semana, su hijo(a) continuará usando el programa *El mundo de los números* para dominar estos conceptos.

Count the things in each row.
Write the number.

 _____

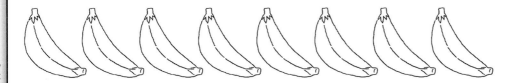 _____

 _____

Letter to Home

Dear Family,

This week, your child will continue to see numbers represented as groups of objects—the way children naturally learn about numbers. Below, your child applied this concept by drawing the number of items needed to make the two boxes equal.

Discuss the activity with your child by asking questions such as the following:

- **How many items are in each box?**
- **If we added one flower to the left box, would the boxes still be equal?**

Assisting Your Child

If . . . your child has difficulty drawing the correct number of items,

Then . . . use real objects such as pennies to help him or her master this concept. Arrange a number of items in a group. Ask your child to place the same amount in another group.

This week, your child will continue using the *Number Worlds* program to master these concepts.

Estimada familia:

Su hijo(a) seguirá esta semana viendo los números representados como grupos de objetos, o sea, la forma en que los niños aprenden naturalmente los números. En el área de abajo, su hijo(a) aplicó este concepto dibujando el número de objetos necesarios para que las dos casillas fueran iguales.

Hable con su hijo(a) sobre esta actividad haciéndole las siguientes preguntas:

- **¿Cuántos objetos hay en cada casilla?**
- **Si agregáramos una flor a la casilla de la izquierda, ¿seguirán siendo iguales las casillas?**

Ayude a su hijo(a)

Si . . . su hijo(a) tiene problemas dibujando el número de objetos correcto,

Entonces . . . use objetos reales como *pennies* a su hijo(a) a dominar este concepto. Ordene un número de objetos en un grupo. Pídale a su hijo que coloque la misma cantidad en otro grupo.

Esta semana, su hijo(a) continuará usando el programa *El mundo de los números* para dominar estos conceptos.

Write how many. Then draw that many in the empty box to make it equal.

How many? _____

Letter to Home

Dear Family,

This week, your child will see numbers as an ordered series and will recognize when one number is missing. Below, your child applied this concept by discovering which number in a series was missing and writing that number in the correct place.

Discuss the activity with your child by asking questions such as the following:

- **Which number was missing?**
- **How did you figure that out?**
- **What number comes next (after 10)?**

Assisting Your Child

If . . . your child has difficulty determining the missing number,
Then . . . practice counting with your child at home.

This week, your child will continue using the **Number Worlds** program to master these concepts.

Estimada familia:

Esta semana, su hijo(a) verá números como una serie ordenada y reconocerá cuando falte un número. En el área de abajo, su hijo(a) aplicó este concepto descubriendo qué número faltaba en una serie y escribiendo ese número en el lugar correcto.

Hable con su hijo(a) sobre esta actividad haciéndole las siguientes preguntas:

- ¿Qué número faltaba?
- ¿Cómo lo determinaste?
- ¿Qué número sigue (después de 10)?

Ayude a su hijo(a)

Si . . . su hijo(a) tiene problemas para determinar el número que falta,
Entonces . . . practique contando con su hijo(a) en la casa.

Esta semana, su hijo(a) continuará usando el programa *El mundo de los números* para dominar estos conceptos.

Write the missing numbers where they belong.

1	2	3		5	6	7	8	9	10
11	12	13	14	15	16		18	19	20
10	9	8	7		5	4	3	2	1

Letter to Home

Dear Family,

This week, your child will see that numbers can be represented by patterns or by numerals. This helps children connect the world of symbols to the world of quantities. Below, your child applied this concept by determining how many dots are on each card and writing that number.

Discuss the activity with your child by asking questions such as the following:

- How did you know what number to write?
- Which cards have similar patterns?

Assisting Your Child

If . . . your child has difficulty determining the correct number,

Then . . . say a number aloud, and have your child draw that many dots.

This week, your child will continue using the *Number Worlds* program to master these concepts.

Estimada familia:

Esta semana, su hijo(a) verá que los números se pueden representar con números o patrones. Esto ayuda a los niños a vincular el mundo de los símbolos con el mundo de las cantidades. En el área de abajo, su hijo(a) aplicó este concepto determinando cuántos puntos hay en cada tarjeta y escribiendo ese número.

Hable con su hijo(a) sobre esta actividad haciéndole las siguientes preguntas:

- ¿Cómo sabías que número ibas a escribir?
- ¿Qué tarjetas tienen patrones parecidos?

Ayude a su hijo(a)

Si . . . su hijo(a) tiene problemas para determinar el número correcto,

Entonces . . . diga ese número en voz alta y pida a su hijo(a) que dibuje ese número de puntos.

Esta semana, su hijo(a) continuará usando el programa *El mundo de los números* para dominar estos conceptos.

Write the number that matches each card.

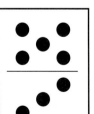 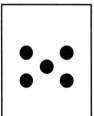 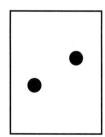 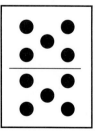

_____   _____   _____   _____

Letter to Home

Dear Family,

This week, your child will see that numbers can be represented as positions on a line. Below, your child applied this concept by identifying the correct position on the number line.

Discuss the activity with your child by asking questions such as the following:
- **How did you figure out which number to write?**
- **What number comes before that number?**
- **What number comes after that number?**

Assisting Your Child

If . . . your child has difficulty determining the correct number,

Then . . . have your child complete each number line below by writing the correct numbers over all the marks.

This week, your child will continue using the **Number Worlds** program to master these concepts.

Estimada familia:

Esta semana, su hijo(a) verá que los números se pueden representar como posiciones en una línea. En el área de abajo, su hijo(a) aplicó este concepto identificando la posición correcta de un número en la recta numérica.

Hable con su hijo(a) sobre esta actividad haciéndole las siguientes preguntas:
- **¿Cómo sabías que número ibas a escribir?**
- **¿Qué número viene antes de ese número?**
- **¿Qué número viene después de ese número?**

Ayude a su hijo(a)

Si . . . su hijo(a) tiene problemas para determinar el número correcto,

Entonces . . . pida a su hijo(a) que complete cada recta numérica de abajo escribiendo los números correctos sobre todas las marcas.

Esta semana, su hijo(a) continuará usando el programa **El mundo de los números** para dominar estos conceptos.

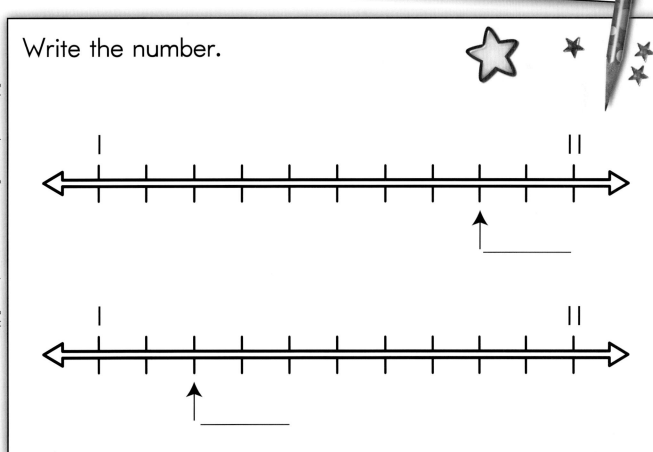

Write the number.

Letter to Home

Dear Family,

This week, your child will continue to see numbers as positions on a line and will learn that a change in position on the line represents a change in quantity. Below, your child applied this concept by figuring out how many spaces up or down the number line to move.

Discuss the activity with your child by asking questions such as the following:

- What number did you start on?
- What number did you land on?
- Is that number bigger or smaller than your starting number?

Assisting Your Child

If . . . your child has difficulty determining the correct number,

Then . . . have your child write the correct numbers above all the marks before working on another problem.

This week, your child will continue using the *Number Worlds* program to master these concepts.

Estimada familia:

Su hijo(a) seguirá esta semana viendo los números como posiciones en una línea y aprenderá que cambiar la posición de un número en la línea representa un cambio en la cantidad. En el área de abajo, su hijo(a) aplicó este concepto determinando cuántos espacios hacia arriba o hacia abajo moverse en la recta numérica.

Hable con su hijo(a) sobre esta actividad haciéndole las siguientes preguntas:

- ¿Con qué número comenzaste?
- ¿En qué número caíste?
- ¿Es ese número mayor o menor que el número con el que comenzaste?

Ayude a su hijo(a)

Si . . . su hijo(a) tiene problemas para determinar el número correcto,

Entonces . . . pida a su hijo(a) que escriba todos los números correctos sobre las marcas antes de comenzar a trabajar en otro problema.

Esta semana, su hijo(a) continuará usando el programa *El mundo de los números* para dominar estos conceptos.

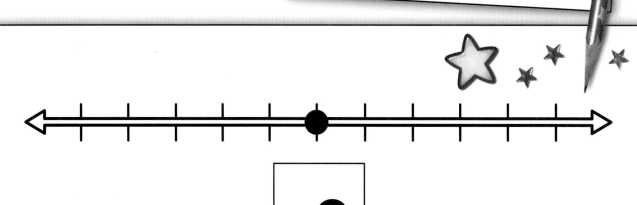

+2

Where did you start? _____

Where did you end? _____

Letter to Home

Dear Family,

This week, your child will begin to work with vertical number lines. Below, your child applied this concept by determining the correct location of a number on the number line.

Discuss the activity with your child by asking questions such as the following:
- **How did you know how many blocks to color?**
- **What number is one above that block?**
- **What number is one below that block?**

Assisting Your Child

If . . . your child has difficulty determining the correct number,

Then . . . have him or her practice with numbers from 1 through 5, and gradually work up to 10.

This week, your child will continue using the **Number Worlds** program to master these concepts.

Estimada familia:

Su hijo(a) empezará esta semana a trabajar con rectas numéricas verticales. En el área de abajo, su hijo(a) aplicó este concepto determinando la ubicación correcta de un número en la recta numérica.

Hable con su hijo(a) sobre esta actividad haciéndole las siguientes preguntas:
- **¿Cómo sabías cuántos bloques debías colorear?**
- **¿Qué número está sobre ese bloque?**
- **¿Qué número está debajo de ese bloque?**

Ayude a su hijo(a)

Si . . . su hijo(a) tiene dificultad para determinar el número correcto,

Entonces . . . pida a su hijo(a) que practique con los números del 1 al 5 y luego que siga gradualmente hasta el número 10.

Esta semana, su hijo(a) continuará usando el programa **El mundo de los números** para dominar estos conceptos.

Color the spaces up to the number shown.

8   3   6

Letter to Home

Dear Family,

This week, your child will continue to add and subtract whole numbers. Below, your child applied this concept by adding to and subtracting from 7.

Discuss the activity with your child by asking questions such as the following:

- How did you know which numbers to ring?
- How did you know which numbers to box?

Assisting Your Child

If . . . your child has difficulty determining the correct number,

Then . . . have your child place his or her finger on a number on the number line, and ask him or her to "move plus one." Repeat this activity until your child is comfortable with plus one/minus one and plus two/minus two.

This week, your child will continue using the *Number Worlds* program to master these concepts.

Estimada familia:

Su hijo(a) seguirá esta semana sumando y restando números enteros. En el área de abajo, su hijo(a) aplicó este concepto sumando 7 y restando de 7.

Hable con su hijo(a) sobre esta actividad haciéndole las siguientes preguntas:

- ¿Cómo sabías qué números ibas a encerrar en un círculo?
- ¿Cómo sabías qué números colocar en la casilla?

Ayude a su hijo(a)

Si . . . su hijo(a) tiene problemas para determinar el número correcto,

Entonces . . . pida a su hijo(a) que coloque su dedo en un número en la recta numérica y pídale que "mueva uno más". Repita esta actividad hasta que su hijo(a) pueda sumar uno y restar uno, y sumar dos y restar dos.

Esta semana, su hijo(a) continuará usando el programa *El mundo de los números* para dominar estos conceptos.

Find 7 on the number line.

Ring the next-door neighbors (+1 and −1).
Write the numbers.

_____   _____

Box the close neighbors (+2 and −2).
Write the numbers.

_____   _____

Letter to Home

Dear Family,

This week, your child will continue to add and subtract whole numbers. Below, your child applied this concept by determining, without the use of a number line, what numbers are before and after a starting point.

Discuss the activity with your child by asking questions such as the following:
- How did you know what number to write?
- How did you add 2?
- How did you figure out what was 1 less?

Assisting Your Child

If . . . your child has difficulty determining the correct number,
Then . . . help your child use a number line or actual objects to answer the questions.

This week, your child will continue using the *Number Worlds* program to master these concepts.

Estimada familia:

Su hijo(a) seguirá esta semana sumando y restando números enteros. En el área de abajo, su hijo(a) aplicó este concepto determinando, sin usar una recta numérica, qué números están antes y después de un punto de partida.

Hable con su hijo(a) sobre esta actividad haciéndole las siguientes preguntas:
- ¿Cómo sabías que número ibas a escribir?
- ¿Cómo sumaste 2?
- ¿Cómo determinaste cuánto era menos 1?

Ayude a su hijo(a)

Si . . . su hijo(a) tiene problemas para determinar el número correcto,
Entonces . . . ayude a su hijo(a) a usar una recta numérica u objetos reales para contestar las preguntas.

Esta semana, su hijo(a) continuará usando el programa *El mundo de los números* para dominar estos conceptos.

Write the number.

1. It is 3 more than 6. _____

2. It is 1 less than 9. _____

3. It is 2 more than 2. _____

Letter to Home

Dear Family,

This week, your child will continue to add and subtract whole numbers. Below, your child applied this concept by adding sets of objects.

Discuss the activity with your child by asking questions such as the following:

- How did you figure out each answer?
- What does the equal sign mean?

Assisting Your Child

If . . . your child has difficulty determining the correct answer,

Then . . . help your child draw a number line at the bottom of this paper, and use the number line to solve the problem.

This week, your child will continue using the **Number Worlds** program to master these concepts.

Estimada familia:

Su hijo(a) seguirá esta semana sumando y restando números enteros. En el área de abajo, su hijo(a) aplicó este concepto sumando grupos de objetos.

Hable con su hijo(a) sobre esta actividad haciéndole las siguientes preguntas:

- ¿Cómo determinaste cada respuesta?
- ¿Qué significa el signo de igualdad?

Ayude a su hijo(a)

Si . . . su hijo(a) tiene problemas para determinar la respuesta correcta,

Entonces . . . ayude a su hijo(a) a dibujar una recta numérica al final de esta hoja y use la recta numérica para resolver el problema.

Esta semana, su hijo(a) continuará usando el programa **El mundo de los números** para dominar estos conceptos.

Write the answer.

1. ◯◯◯ + ◯◯ = _____

2. ◯◯◯◯ + ◯◯◯ = _____

3. ◯◯◯◯◯◯ + ◯ = _____

Letter to Home

Dear Family,

This week, your child will continue to add and subtract whole numbers. Below, your child applied this concept by matching set size to a numeral and increasing the number by 1.

Discuss the activity with your child by asking questions such as the following:

- **How did you know what number to write?**
- **What is one more than four?**
- **What is one more than seven?**
- **What does the plus sign mean?**

Assisting Your Child

If . . . your child has trouble recognizing the +1 pattern,

Then . . . have him or her use a number line to reinforce this concept.

This week, your child will continue using the **Number Worlds** program to master these concepts.

Estimada familia:

Su hijo(a) seguirá esta semana sumando y restando números enteros. En el área de abajo, su hijo(a) aplicó este concepto emparejando el tamaño de un grupo y un número, y aumentando el número en 1.

Hable con su hijo(a) sobre esta actividad haciéndole las siguientes preguntas:

- **¿Cómo sabías que número ibas a escribir?**
- **¿Cuánto es uno más cuatro?**
- **¿Cuánto es uno más siete?**
- **¿Qué significa el signo de suma?**

Ayude a su hijo(a)

Si . . . su hijo(a) tiene problemas para reconocer el patrón +1,

Entonces . . . pida a su hijo(a) que use una recta numérica para reforzar este concepto.

Esta semana, su hijo(a) continuará usando el programa **El mundo de los números** para dominar estos conceptos.

Write the answer.

1.

2.

3. ⬭⬭ + ⬭ = _____

Letter to Home

Dear Family,

This week, your child will continue to add and subtract whole numbers. Below, your child applied this concept by matching set size to a numeral and decreasing that number by different amounts.

Discuss the activity with your child by asking questions such as the following:

- How did you know what number to write?
- What does the minus sign mean?

Assisting Your Child

If . . . your child has difficulty determining the correct number,

Then . . . have him or her use a number line for help. You can also practice subtraction with your child at home. For example, ask, "If I have 3 spoons and I take away 1 spoon, how many spoons do I have left?"

This week, your child will continue using the **Number Worlds** program to master these concepts.

Estimada familia:

Su hijo(a) seguirá esta semana sumando y restando números enteros. En el área de abajo, su hijo(a) aplicó este concepto emparejando el tamaño de un grupo y un número, y disminuyendo ese número en diferentes cantidades.

Hable con su hijo(a) sobre esta actividad haciéndole las siguientes preguntas:

- ¿Cómo sabías que número ibas a escribir?
- ¿Qué significa el signo de resta?

Ayude a su hijo(a)

Si . . . su hijo(a) tiene problemas para determinar el número correcto,

Entonces . . . pida a su hijo(a) que use una recta numérica. También puede practicar con su hijo(a) la resta en la casa. Pregunte por ejemplo: "Si tengo 3 cucharas y quito una, ¿cuántas cucharas me quedan?"

Esta semana, su hijo(a) continuará usando el programa **El mundo de los números** para dominar estos conceptos.

Write the answer.

1.  = _____

2.  = _____

3.  = _____

Letter to Home

Dear Family,

This week, your child will realize that addition is equivalent to forward movement along the number line. Below, your child applied this concept by writing an equation that describes the action shown on the number line.

Discuss the activity with your child by asking questions such as the following:
- How did you know what numbers to write?
- In which question did you move the farthest?

Assisting Your Child

If . . . your child forgets to write the plus and equal signs,
Then . . . ask him or her, "What do we write to show that we added? What do we write before the answer?"

This week, your child will continue using the **Number Worlds** program to master these concepts.

Estimada familia:

Esta semana, su hijo(a) se dará cuenta que la suma es equivalente a desplazarse hacia adelante en una recta numérica. En el área de abajo, su hijo(a) aplicó este concepto escribiendo una ecuación que describe la acción que se muestra en la recta numérica.

Hable con su hijo(a) sobre esta actividad haciéndole las siguientes preguntas:
- ¿Cómo sabías que números ibas a escribir?
- ¿En qué pregunta te desplazaste más?

Ayude a su hijo(a)

Si . . . su hijo(a) olvida cómo escribir los signos de suma e igualdad,
Entonces . . . pregunte a su hijo(a), "¿Qué escribimos para mostrar que sumamos? ¿Qué escribimos antes de la respuesta?"

Esta semana, su hijo(a) continuará usando el programa **El mundo de los números** para dominar estos conceptos.

Write the number story.

1.

   _____

2.

   _____

Letter to Home

Dear Family,

This week, your child will continue to write equations. Below, your child applied this concept by adding Dot Set Cards and writing the corresponding equation.

Discuss the activity with your child by asking questions such as the following:

- **What does the plus sign mean?**
- **What does the equal sign mean?**
- **What does the minus sign mean?**

Assisting Your Child

If . . . your child has trouble remembering which symbol to use,

Then . . . discuss the plus sign and minus sign at home. For example, say, "I had 2 plates on the table, and I just put down 2 more. What sign tells what I did?"

This week, your child will continue using the *Number Worlds* program to master these concepts.

Estimada familia:

Su hijo(a) seguirá esta semana escribiendo ecuaciones. En el área de abajo, su hijo(a) aplicó este concepto sumando Tarjetas con puntos y escribiendo la ecuación correspondiente.

Hable con su hijo(a) sobre esta actividad haciéndole las siguientes preguntas:

- **¿Qué significa el signo de suma?**
- **¿Qué significa el signo de igualdad?**
- **¿Qué significa el signo de resta?**

Ayude a su hijo(a)

Si . . . su hijo(a) tiene problemas para recordar qué símbolo usar,

Entonces . . . hable en casa sobre los signos de suma y resta. Diga por ejemplo: "Tengo 2 platos en la mesa y acabo de poner 2 más. ¿Qué signo indica lo que hice?"

Esta semana, su hijo(a) continuará usando el programa *El mundo de los números* para dominar estos conceptos.

Write the number story and the answer.

1. 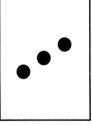 + 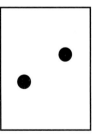 = _____

2. 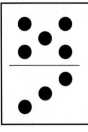 − 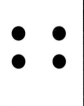 = _____

Letter to Home

Dear Family,

This week, your child will learn about graphs. Below, your child applied this concept by graphing numbers of pets.

Discuss the activity with your child by asking questions such as the following:

- **How many spaces did you color for this pet? Why?**
- **There are the most of which pet? How do you know?**

Assisting Your Child

If . . . your child has trouble transferring the quantities to the graph,

Then . . . have him or her cut out or draw pictures of animals and arrange them to match the graph format before counting the pictures and coloring the graph.

This week, your child will continue using the **Number Worlds** program to master these concepts.

Estimada familia:

Su hijo(a) aprenderá esta semana sobre gráficos. En el área de abajo, su hijo(a) aplicó este concepto graficando el número de mascotas.

Hable con su hijo(a) sobre esta actividad haciéndole las siguientes preguntas:

- **¿Cuántos espacios coloreaste para esta mascota? ¿Por qué?**
- **¿Cuál mascota aparece más frecuente? ¿Cómo lo sabes?**

Ayude a su hijo(a)

Si . . . su hijo tiene problemas transfiriendo las cantidades al gráfico,

Entonces . . . pida a su hijo(a) que recorte o haga dibujos de animales y los ordene para corresponder el formato del gráfico antes de contar los dibujos y colorear el gráfico.

Esta semana, su hijo(a) continuará usando el programa *El mundo de los números* para dominar estos conceptos.

Color the spaces on the graph to show how many.

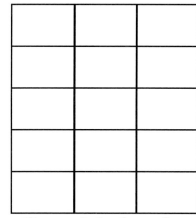

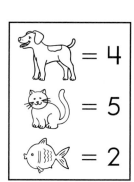

dog = 4
cat = 5
fish = 2

Letter to Home

Dear Family,

This week, your child will connect counting to addition. Below, your child applied this concept by using different colors for different addends on a number line and then completing the equation.

Discuss the activity with your child by asking questions such as the following:

- How did you know how many circles to color red (or blue)?
- Do you see a pattern in the colors?

Assisting Your Child

If . . . your child has difficulty coloring or filling in the blank,

Then . . . help your child by writing the numbers 1–5 above the top row of circles and having him or her count each addend aloud.

This week, your child will continue using the **Number Worlds** program to master these concepts.

Estimada familia:

Esta semana su hijo(a) va a conectar el conteo con la suma. En el área de abajo, su hijo(a) aplicó este concepto a continuación usando diferentes colores para diferenctes sumandos en una recta numérica y luego completando la ecuación.

Hable con su hijo(a) sobre esta actividad haciéndole las siguientes preguntas:

- ¿Cómo supiste cuántos círculos debías pintar de rojo (o azul)?
- ¿Ves un patrón en los colores?

Ayude a su hijo(a)

Si . . . su hijo(a) tiene problemas para colorear o llenar el espacio en blanco,

Entonces . . . ayude a su hijo(a) a escribir los números 1–5 arriba de la fila superior de círculos y pídale que cuente cada sumando en voz alta.

Esta semana, su hijo(a) va a seguir usando el programa **El mundo de los números** para dominar estos conceptos.

Color the circles equal to the first number red. Color the circles equal to the second number blue. Write the sum.

   1 + 4 = _____

       2 + 3 = _____

       3 + 2 = _____

       4 + 1 = _____

Letter to Home

Dear Family,

This week, your child will connect counting to adding. Below, your child applied this concept by using different shapes to write an equation, or "number story," with three addends.

Discuss the activity with your child by asking questions such as the following:

- **How did you know which numbers to use in the number stories?**
- **Do you see a pattern in the numbers?**

Assisting Your Child

If . . . your child has difficulty writing the equations,

Then . . . help your child by writing a blank equation (____ + ____ + ____ = ____) as a model.

This week, your child will continue using the *Number Worlds* program to master these concepts.

Estimada familia:

Esta semana, su hijo(a) va a conectar el conteo con la suma. En el área de abajo, su hijo(a) aplicó este concepto a continuación usando diferentes figuras para escribir una ecuación, o una "historia con números", con tres sumandos.

Hable con su hijo(a) sobre esta actividad haciéndole las siguientes preguntas:

- **¿Cómo supiste qué números debías usar en las historias con números?**
- **¿Ves un patrón en los números?**

Ayude a su hijo(a)

Si . . . su hijo(a) tiene problemas escribiendo las ecuaciones,

Entonces . . . ayúdelo(a) escribiendo una ecuación en blanco (____ + ____ + ____ = ____) como modelo.

Esta semana, su hijo(a) va a seguir usando el programa *El mundo de los números* para dominar estos conceptos.

Add the shapes. Write the number story.

_____

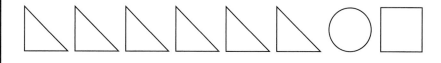

_____

Letter to Home

Dear Family,

This week, your child will use addition and subtraction in successive operations. Below, your child applied this concept by solving the equations.

Discuss the activity with your child by asking questions such as the following:

- How did you figure out what numbers to write in the blanks?
- What do you notice about the first and last problems?

Assisting Your Child

If . . . your child has difficulty completing the equations,

Then . . . help your child draw a number line at the bottom of this paper. Have your child place his or her finger on the starting point and move right for addition and left for subtraction.

This week, your child will continue using the *Number Worlds* program to master these concepts.

Estimada familia:

Esta semana, su hijo(a) va a usar la suma y la resta en operaciones sucesivas. En el área de abajo, su hijo(a) aplicó este concepto resolviendo las siguientes ecuaciones.

Hable con su hijo(a) sobre esta actividad haciéndole las siguientes preguntas:

- ¿Cómo determinaste qué números debías escribir en los espacios en blanco?
- ¿Qué has notado acerca del primer y último problema?

Ayude a su hijo(a)

Si . . . su hijo(a) tiene problemas completando las ecuaciones,

Entonces . . . ayúdelo(a) a trazar una recta numérica en la parte inferior de esta hoja. Pídale que coloque su dedo en el punto de comienzo y lo mueva a la derecha para sumar y a la izquierda para restar.

Esta semana, su hijo(a) va a seguir usando el programa *El mundo de los números* para dominar estos conceptos.

Write the answer.

1. $1 + 5 + 1 + 3 =$ _____

2. $2 + 5 - 3 + 5 =$ _____

3. $4 + 8 - 3 + 2 =$ _____

4. $1 + 3 + 1 + 5 =$ _____

Letter to Home

Dear Family,

This week, your child will connect counting and subtraction. Below, your child applied this concept by eliminating amounts being taken away and then completing the equation.

Discuss the activity with your child by asking questions such as the following:
- **How did you know how many circles to cross out?**
- **How did you know what number to write in the blank?**

Assisting Your Child
If . . . your child has difficulty crossing out the correct number of circles,
Then . . . help your child by writing the numbers 1–5 above the top row of circles and having him or her count each circle aloud.

This week, your child will continue using the *Number Worlds* program to master these concepts.

Estimada familia:

Esta semana su hijo(a) va a conectar el conteo con la resta. En el área de abajo, su hijo(a) aplicó este concepto a continuación eliminando cantidades que se quitan y luego completando la ecuación.

Hable con su hijo(a) sobre esta actividad haciéndole las siguientes preguntas:
- **¿Cómo supiste cuántos círculos debías tachar?**
- **¿Cómo supiste qué número debías escribir en el espacio en blanco?**

Ayude a su hijo(a)
Si . . . su hijo(a) tiene problemas tachando el número correcto de círculos,
Entonces . . . ayúdelo(a) escribiendo los números 1–5 arriba de la fila superior de círculos y pidiendo a él o ella que cuente cada círculo en voz alta.

Esta semana, su hijo(a) va a seguir usando el programa *El mundo de los números* para dominar estos conceptos.

Cross out circles that equal the number being taken away. How many are left?

        5 − 1 = _____

        5 − 2 = _____

        5 − 5 = _____

Letter to Home • Week 19

Letter to Home

Dear Family,

This week, your child will connect counting and subtraction. Below, your child applied this concept by writing subtraction equations that describe the pictures.

Discuss the activity with your child by asking questions such as the following:
- Which number did you write first? Why?
- Which number did you write second? Why?

Assisting Your Child

If . . . your child has difficulty determining which number to write first,
Then . . . remind him or her to start with the total number of fish, then write how many have been taken away. Model subtraction at home using a group of books or toys.

This week, your child will continue using the *Number Worlds* program to master these concepts.

Estimada familia:

Esta semana su hijo(a) va a conectar el conteo con la resta. Su hijo(a) aplicó este concepto a continuación escribiendo ecuaciónes con restas que describen los dibujos.

Hable con su hijo(a) sobre esta actividad haciéndole las siguientes preguntas:
- ¿Qué número escribiste primero? ¿Por qué?
- ¿Qué número escribiste de segundo? ¿Por qué?

Ayude a su hijo(a)

Si . . . su hijo(a) tiene problemas determinando qué número debe escribir primero,
Entonces . . . recuérdele que comience con el número total de peces y que luego escriba cuántos se quitan. Haga un modelo de la resta en su casa usando un grupo de libros o juguetes.

Esta semana, su hijo(a) va a seguir usando el programa *El mundo de los números* para dominar estos conceptos.

Write the number story.

 ___ − ___ = ___

 ___ − ___ = ___

 ___ − ___ = ___

Letter to Home

Dear Family,

This week your child will write equations that may include double-digit numbers. Your child applied this concept below by creating and solving equations.

Discuss the activity with your child by asking questions such as the following:

- How did you figure out the answer?
- What does the equal sign mean?

Assisting Your Child

If . . . your child has trouble solving the problems,

Then . . . help him or her use a number line or a group of objects to figure out the answer.

This week your child will continue using the **Number Worlds** program to master these concepts.

Estimada familia:

Esta semana su hijo(a) va a escribir ecuaciones con números de dos dígitos. Él o ella aplicó este concepto a continuación creando y resolviendo ecuaciones. Hable con su hijo(a) sobre esta actividad haciéndole las siguientes preguntas:

- ¿Cómo averiguaste la respuesta?
- ¿Qué significa el signo de igualdad?

Ayude a su hijo(a)

Si . . . su hijo(a) tiene dificultad resolviendo los problemas,

Entonces . . . ayúdelo(a) a usar una recta numérica o un grupo de objetos para determinar la respuesta.

Esta semana su hijo(a) seguirá usando el programa *El mundo de los números* para dominar estos conceptos.

Write the number story.

___ + ___ = ___

___ + ___ = ___

Letter to Home

Dear Family,

This week your child will solve subtraction problems involving double-digit numbers. Your child applied this concept below by completing the subtraction equations.

Discuss the activity with your child by asking questions such as the following:

- **How many did you start with? How many did you subtract?**
- **Is the number you have now bigger or smaller than the one you started with?**

Assisting Your Child

If . . . your child has trouble solving the problems,

Then . . . practice subtraction at home as your child picks up toys, eats snacks, and so on. (For example, "How many crackers do you have? How many will you have if you eat 4 crackers?")

This week your child will continue using the **Number Worlds** program to master these concepts.

Estimada familia:

Esta semana su hijo(a) va a resolver problemas de resta con números de dos dígitos. Él o ella aplicó este concepto a continuación completando las ecuaciones de resta.

Hable con su hijo(a) sobre esta actividad haciéndole las siguientes preguntas:

- **¿Con cuántos comenzaste? ¿Cuántos restaste?**
- **El número que tienes ahora, ¿es mayor o menor que el que tenías al principio?**

Ayude a su hijo(a)

Si . . . su hijo(a) tiene dificultad resolviendo los problemas,

Entonces . . . practique la resta en su casa cuando su hijo(a) recoja los juguetes, coma meriendas y así sucesivamente. (Por ejemplo: "¿Cuántas galletas tienes? Si te comes 4, ¿cuántas te quedarán?")

Esta semana su hijo(a) seguirá usando el programa *El mundo de los números* para dominar estos conceptos.

Write the answer.

1. 15 − 5 = _____

2. 12 − 6 = _____

3. 14 − 8 = _____

Letter to Home

Dear Family,

This week your child will write and solve addition and subtraction problems. Your child applied this concept by solving the equations below.

Discuss the activity with your child by asking questions such as the following:

- What number did you start with?
- Did you add or subtract? How did you know which one to do?
- Is your answer higher or lower than the number you started with?

Assisting Your Child

If . . . your child has trouble solving the problems,

Then . . . encourage him or her to use the number line or groups of objects for reference.

This week your child will continue using the **Number Worlds** program to master these concepts.

Estimada familia:

Esta semana su hijo(a) va a escribir y resolver problemas de suma y resta. Él o ella aplicó este concepto resolviendo las ecuaciones a continuación.

Hable con su hijo(a) sobre esta actividad haciéndole las siguientes preguntas:

- ¿Con qué número comenzaste?
- ¿Sumaste o restaste? ¿Cómo sabes qué operación debes usar?
- Tu respuesta, ¿es mayor o menor que el número con que comenzaste?

Ayude a su hijo(a)

Si . . . su hijo(a) tiene dificultad resolviendo los problemas,

Entonces . . . anímelo(a) a usar una recta numérica o grupos de objetos como referencia.

Esta semana su hijo(a) va a seguir usando el programa **El mundo de los números.**

Write the answer.

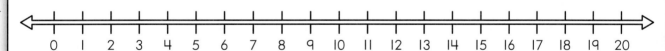

1. 18 − 6 = _____

2. 9 + 5 = _____

3. 13 − 7 = _____

Letter to Home

Dear Family,

This week your child will gain further practice with solving subtraction problems and exploring the number sequence. Your child applied this concept below by filling in the missing numbers.

Discuss the activity with your child by asking questions such as the following:

- How did you figure out the answer?
- Which number is higher when we count up? How do you know?

Assisting Your Child

If . . . your child has trouble solving the subtraction problems,
Then . . . allow him or her to use a number line for reference.

This week your child will continue using the **Number Worlds** program to master these concepts.

Estimada familia:

Esta semana su hijo(a) va a practicar más la resolución de problemas de resta y va a examinar la secuencia numérica. Él o ella aplicó este concepto a continuación escribiendo los números que faltan.

Hable con su hijo(a) sobre esta actividad haciéndole las siguientes preguntas:

- ¿Cómo averiguaste la respuesta?
- ¿Qué número es mayor cuando contamos en forma ascendente? ¿Cómo lo sabes?

Ayude a su hijo(a)

Si . . . su hijo(a) tiene dificultad resolviendo los problemas de resta,
Entonces . . . dígale que use una recta numérica como referencia.

Esta semana su hijo(a) seguirá usando el programa *El mundo de los números* para dominar estos conceptos.

Write the missing number.

1. 7 − 5 = _____

2. 10 − 7 = _____

3. 8 − 4 = _____

Letter to Home

Dear Family,

This week your child will strengthen his or her understanding of numbers to 100 and determine the location of each number in the number sequence. Your child applied this concept below by filling in the missing numbers.

Discuss the activity with your child by asking questions such as the following:

- What number comes next? How did you figure that out?
- How many tens does this number have? How many ones does it have?

Assisting Your Child

If . . . your child has trouble remembering the number sequence,
Then . . . help him or her identify the pattern in the ones and tens places.

This week your child will continue using the **Number Worlds** program to master these concepts.

Estimada familia:

Esta semana su hijo(a) va a fortalecer su conocimiento sobre los números hasta el 100, y va a determinar el lugar de cada número en la secuencia numérica. Él o ella aplicó este concepto a continuación escribiendo los números que faltan.

Hable con su hijo(a) sobre esta actividad haciéndole las siguientes preguntas:

- ¿Qué número va después? ¿Cómo lo determinaste?
- ¿Cuántas decenas tiene este número? ¿Cuántas unidades tiene?

Ayude a su hijo(a)

Si . . . su hijo(a) tiene problemas recordando la secuencia numérica,
Entonces . . . ayúdelo(a) a identificar el patrón en los lugares de decenas y unidades.

Esta semana su hijo(a) seguirá usando el programa *El mundo de los números* para dominar estos conceptos.

Write the missing number.

1. 23 _____ 25 26 27

2. 46 47 48 49 _____

3. 69 70 _____ 72 73

Letter to Home

Dear Family,

This week your child will reinforce counting and number-sequence skills. Your child applied this concept below by solving an addition word problem.

Discuss the activity with your child by asking questions such as the following:

- **What facts do you know? What do you want to find out?**
- **Will you add or subtract to find the answer? Why?**

Assisting Your Child

If . . . your child has trouble identifying which numbers should be added to find the answer, **Then . . .** help your child look for clue words such as *how many* and *altogether*. These words are clues that prompt your child to add the numbers they know from the story problem to find the answer.

This week your child will continue using the *Number Worlds* program to master these concepts.

Estimada familia:

Esta semana su hijo(a) va a reforzar las destrezas de conteo y secuencia numérica. Él o ella aplicó este concepto a continuación resolviendo un problema de suma.

Hable con su hijo(a) sobre esta actividad haciéndole las siguientes preguntas:

- **¿Qué operaciones conoces? ¿Qué quieres averiguar?**
- **¿Vas a sumar o restar para buscar la respuesta? ¿Por qué?**

Ayude a su hijo(a)

Si . . . su hijo(a) tiene problemas identificando los números que deben sumarse para hallar la respuesta, **Entonces . . .** ayúdelo(a) a buscar palabras que den una pista como: *cuántos* y *en total*. Estas palabras son pistas que inducen a su hijo(a) a sumar los números que contiene el problema para hallar la respuesta.

Esta semana su hijo(a) seguirá usando el programa *El mundo de los números* para dominar estos conceptos.

Write the number story.

I had 2 kittens. A friend gave me 1 more. How many kittens do I have altogether?

_____

Letter to Home

Dear Family,

This week your child will continue to explore the number sequence and will be able to describe the position of a number in terms of tens and ones. Your child applied this concept below by identifying the tens and ones in each number.

Discuss the activity with your child by asking questions such as the following:

- Is this number close to or far from 0?
- What other numbers would you pass to get to this number?

Assisting Your Child

If . . . your child has trouble remembering which digit shows how many tens,

Then . . . remind him or her that when there are two digits, the left digit shows how many tens. Practice identifying tens and ones in 10, 20, 30, and so on.

This week your child will continue using the **Number Worlds** program to master these concepts.

Estimada familia:

Esta semana su hijo(a) va a seguir examinando la secuencia numérica y podrá describir la posición de un número desde el punto de vista de decenas y unidades. Él o ella aplicó este concepto a continuación identificando las decenas y unidades en cada número.

Hable con su hijo(a) sobre esta actividad haciéndole las siguientes preguntas:

- Este número, ¿está cerca o lejos de 0?
- ¿Qué otros números pasarías para llegar a este número?

Ayude a su hijo(a)

Si . . . su hijo(a) tiene problemas recordando qué dígito muestra el número de decenas,

Entonces . . . recuérdele que cuando hay dos dígitos, el dígito izquierdo muestra el número de decenas. Practique con él o ella la manera como identificar las decenas y unidades en 10, 20, 30 y así sucesivamente.

Esta semana su hijo(a) seguirá usando el programa **El mundo de los números** para dominar estos conceptos.

Write how many tens and ones.

1. 64 _____ tens _____ ones

2. 27 _____ tens _____ ones

3. 35 _____ tens _____ ones

Letter to Home

Dear Family,

This week your child will gain further practice with addition and subtraction. Your child applied these concepts by solving the multiple-addend problems below.

Discuss the activity with your child by asking questions such as the following:

- How did you figure out the answer?
- What would happen if you changed the order of the numbers? Would the answer change?

Assisting Your Child

If . . . your child has trouble solving the problems,

Then . . . encourage him or her to break the problems into smaller parts. For example, 3 + 5 = 8; 8 + 2 = 10.

This week your child will continue using the *Number Worlds* program to master these concepts.

Estimada familia:

Esta semana su hijo(a) va a practicar más la suma y la resta. Él o ella aplicó estos conceptos resolviendo problemas con sumandos múltiples a continuación.

Hable con su hijo(a) sobre esta actividad haciéndole las siguientes preguntas:

- ¿Cómo averiguaste la respuesta?
- ¿Qué pasaría si cambiaras el orden de los números? ¿Cambiaría la respuesta?

Ayude a su hijo(a)

Si . . . su hijo(a) tiene dificultad resolviendo los problemas,

Entonces . . . anímelo(a) a dividir los problemas en partes más pequeñas. Por ejemplo: 3 + 5 = 8; 8 + 2 = 10.

Esta semana su hijo(a) seguirá usando el programa *El mundo de los números* para dominar estos conceptos.

Write the answer.

1. 3 + 5 + 2 = _____

2. 1 + 4 + 3 = _____

3. 5 + 4 + 3 = _____

Letter to Home

Dear Family,

This week your child will practice solving subtraction word problems. Your child applied this concept by writing and solving the equations below.

Discuss the activity with your child by asking questions such as the following:

- **What are we trying to find out? What facts do we know?**
- **Which number did you put first in your number story? Why?**
- **Did you add or subtract? Why?**

Assisting Your Child

If ... your child has trouble solving the problems,
Then ... allow him or her to use objects to model the problems.

This week your child will continue using the *Number Worlds* program to master these concepts.

Estimada familia:

Esta semana su hijo(a) va a practicar la resolución de problemas de resta. Él o ella este concepto a continuación escribiendo y resolviendo las ecuaciones a continuación.

Hable con su hijo(a) sobre esta actividad haciéndole las siguientes preguntas:

- **¿Qué estamos tratando de averiguar? ¿Qué operaciones conocemos?**
- **¿Qué número pusiste primero en el problema? ¿Por qué?**
- **¿Sumaste o restaste? ¿Por qué?**

Ayude a su hijo(a)

Si ... su hijo(a) tiene dificultad resolviendo los problemas,
Entonces ... dígale que use objetos para hacer un modelo de los problemas.

Esta semana su hijo(a) seguirá usando el programa *El mundo de los números* para dominar estos conceptos.

Write the number story.

1. I had 8 toys. I gave 3 to my friend. How many toys do I have left?

   _____

2. I had 5 carrots. Then I ate 2. How many carrots are left?

   _____

Letter to Home

Dear Family,

This week your child will gain further practice solving word problems. Your child applied this concept by solving the problems below.

Discuss the activity with your child by asking questions such as the following:

- What facts do we know?
- What are we trying to find out?

Assisting Your Child

If . . . your child has trouble solving the problem,
Then . . . work with him or her to make a chart or draw pictures that illustrate the problem.

This week your child will continue using the *Number Worlds* program to master these concepts.

Estimada familia:

Esta semana su hijo(a) va a practicar más la resolución de problemas narrativos. Él o ella aplicó este concepto resolviendo los problemas a continuación.

Hable con su hijo(a) sobre esta actividad haciéndole las siguientes preguntas.

- ¿Qué operaciones conocemos?
- ¿Qué estamos tratando de averiguar?

Ayude a su hijo(a)

Si . . . su hijo(a) tiene dificultad resolviendo el problema,
Entonces . . . trabaje con él o ella para hacer una tabla o dibujos para ilustrar el problema.

Esta semana su hijo(a) seguirá usando el programa *El mundo de los números* para dominar estos conceptos.

Write the answer.

1. I have 2 cats. They each eat 1 scoop of food a day. How much do they eat in 2 days?

   _____

2. I have 5 scoops of food. How much extra food do I have?

   _____

Count and Compare

Name _____

Object Land • Activity Sheet 1   **B1**

Name _____

Party!

B2 **Object Land** • Activity Sheet 2

Subtracting Counters

Name _____

Take away the red.

5 −

10 −

··

Take away the yellow.

5 −

10 −

··

Take away the green.

5

10

··

Take away the blue.

5

10

··

Object Land • Activity Sheet 3

Dragon Quest 1
Record Form

Name _____

☐ + _____ = ☐

☐ + _____ = ☐

☐ + _____ = ☐

☐ + _____ = ☐

☐ + _____ = ☐

☐ + _____ = ☐

☐ + _____ = ☐

☐ + _____ = ☐

☐ + _____ = ☐

Building Numbers

Name _____

① + ① = _____

① + ① + ① = _____

① + ① + ① + ① = _____

① + ① + ① + ① + ① = _____

① + ① + ① + ① + ① + ① = _____

① + ① + ① + ① + ① + ① + ① = _____

① + ① + ① + ① + ① + ① + ① + ① = _____

① + ① + ① + ① + ① + ① + ① + ① + ① = _____

① + ① + ① + ① + ① + ① + ① + ① + ① + ① = _____

Picture Land • Activity Sheet 2   **B5**

Five   Name _____

(Blue) + (Red) = 5

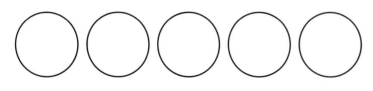 ____ + ____ = 5

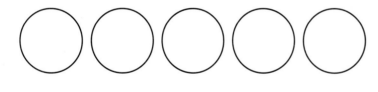 ____ + ____ = 5

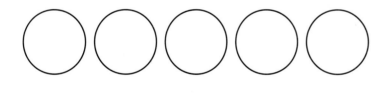 ____ + ____ = 5

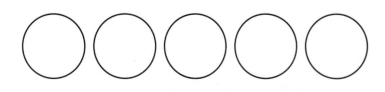 ____ + ____ = 5

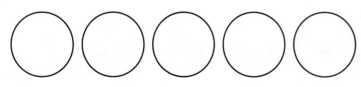 ____ + ____ = 5

____ + ____ = 5

Six

Name _____

$\bigcirc$ Blue $+$ $\bigcirc$ Red $= 6$

$\bigcirc\bigcirc\bigcirc\bigcirc\bigcirc\bigcirc$ _____ $+$ _____ $= 6$

$\bigcirc\bigcirc\bigcirc\bigcirc\bigcirc\bigcirc$ _____ $+$ _____ $= 6$

$\bigcirc\bigcirc\bigcirc\bigcirc\bigcirc\bigcirc$ _____ $+$ _____ $= 6$

$\bigcirc\bigcirc\bigcirc\bigcirc\bigcirc\bigcirc$ _____ $+$ _____ $= 6$

$\bigcirc\bigcirc\bigcirc\bigcirc\bigcirc\bigcirc$ _____ $+$ _____ $= 6$

$\bigcirc\bigcirc\bigcirc\bigcirc\bigcirc\bigcirc$ _____ $+$ _____ $= 6$

$\bigcirc\bigcirc\bigcirc\bigcirc\bigcirc\bigcirc$ _____ $+$ _____ $= 6$

Picture Land • Activity Sheet 4

Seven   Name _____

$\bigcirc$ Blue + $\bigcirc$ Red = 7

○ ○ ○ ○ ○ ○ ○   ____ + ____ = 7

○ ○ ○ ○ ○ ○ ○   ____ + ____ = 7

○ ○ ○ ○ ○ ○ ○   ____ + ____ = 7

○ ○ ○ ○ ○ ○ ○   ____ + ____ = 7

○ ○ ○ ○ ○ ○ ○   ____ + ____ = 7

○ ○ ○ ○ ○ ○ ○   ____ + ____ = 7

○ ○ ○ ○ ○ ○ ○   ____ + ____ = 7

○ ○ ○ ○ ○ ○ ○   ____ + ____ = 7

B8 Picture Land • Activity Sheet 5

Eight

Name _____

(Blue) + (Red) = 8

○○○○○○○○ ____ + ____ = 8

○○○○○○○○ ____ + ____ = 8

○○○○○○○○ ____ + ____ = 8

○○○○○○○○ ____ + ____ = 8

○○○○○○○○ ____ + ____ = 8

○○○○○○○○ ____ + ____ = 8

○○○○○○○○ ____ + ____ = 8

○○○○○○○○ ____ + ____ = 8

○○○○○○○○ ____ + ____ = 8

Picture Land • Activity Sheet 6

Nine   Name _____

$\bigcirc$ Blue + $\bigcirc$ Red = 9

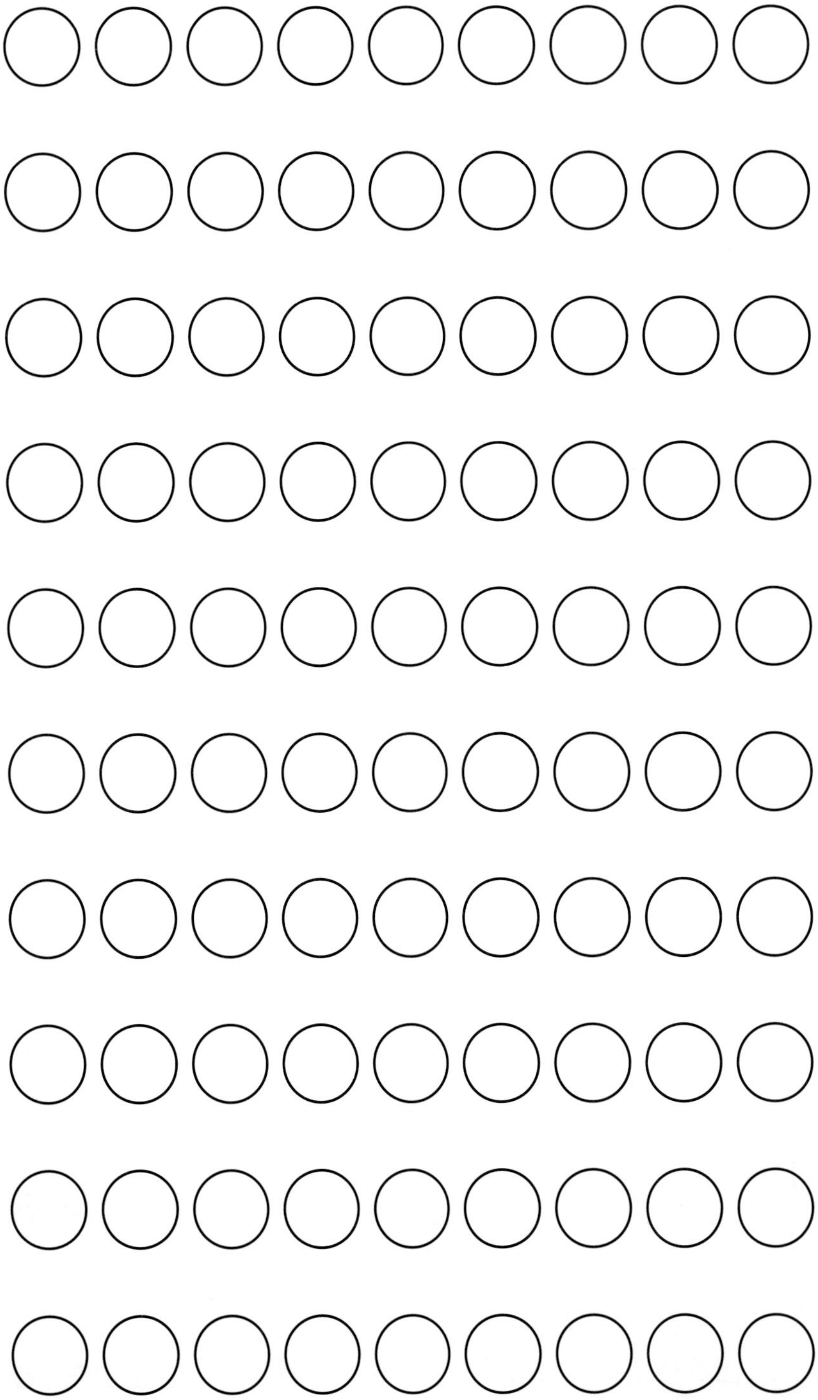

○○○○○○○○○   ____ + ____ = 9

○○○○○○○○○   ____ + ____ = 9

○○○○○○○○○   ____ + ____ = 9

○○○○○○○○○   ____ + ____ = 9

○○○○○○○○○   ____ + ____ = 9

○○○○○○○○○   ____ + ____ = 9

○○○○○○○○○   ____ + ____ = 9

○○○○○○○○○   ____ + ____ = 9

○○○○○○○○○   ____ + ____ = 9

○○○○○○○○○   ____ + ____ = 9

Picture Land • Activity Sheet 7

Ten   Name _____

$\text{Blue} + \text{Red} = 10$

○○○○○○○○○○   ___ + ___ = 10

○○○○○○○○○○   ___ + ___ = 10

○○○○○○○○○○   ___ + ___ = 10

○○○○○○○○○○   ___ + ___ = 10

○○○○○○○○○○   ___ + ___ = 10

○○○○○○○○○○   ___ + ___ = 10

○○○○○○○○○○   ___ + ___ = 10

○○○○○○○○○○   ___ + ___ = 10

○○○○○○○○○○   ___ + ___ = 10

○○○○○○○○○○   ___ + ___ = 10

○○○○○○○○○○   ___ + ___ = 10

Picture Land • Activity Sheet 8

Counter Dropping Record Form

Name _____

| Blue | Red |

____ + ____ = ____

____ + ____ = ____

____ + ____ = ____

____ + ____ = ____

____ + ____ = ____

____ + ____ = ____

____ + ____ = ____

____ + ____ = ____

____ + ____ = ____

____ + ____ = ____

____ + ____ = ____

B12 **Picture Land** • Activity Sheet 9

Dragon Quest 2 and 3
Record Form

Name _____

☐ _____ = ☐

☐ _____ = ☐

☐ _____ = ☐

☐ _____ = ☐

☐ _____ = ☐

☐ _____ = ☐

☐ _____ = ☐

☐ _____ = ☐

☐ _____ = ☐

Picture Land • Activity Sheet 10

Fish Pond

Name _____

B14 **Picture Land** • Activity Sheet 11

Fish Pond Record Form

Name _____

How many fish are left in the pond?

20 − _____ = _____

_____ − _____ = _____

_____ − _____ = _____

_____ − _____ = _____

_____ − _____ = _____

_____ − _____ = _____

_____ − _____ = _____

_____ − _____ = _____

Picture Land • Activity Sheet 12

Two-Team Record Form   Name _____

Team 1	Team 2
___ + ___ = ___	___ + ___ = ___
___ + ___ = ___	___ + ___ = ___
___ + ___ = ___	___ + ___ = ___
___ + ___ = ___	___ + ___ = ___
___ + ___ = ___	___ + ___ = ___
___ + ___ = ___	___ + ___ = ___
___ + ___ = ___	___ + ___ = ___
___ + ___ = ___	___ + ___ = ___
___ + ___ = ___	___ + ___ = ___
___ + ___ = ___	___ + ___ = ___

Picture Land • Activity Sheet 13

Name _____

Game Story for Addition

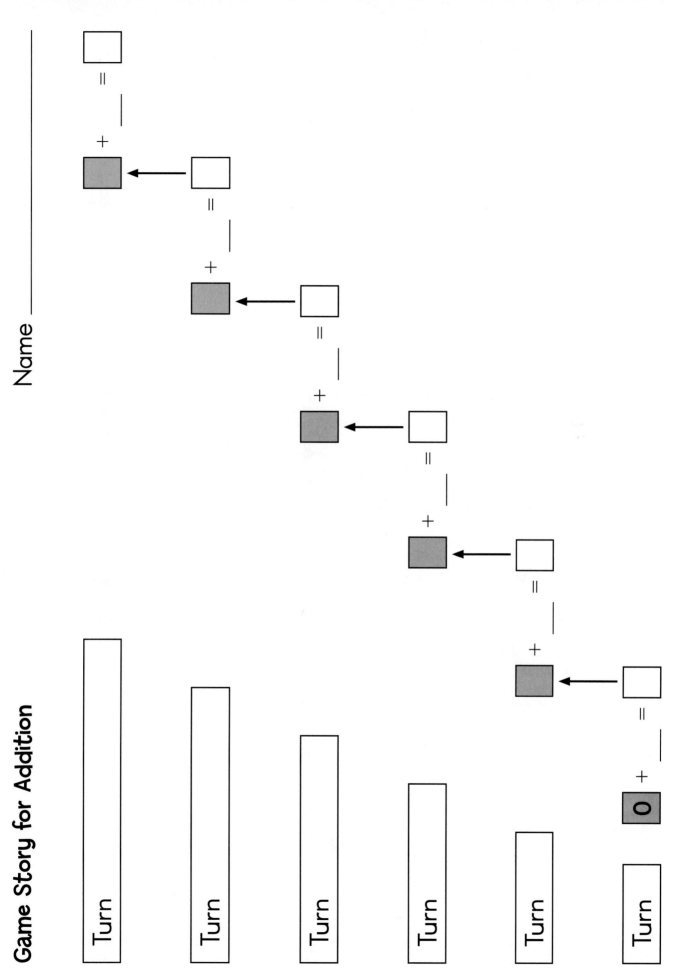

Line Land • Activity Sheet 1 **B17**

Number Lines to 20

Name _____

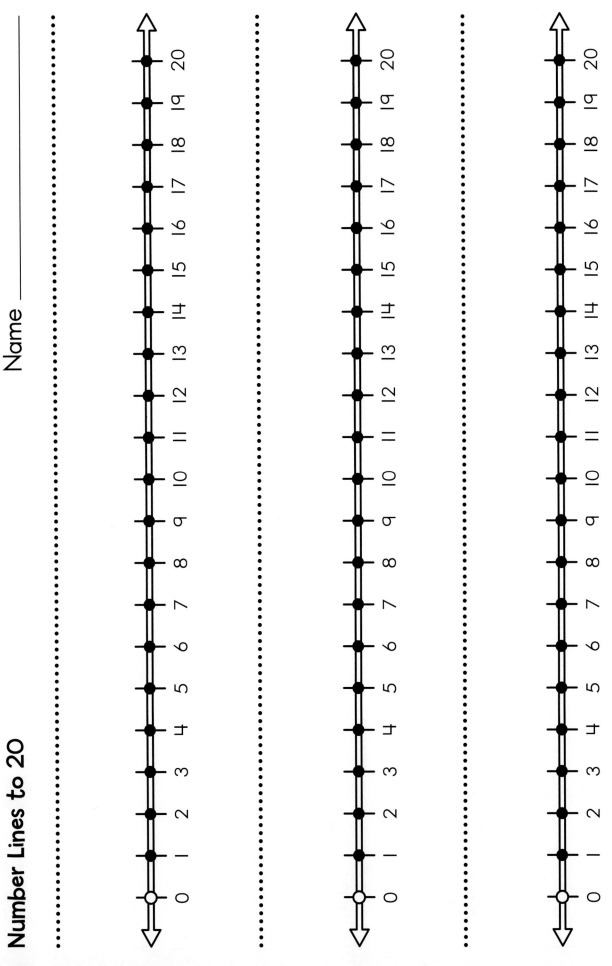

B18 **Line Land** • Activity Sheet 2

Game Story for Subtraction

Name _____

Turn Turn Turn Turn Turn Turn

Line Land • Activity Sheet 3   **B19**

Multiples of 10 Number Line

Name _____

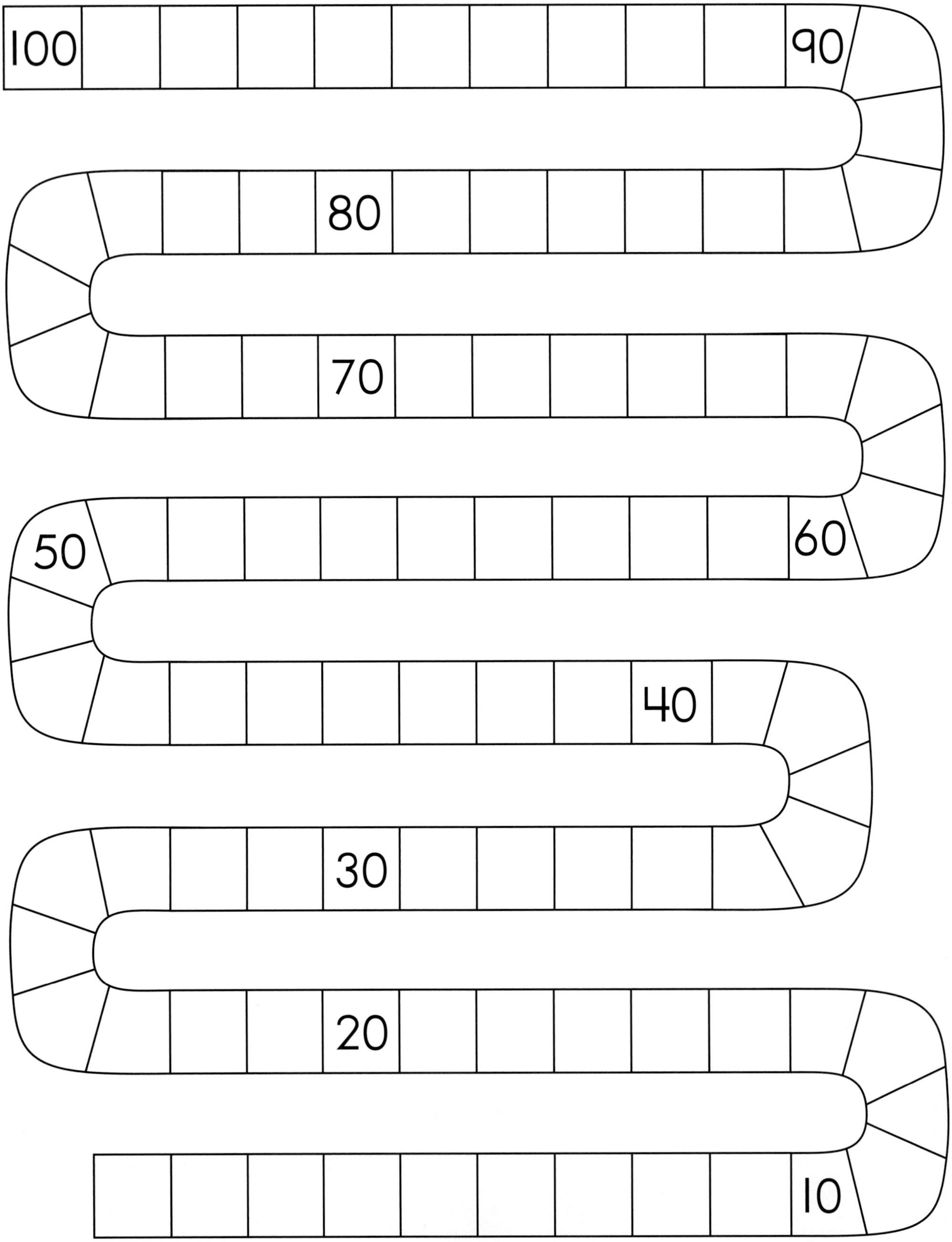

100 ... 90
80
70
50 ... 60
40
30
20
10

B20 Line Land • Activity Sheet 4

Graphing and Counting

Name _____

15				
14				
13				
12				
11				
10				
9				
8				
7				
6				
5				
4				
3				
2				
1				
	Red	Yellow	Blue	Green

Color	How many?
Red	
Yellow	
Blue	
Green	

Sky Land • Activity Sheet 1

Shopping Trip   Name _____

pencil 2¢ | ring 3¢ | crayons 3¢ | ball 4¢ | apple 2¢

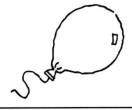 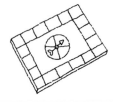  

balloon 3¢ | game 4¢ | goldfish 3¢ | sunglasses 5¢

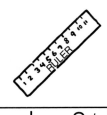 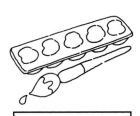 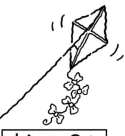 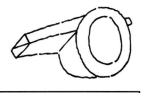

ruler 2¢ | paint 2¢ | kite 2¢ | whistle 5¢

notebook 4¢ | piggybank 1¢ | hairbrush 5¢ | flower 1¢

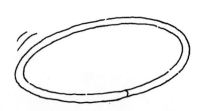 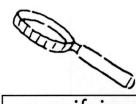 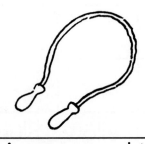 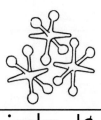

hulahoop 3¢ | magnifying glass 4¢ | jumprope 1¢ | jacks 1¢

B22 Multi-Land • Activity Sheet 1

Multiple-Players Record Form

Name _____

	Player 1	Player 2	Player 3	Player 4	Player 5
Turn 1					
Turn 2					
Turn 3					
Turn 4					
Turn 5					
Turn 6					
Turn 7					
Turn 8					
Turn 9					
Turn 10					

Multi-Land • Activity Sheet 2

Shape Blocks

Instructions:
Enlarge shapes 200% on a photocopier. Transfer the shapes to different-colored foam sheets and cut out several of each shape.

Suggested activities:
1. Create shape "flash cards" by labeling the shapes on one side.
2. Encourage children to sort the shapes by different attributes (size, color, number of sides, and so on).
3. Have children make patterns with the shapes. You may create a repeating pattern and ask children to continue it, or you may have children create their own patterns.
4. Encourage children to solve tangram puzzles by using the tangram shapes to match puzzle outlines. Visit **SRAonline.com** for printable tangram puzzle outlines.
5. Encourage children to create their own shape pictures using the shape blocks. Children may then challenge a partner to create a matching picture.

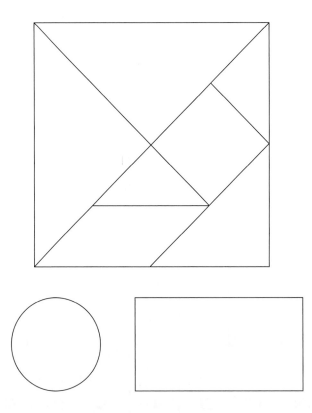

About Mathematics Intervention

What is an intervention?

An intervention is any instructional or practice activity designed to help students who are not making adequate progress.
- For struggling students, this requires an acceleration of development over a sufficient period of time.
- If the problem is small, the intervention might be brief.
- If the problem or lag in development is large, the intervention may last for weeks, months, or all year.

What is the three-tier approach?

Layers of intervention are organized in tiers to enable schools to respond to student needs.

Tier 1
Core Instruction: high quality, comprehensive instruction for all students. It reduces the number of students who later become at-risk for academic problems.
Tier 2
Supplemental Intervention: Addresses essential content for students who are not making adequate progress in the core program. It reduces the need for more intensive intervention.
Tier 3
Intensive Intervention: Increases in intensity and duration for students with low-level skills and a sustained lack of adequate progress within Tiers 1 and 2.

How is *Number Worlds* mathematics intervention different?

Number Worlds instruction is:
- More explicit
- More systematic
- More intense
- More supportive

Number Worlds is designed for both Tier 2 and Tier 3 students. Tier 2 students may spend a brief time in the program learning a key concept and then be quickly reintegrated back into the core instructional program. Levels A–C are also appropriate for Tier 1 students.

Students in Tier 3 will most likely need to complete the entire *Number Worlds* curriculum at their learning levels. Tier 3 requires intensive intervention for students with low skills and a sustained lack of adequate progress within Tiers 1 and 2. Teaching at this level is more intensive and includes more explicit instruction that is designed to meet the individual needs of struggling students. Group size is smaller, and the duration of daily instruction is longer.

Program Research

Building Number Sense with *Number Worlds:*
A Mathematics Program for Young Children
Sharon Griffin

What is number sense? We all know number sense when we see it but, if asked to define what it is and what it consists of, most of us, including the teachers among us, would have a much more difficult time. Yet this is precisely what we need to know to teach number sense effectively. Consider the answers three kindergarten children provide when asked the following question from the Number Knowledge Test (Griffin & Case, 1997): "Which is bigger: seven or nine?"

> Brie responds quickly, saying "Nine." When asked how she figured it out, she says, "Well, you go, 'seven' (pause) 'eight', 'nine' (putting up two fingers while saying the last two numbers). That means nine has two more than seven. So it's bigger."
>
> Leah says, hesitantly, "Nine?" When asked how she figured it out, she says, "Because nine's a big number."
>
> Caitlin looks genuinely perplexed, as if the question was not a sensible thing to ask, and says, "I don't know."

Kindergarten teachers will immediately recognize that Brie's answer provides evidence of a well-developed number sense for this age level and Leah's answer, a more fragile and less-developed number sense. The knowledge that lies behind this "sense" may be much less apparent. What knowledge does Brie have that enables her to come up with the answer in the first place and to demonstrate good number sense in the process?

1. Knowledge that underlies number sense
Research conducted with the Number Knowledge Test and several other cognitive developmental measures (see Griffin, 2002; Griffin & Case, 1997 for a summary of this research) suggests that the following understandings lie at the heart of the number sense that 5-year-olds like Brie are able to demonstrate on this problem. They know (a) that numbers indicate quantity and therefore, that numbers, themselves, have magnitude; (b) that the word "bigger" or "more" is sensible in this context; (c) that the numbers 7 and 9, like every other number from 1 to 10, occupy fixed positions in the counting sequence; (d) that 7 comes before 9 when you are counting up; (e) that numbers that come later in the sequence—that are higher up— indicate larger quantities and therefore, that 9 is bigger (or more) than 7.

Brie provided evidence of an additional component of number sense in the explanation she provided for her answer. By using the Count-On strategy to show that nine comes two numbers after seven and by suggesting that this means "it has two more than seven," Brie demonstrated that she also knows (f) that each counting number up in the sequence corresponds precisely to an increase of one unit in the size of a set. This understanding, possibly more than any of the others listed above, enables children to use the counting numbers alone, without the need for real objects, to solve quantitative problems involving the joining of two sets. In so doing, it transforms mathematics from something that can only be done out there (e.g., by manipulating real objects) to something that can be done in their own heads, and under their own control.

This set of understandings, the core of *number sense*, forms a knowledge network that Case and Griffin (1990), see also Griffin and Case (1997), have called a *central conceptual structure for number.* Research conducted by these investigators has shown that this structure is central in at least two ways (see Griffin, Case, & Siegler, 1994). First, it enables children to make sense of a broad range of quantitative problems across contexts and to answer questions, for example, about two times on a clock (Which is longer?), two positions on a path (Which is farther?), and two sets of coins (Which is worth more?). Second, it provides the foundation on which children's learning of more complex number concepts, such as those involving double-digit numbers, is built. For this reason, this network of knowledge is an important set of understandings that should be taught in the preschool years, to all children who do not spontaneously acquire them.

2 How can this knowledge be taught?

Number Worlds, a mathematics program for young children (formerly called *Rightstart*), was specifically developed to teach this knowledge and to provide a test for the cognitive developmental theory (i.e., Central Conceptual Structure theory; see Case & Griffin, 1990) on which the program was based. Originally developed for kindergarten, the program (see Griffin & Case, 1995) was expanded to teach a broader range of understandings when research findings provided strong evidence that (a) children who were exposed to the program acquired the knowledge it was designed to teach (i.e., the central conceptual structure for number), and (b) the theoretical postulates on which the program was based were valid (see Griffin & Case, 1996; Griffin, Case, & Capodilupo, 1995; Griffin et al., 1994). Programs for grades one and two were developed to teach the more complex central conceptual structures that underlie base-ten understandings (see Griffin, 1997, 1998) and a program for preschool was developed (see Griffin, 2000) to teach the "precursor" understandings that lay the foundation for the development of the central conceptual structure for number.

Because the four levels of the program are based on a well-developed theory of cognitive development, they provided a finely graded sequence of activities (and associated knowledge objectives) that recapitulates the natural developmental progression for the age range of 3–9 years and that allows each child to enter the program at a point that is appropriate for his or her own development, and to progress through the program to teach 20 or more children at any one time and every effort has been made, in the construction of the **Number Worlds** program, to make it as easy as possible for teachers to accommodate the developmental needs of individual children (or groups of children) in their classroom. Five instructional principles that lie at the heart of the program are described below and are used to illustrate several features of the program that have already been mentioned and several that have not yet been introduced.

2.1 Principle 1: Build upon children's current knowledge

Each new idea that is presented to children must connect to their existing knowledge if it is going to make any sense at all. Children must also be allowed to use their existing knowledge to construct new knowledge that is within reach—that is one step beyond where they are now—and a set of bridging contexts and other instructional supports should be in place to enable them to do so.

In the examples of children's thinking presented earlier, three different levels of knowledge are apparent. Brie appears to have acquired the knowledge network that underlies number sense and to be ready, therefore, to move on to the next developmental level: to connect this set of understandings to the written numerals (i.e., the formal symbols) associated with each counting word. Leah appears to have some understanding of some of the components of this network (i.e., that number have magnitude) and to be ready to use this understanding as a base to acquire the remaining understandings (e.g., that a number's magnitude and its position in the counting sequence are directly related). Caitlin demonstrated little understanding of any element of this knowledge network and she might benefit, therefore, from exposure to activities that will help her acquire the "precursor" knowledge needed to build this network, namely knowledge of counting (e.g., the one-to-one correspondence rule) and knowledge of quantity (e.g., an intuitive understanding of relative amount). Although all three children are in kindergarten, each child appears to be at a different point in the developmental trajectory and to require a different set of learning opportunities; ones that will enable each child to use her existing knowledge to construct new knowledge at the next level up.

To meet these individual needs, teachers need (a) a way to assess children's current knowledge, (b) activities that are multi-leveled so children with different entering knowledge can all benefit from exposure to them, and (c) activities that are carefully sequenced and that span several developmental levels so children with different entering knowledge can be exposed to activities that are appropriate for their level of understanding. These are all available in the **Number Worlds** program and are illustrated in various sections of this paper.

Program Research

2.2 Principle 2: Follow the natural developmental progression when selecting new knowledge to be taught

Researchers who have investigated the manner in which children construct number knowledge between the ages of 3 and 9 years have identified a common progression that most, if not all, children follow (see Griffin, 2002; Griffin and Case, 1997 for a summary of this research). As suggested earlier, by the age of 4 years, most children have constructed two "precursor" knowledge networks—knowledge of counting and knowledge of quantity—that are separate in this stage and that provide the base for the next developmental stage. Sometime in kindergarten, children become able to integrate these knowledge networks—to connect the world of counting numbers to the world of quantity—and to construct the central conceptual understandings that were described earlier. Around the age of 6 or 7 years, children connect this integrated knowledge network to the world of formal symbols and, by the age of 8 or 9 years, most children become capable of expanding this knowledge network to deal with double-digit numbers and the base-ten system. A mathematics program that provides opportunities for children to use their current knowledge to construct new knowledge that is a natural next step, and that fits their spontaneous development, will have the best chance of helping children make maximum progress in their mathematics learning and development.

Because there are limits in development on the complexity of information children can handle at any particular age/stage (see Case, 1992), it makes no sense to attempt to speed up the developmental process by accelerating children through the curriculum. However, for children who are at an age when they should have acquired the developmental milestones but for some reason haven't, exposure to a curriculum that will give them ample opportunities to do so makes tremendous sense. It will enable them to catch up to their peers and thus, to benefit from the formal mathematics instruction that is provided in school. Children who are developing normally also benefit from opportunities to broaden and deepen the knowledge networks they are constructing, to strengthen these understandings, and to use them in a variety of contexts.

2.3 Principle 3: Teach computational fluency as well as conceptual understanding

Because computational fluency and conceptual understanding have been found to go hand in hand in children's mathematical development (see Griffin, 2003; Griffin et al., 1994), opportunities to acquire computational fluency, as well as conceptual understanding, are built into every **Number Worlds** activity. This is nicely illustrated in the following activities, drawn from different levels of the program.

In The Mouse and the Cookie Jar Game (created for the preschool program and designed to give 3- to 4-year-olds an intuitive understanding of subtraction), children are given a certain number of counting chips (with each child receiving the same number but a different color) and told to pretend their chips are cookies. They are asked to count their cookies and, making sure they remember how many they have and what their color is, to deposit them in the cookie jar for safe keeping. While the children sleep, a little mouse comes along and takes one (or two) cookies from the jar. The problem that is then posed to the children is "How can we figure out whose cookie(s) the mouse took?"

Although children quickly learn that emptying the jar and counting the set of cookies that bears their own color is a useful strategy to use to solve this problem, it takes considerably longer for many children to realize that, if they now have four cookies (and originally had five), it means that they have one fewer and the mouse has probably taken one of their cookies. Children explore this problem by counting and recounting the remaining sets, comparing them to each other (e.g., by aligning them) to see who has the most or least, and ultimately coming up with a prediction. When a prediction is made, children search the mouse's hole to see whose cookie had been taken and to verify or revise their prediction. As well as providing opportunities to perfect their counting skills, this activity gives children concrete opportunities to experience simple quantity transformations and to discover how the counting numbers can be used to predict and explain differences in amount.

The *Dragon Quest Game* that was developed for the Grade 1 program teaches a much more sophisticated set of understandings. Children are introduced to Phase 1 activity by being told a story about a fire-breathing dragon that has been terrorizing the village where children live. The children playing the game are heroes who have been chosen to seek out the dragon and put out his fire. To extinguish this dragon's fire (as opposed to the other, more powerful dragons they will encounter in later phases) a hero will need at least 10 pails of water. If a hero enters into the dragon's area with less than 10 pails of water, he or she will become the dragon's prisoner and can only be rescued by one of the other players.

To play the game, children take turns rolling a die and moving their playing piece along the colored game board. If they land on a well pile (indicated by a star), they can pick a card from the face-down deck of cards, which illustrate, with images and symbols (e.g., +4) a certain number of pails of water. Children are encouraged to add up their pails of water as they receive them and they are allowed to use a variety of strategies to do so, ranging from mental math (which is encouraged) to the use of tokens to keep track of the quantity accumulated. The first child to reach the dragon's lair with at least 10 pails of water can put out the dragon's fire and free any teammates who have become prisoners.

As children play this game and talk about their progress, they have ample opportunity to connect numbers to several different quantity representations (e.g., dot patterns on the die; distance of their pawn along the path; sets of buckets illustrated on the cards; written numerals also provided on the cards) and to acquire an appreciation of numerical magnitude across these contexts. With repeated play, they also become capable of performing a series of successive addition operations *in their heads* and of expanding the well pile. When they are required to submit formal proof to the mayor of the village that they have amassed sufficient pails of water to put out the dragon's fire before they are allowed to do so, they become capable of writing a series of formal expressions to record the number of pails received and spilled over the course of the game. In contexts such as these children receive ample opportunity to use the formal symbol system in increasingly efficient ways to make sense of quantitative problems they encounter in the course of their own activity.

2.4 Principle 4: provide plenty of opportunity for hands-on exploration, problem-solving, and communication

Like the *Dragon Quest Game* that was just described, many of the activities created for the **Number Worlds** program are set in a game format that provides plenty of opportunity for hands-on exploration of number concepts, for problem-solving and for communication. Communication is explicitly encouraged in a set of question prompts that are included with each small group game (e.g., How far are you now? How many more buckets do you need to put out the dragon's fire? How do you know?) as well as in a more general set of dialogue prompts that are included in the teacher's guide. Opportunities for children to discuss what they learned during game play each day, to share their knowledge with their peers, and to make their reasoning explicit are also provided in a Wrap-Up session that is included at the end of each math lesson.

Finally, in the whole group games and activities that were developed for the Warm-Up portion of each math lesson, children are given ample opportunity to count (e.g., up from 1 and down from 10) and to solve mental math problems, in a variety of contexts. In addition to developing computational fluency, these activities expose children to the language of mathematics and give them practice using it. Although this is valuable for all children, it is especially useful for ESL children, who may know how to count in their native language but not yet in English. Allowing children to take turns in these activities and to perform individually gives teachers opportunities to assess each child's current level of functioning, important for instructional planning, and gives children opportunities to learn from each other.

2.5 Principle 5: Expose children to the major ways number is represented and talked about in developed societies

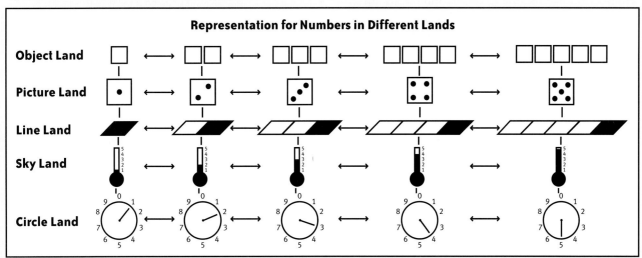

Program Research

Number is represented in our culture in five major ways: as a group of objects, a dot-set pattern, a position on a line, a position on a scale (e.g., a thermometer), and a point on a dial. In each of these contexts, number is also talked about in different ways, with a larger number (and quantity) described as "more" in the world of dot-sets, as "further along" in the world of paths and lines, as "higher up" in the world of scale measures, and as "further around" in the world of dials. Children who are familiar with these forms of representation and the language used to talk about number in these contexts have a much easier time making sense of the number problems they encounter inside and outside of school.

In the **Number Worlds** program, children are systematically exposed to these forms of representations as they explore five different "lands." Learning activities developed for each land share a particular form of number representation while they simultaneously address specific knowledge goals for each grade level. Many of the games, like *Dragon Quest*, also expose children to multiple representations of number in one activity so children can gradually come to see the ways they are equivalent.

3 Discussion

Children who have been exposed to the **Number Worlds** program do very well on number questions like the one presented in the introduction and on the Number Knowledge Test (Griffin & Case, 1997) from which this question was drawn. In several evaluation studies conducted with children from low-income communities, children who received the **Number Worlds** program made significant gains in conceptual knowledge of number and in number sense, when compared to matched-control groups who received readiness training of a different sort. These gains enabled them to start their formal schooling in grade one on an equal footing with their more advantaged peers, to perform as well as groups of children from China and Japan on a computation test administered at the end of grade one, and to keep pace with their more advantaged peers (and even outperform them on some measures) as they progressed through the first few years of formal schooling (Griffin & Case, 1997).

Teachers also report positive gains from using the **Number Worlds** program and from exposure to the instructional principles on which it is based. Although all teachers acknowledge that implementing the program and putting the principles into action is not an easy task, many claim that their teaching of all subjects has been transformed in the process. They now facilitate discussion rather than dominating it; they pay much more attention to what children say and do; and they now allow children to take more responsibility for their own learning, with positive and surprising results. Above all, they now look forward to teaching math and they and their students are eager to do more of it.

Griffin, S. "Building Number Sense with Number Worlds: A mathematics program for young children." In *Early Childhood Research Quarterly* 19 (2004) 173–180. Elsevier.

References

Case, R. (1992). *The mind's staircase: Exploring the conceptual underpinnings of children's thought and knowledge.* Hillsdale, NJ: Erlbaum.

Case, R., & Griffin, S. (1990). Child cognitive development: The role of central conceptual structures in the development of scientific and social thought. In E.A. Hauert (Ed.), *Developmental psychology: Cognitive, perceptuo-motor, and neurological perspectives* (pp. 193–230). North-Holland: Elsevier.

Griffin, S. (1997). *Number Worlds: Grade one level.* Durham, NH: Number Worlds Alliance Inc.

Griffin, S. (1998). *Number Worlds: Grade two level.* Durham, NH: Number Worlds Alliance Inc.

Griffin, S. (2000). *Number Worlds: Preschool level.* Durham, NH: Number Worlds Alliance Inc.

Griffin, S. (2002). The development of math competence in the preschool and early school years: Cognitive foundations and instructional strategies. In J. Royer (Ed.), *Mathematical cognition* (pp. 1–32). Greenwich, CT: Information Age Publishing.

Griffin, S. (2003). Laying the foundation for computational fluency in early childhood. *Teaching children mathematics,* 9, 306–309.

Griffin, S. (1997). *Number Worlds: Kindergarten level.* Durham, NH: Number Worlds Alliance Inc.

Griffin, S., & Case, R. (1996). Evaluating the breadth and depth of training effects when central conceptual structures are taught. *Society for research in child development monographs,* 59, 90–113.

Griffin, S., & Case, R. (1997). Re-thinking the primary school math curriculum: An approach based on cognitive science. *Issues in Education,* 3, 1–49.

Griffin, S., Case, R., & Siegler, R. (1994). Rightstart: Providing the central conceptual prerequisites for first formal learning of arithmetic to students at-risk for school failure. In K. McGilly (Ed.), *Classroom lessons: Integrating cognitive theory and classroom practice* (pp. 24-49). Cambridge, MA: Bradford Books MIT Press.

Griffin, S., Case, R., & Capodilupo, A. (1995). Teaching for understanding: The importance of central conceptual structures in the elementary mathematics curriculum. In A. McKeough, I. Lupert, & A. Marini (Eds.), *Teaching for transfer: Fostering generalization in learning* (pp. 121–151). Hillsdale, NJ: Erlbaum.

Math Proficiencies

Mathematical proficiency has five strands. These strands are not independent; they represent different aspects of a complex whole . . . they are interwoven and interdependent in the development of proficiency in mathematics.

—Kilpatrick, J., Swafford, J., and Findell, B., eds. *Adding It Up: Helping Children Learn Mathematics.* Washington D.C.: National Research Council/National Academy Press, 2001, pp. 115–133.

Number Worlds develops all five proficiencies in each lesson, so that students build their skills, conceptual understanding, and reasoning powers as well as their abilities to apply mathematics and see it as useful.

Math Proficiencies

1 UNDERSTANDING (Conceptual Understanding): Comprehending mathematical concepts, operations, and relations—knowing what mathematical symbols, diagrams, and procedures mean. *Conceptual Understanding* refers to an integrated and functional grasp of mathematical ideas. Students with conceptual understanding know more than isolated facts and methods. They understand why a mathematical idea is important and the kinds of contexts in which it is useful. They have organized their knowledge into a coherent whole, which enables them to learn new ideas by connecting those ideas to what they already know. Conceptual understanding also supports retention. Because facts and methods learned with understanding are connected, they are easier to remember and use, and they can be reconstructed when forgotten. If students understand a method, they are unlikely to remember it incorrectly.

A significant indicator of conceptual understanding is being able to represent mathematical situations in different ways and knowing how different representations can be useful for different purposes.

Knowledge that has been learned with understanding provides the bases for generating new knowledge and for solving new and unfamiliar problems. When students have acquired conceptual understanding in an area of mathematics, they see the connections among concepts and procedures and can give arguments to explain why some facts are consequences of others. They gain confidence, which then provides a base from which they can move to another level of understanding.

2 COMPUTING (Procedural Fluency): Carrying out mathematical procedures, such as adding, subtracting, multiplying, and dividing numbers flexibly, accurately, efficiently, and appropriately. *Procedural Fluency* refers to knowledge of procedures, knowledge of when and how to use them appropriately, and skill in performing them flexibly, accurately, and efficiently. In the domain of number, procedural fluency is especially needed to support conceptual understanding of place value and the meaning of rational numbers. It also supports the analysis of similarities and differences among methods of calculating. These methods include written procedures, and mental methods for finding certain sums, differences, products, or quotients, and methods that use calculators, computers, or manipulative materials such as blocks, counters, or beads.

Students need to be efficient and accurate in performing basic computations with whole numbers without always having to refer to tables or other aids. They also need to know reasonably efficient and accurate ways to add, subtract, multiply, and divide multidigit numbers, mentally and with pencil and paper. A good conceptual understanding of place value in the base-ten system supports the development of fluency in multidigit computation. Such understanding also supports simplified but accurate mental arithmetic and more flexible ways of dealing with numbers than many students ultimately achieve.

Math Proficiencies

3 **APPLYING** (Strategic Competence): Being able to formulate problems mathematically and to devise strategies for solving them using concepts and procedures appropriately. *Strategic Competence* refers to the ability to formulate mathematical problems, represent them, and solve them. This strand is similar to what has been called problem solving and problem formulation. Although in school, students are often presented with clearly specified problems to solve, outside of school they encounter situations in which part of the difficulty is to figure out exactly what the problem is. Then they need to formulate the problem so that they can use mathematics to solve it. Consequently, they are likely to need experience and practice in problem formulating as well as in problem solving. They should know a variety of solution strategies as well as which strategies might be useful for solving a specific problem.

To represent a problem accurately, students must first understand the situation, including its key features. They then need to generate a mathematical representation of the problem that captures the core mathematical elements and ignores the irrelevant features.

Students develop procedural fluency as they use their strategic competence to choose among effective procedures. They also learn that solving challenging mathematics problems depends on the ability to carry out procedures readily, and conversely, that problem-solving experience helps them acquire new concepts and skills.

4 **REASONING** (Adaptive Reasoning): Using logic to explain and justify a solution to a problem or to extend from something known to something not yet known. *Adaptive Reasoning* refers to the capacity to think logically about the relationships among concepts and situations. Such reasoning is correct and valid, stems from careful consideration of alternatives, and includes knowledge of how to justify the conclusions. In mathematics, adaptive reasoning is the glue that holds everything together and guides learning. One uses it to navigate through the many facts, procedures, concepts, and solution methods and to see that they all fit together in some way that they make sense. In mathematics, deductive reasoning is used to settle disputes and disagreements. Answers are right because they follow from some agreed-upon assumptions through a series of logical steps. Students who disagree about a mathematical answer need not rely on checking with the teacher, collecting opinions from their classmates, or gathering data from outside the classroom. In principle, they need only check that their reasoning is valid.

Research suggests that students are able to display reasoning ability when three conditions are met: They have a sufficient knowledge base, the task is understandable and motivating, and the context is familiar and comfortable.

5 **ENGAGING** (Productive Disposition): Seeing mathematics as sensible, useful, and doable—if you work at it—and being willing to do the work. *Productive disposition* refers to the tendency to see sense in mathematics, to perceive it as both useful and worthwhile, to believe that steady effort in learning mathematics pays off, and to see oneself as an effective learner and doer of mathematics. If students are to develop conceptual understanding, procedural fluency, strategic competence, and adaptive reasoning abilities, they must believe mathematics is understandable, not arbitrary; that with diligent effort, it can be learned and used; and that they are capable of figuring it out. Developing a productive disposition requires frequent opportunities to make sense of mathematics, to recognize the benefits of perseverance, and to experience the rewards of sense making in mathematics.

Students' disposition toward mathematics is a major factor in determining their educational success. Students who have developed a productive disposition are confident in their knowledge and abilities. They see that mathematics is both reasonable and intelligible and believe that, with appropriate effort and experience, they can learn.

Content Strands

Content Strands of Mathematics

Number Sense and Place Value

Understanding of the significance and use of numbers in counting, measuring, comparing, and ordering.

> "It is very important for teachers to provide children with opportunities to recognize the meaning of mathematical symbols, mathematical operations, and the patterns or relationships represented in the child's work with numbers. For example, the number sense that a child acquires should be based upon an understanding that inverse operations, such as addition and subtraction, undo the operations of the other. Instructionally, teachers must encourage their students to think beyond simply finding the answer and to actually have them think about the numerical relationships that are being represented or modeled by the symbols, words, or materials being used in the lesson."
>
> Kilpatrick, J., Swafford, J. and Findell, B. eds. Adding It Up: Helping Children Learn Mathematics. Washington, D.C.: National Research Council/National Academy Press, 2001, p. 270–271.

Number Worlds and Number Sense

Goal: Firm understanding of the significance and use of numbers in counting, measuring, comparing and ordering. The ability to think intelligently, using numbers. This basic requirement of numeracy includes the ability to recognize given answers as absurd, without doing a precise calculation, by observing that they violate experience, common sense, elementary logic, or familiar arithmetic patterns. It also includes the use of imagination and insight in using numbers to solve problems. Children should be able to recognize when, for example, a trial-and-error method is likely to be easier to use and more manageable than a standard algorithm.

Developing number sense is a primary goal of ***Number Worlds*** in every grade. Numbers are presented in a variety of representations and integrated in many contexts so that students develop thorough understanding of numbers.

Algebra

Algebra is the branch of mathematics that uses symbols to represent arithmetic operations. Algebra extends arithmetic through the use of symbols and other notations such as exponents and variables. Algebraic thinking involves understanding patterns, equations, and relationships and includes concepts of functions and inverse operations. Because algebra uses symbols rather than numbers, it can produce general rules that apply to all numbers. What most people commonly think of as algebra involves the manipulation of equations and the solving of equations. Exposure to algebraic ideas can and should occur well before students first study algebra in middle school or high school. Even primary students are capable of understanding many algebraic concepts. Developing algebraic thinking in the early grades smoothes the transition to algebra in middle school and high school and ensures success in future math and science courses as well as in the workplace.

> "Algebra begins with a search for patterns. Identifying patterns helps bring order, cohesion, and predictability to seemingly unorganized situations and allows one to make generalizations beyond the information directly available. The recognition and analysis of patterns are important components of the young child's intellectual development because they provide a foundation for the development of algebraic thinking."
>
> Clements, Douglas and Sarama, J., eds. *Engaging Young Children in Mathematics: Standards for Early Childhood Mathematics Education.* Mahwah, New Jersey: Lawrence Erlbaum Associates, Publishers, 2004., p. 52.

Number Worlds and Algebra

Goal: Understanding of functional relationships between variables that represent real world phenomena that are in a constant state of change. Children should be able to draw the graphs of functions and to derive information about functions from their graphs. They should understand the special importance of linear functions and the connection between the study of functions and the solutions of equations and inequalities.

The algebra readiness instruction that begins in the Pre-K level is designed to prepare students for future work in algebra by exposing them to algebraic thinking, including looking for patterns, using variables, working with functions, using integers and exponents, and being aware that mathematics is far more than just arithmetic.

Content Strands

Arithmetic

Arithmetic, one of the oldest branches of mathematics, arises from the most fundamental of mathematical operations: counting. The arithmetic operations—addition, subtraction, multiplication, division, and place holding—form from the basis of the mathematics we use regularly. Mastery of the basic operations with whole numbers (addition, subtraction, multiplication, and division)

> "Although some educators once believed that children memorized their "basic facts" as conditioned responses, research now shows that children do not move from knowing nothing about sums and differences of numbers to having the basic number combinations memorized. Instead, they move through a series of progressively more advance and abstract methods for working out the answers to simple arithmetic problems. Furthermore, as children get older, they use the procedures more and more efficiently."
>
> Kilpatrick, J., Swafford, J., and Findell, B., eds. *Adding It Up: Helping Children Learn Mathematics.* Washington, D.C.: National Research Council/National Academy Press, 2001, pp. 182–183.

Number Worlds and Arithmetic

Goal: Mastery of the basic operations with whole numbers (addition, subtraction, multiplication, and division). Whatever other skills and understandings children acquire, they must have the ability to calculate a precise answer when necessary. This fundamental skill includes not only knowledge of the appropriate arithmetic algorithms but also mastery of the basic addition, subtraction, multiplication, and division facts and understanding of the positional notation (base ten) of the whole numbers.

Cumulative assessment occurs throughout the program to indicate when mastery of concepts and skills is expected. Once taught, arithmetic skills are also integrated into other topics such as data analysis.

Fractions, Decimals, and Percents

Understand rational numbers (fractions, decimals, and percents) and their relationships to each other, including the ability to perform calculations and to use rational numbers in measurement.

> "Children need to learn that rational numbers are numbers in the same way that whole numbers are numbers. For children to use rational numbers to solve problems, they need to learn that the same rational number may be represented in different ways, as a fraction, a decimal, or a percent. Fraction concepts and representations need to be related to those of division, measurement, and ratio. Decimal and fractional representations need to be connected and understood. Building these connections takes extensive experience with rational numbers over a substantial period of time. Researchers have documented that difficulties in working with rational numbers can often be traced to weak conceptual understanding. . . . Instructional sequences in which more time is spent at the outset on developing meaning for the various representations of rational numbers and the concept of unit have been shown to promote mathematical proficiency."
>
> Kilpatrick, J., Swafford, J., and Findell, B., eds. *Adding It Up: Helping Children Learn Mathematics.* Washington, D.C.: National Research Council/National Academy Press, 2001, pp. 415–416.

Number Worlds and Rational Numbers

Goal: Understanding of rational numbers and of the relationship of fractions to decimals. Included here are the ability to do appropriate calculations with fractions or decimals (or both, as in fractions of decimals), the use of decimals in (metric unit) measurements, the multiplication of fractions as a model for the "of" relation and as a model for areas of rectangles.

Goal: Understanding of the meaning of rates and of their relationship to the arithmetic concept of ratio. Children should be able to calculate ratios, proportions, and percentages; understand how to use them intelligently in real-life situations; understand the common units in which rates occur (such as kilometers per hour, cents per gram); understand the meaning of per; and be able to express ratios as fractions.

In ***Number Worlds*** understanding of rational number begins at the earliest grades with sharing activities and develops understanding of rational numbers with increasing sophistication at each grade.

Content Strands

Geometry

Geometry is the branch of mathematics that deals with the properties of space. Plane geometry is the geometry of flat surfaces and solid geometry is the geometry of three-dimensional solids. Geometry has many more fields, including the study of spaces with four or more dimensions.

> "Geometry can be used to understand and to represent the objects, directions, and locations in our world, and the relationships between them. Geometric shapes can be described, analyzed, transformed, and composed and decomposed into other shapes."
>
> Clements, Douglas and Sarama, J., eds. *Engaging Young Children in Mathematics: Standards for Early Childhood Mathematics Education.* Mahwah, New Jersey: Lawrence Erlbaum Associates, Publishers, 2004., p. 39.

Number Worlds and Geometry

Goal: Understanding of an ability to use the geometric concepts of perimeter, area, volume, and congruency as applied to simple figures.

Data Analysis and Applications

Data Analysis encompasses the ability to organize information to make it easier to use and the ability to interpret data and graphs.

> "Describing data involves reading displays of data (e.g., tables, lists, graphs); that is, finding information explicitly stated in the display, recognizing graphical conventions, and making direct connections between the original data and the display. The process is essentially what has been called reading the data. . . . The process of organizing and reducing data incorporates mental actions such as ordering, grouping, and summarizing. Data reduction also includes the use of representative measures of center (often termed *measures of central tendency*) such as mean, mode, or media, and measures of spread such as range and standard deviation."
>
> Kilpatrick, J., Swafford, J., and Findell, B., eds. *Adding It Up: Helping Children Learn Mathematics.* Washington, D.C.: National Research Council/National Academy Press, 2001, pp. 289.

Number Worlds and Data Analysis

Goal: Ability to organize and arrange data for greater intelligibility. Children should develop not only the routine skills of tabulating and graphing results but also, at a higher level, the ability to detect patterns and trends in poorly organized data either before or after reorganization. In addition, children need to develop the ability to extrapolate and interpolate from data and from graphic representations. Children should also know when extrapolation or interpolation is justified and when it is not.

In *Number Worlds* students work with graphs beginning in Pre-K. In each grade the program emphasizes understanding what data show.

Problem Solving

Number Worlds and Problem Solving

Goal: Students must develop the critical thinking skills useful for solving problems. Computational skills are, of course, important, but they are not enough. Students also need an arsenal of critical-thinking skills (sometimes called problem-solving strategies and sometimes called heuristics) that they can call upon to solve particular problems. These skills should not be taught in isolation—students should learn to use them in different contexts. By doing so students are more likely to recognize in which situations a particular skill will be useful and when it is not likely to be useful. We can group critical-thinking skills into two categories—those that are useful in virtually all situations and those that are useful in specific contexts.

In *Number Worlds* students solve problems throughout the daily lessons in all levels of the program.

Technology Overview

Technology

Technology has changed the world of mathematics. Technological tools have eliminated the need for tedious calculations and have enabled significant advances in applications of mathematics. Technology can also help to make teaching more effective and efficient. Well-designed math software activities have proven effective in advancing children's math achievements. Technology can also help teachers organize planning and instruction and manage record keeping.

Number Worlds integrates two key software programs throughout the lessons to help students develop richer mathematical skills and conceptual understanding: **eMathTools** and **Building Blocks,** which are explained on the following pages.

Building Blocks Software Activities

Building Blocks software provides computer math activities that address specific developmental levels of the math learning trajectories. **Building Blocks** software is critical to **Number Worlds**. The engaging research-based activities provide motivating development and support of concepts.

Some **Building Blocks** activities have different levels of difficulty indicated by ranges in the Activity Names below. The list provides an overview of all of the **Building Blocks** activities along with the domains, descriptions, and appropriate age ranges.

Domain: Trajectory	Activity Name	Description	Age Range (in years)
Geometry: Composition/Decomposition	Create a Scene	Students explore shapes by moving and manipulating them to make pictures.	4–12
Geometry: Composition/Decomposition	Piece Puzzler 1–5, Piece Puzzler Free Explore, and Super Shape 1–7	Students complete pictures using pattern or tangram shapes.	4–12
Geometry: Imagery	Geometry Snapshots 1–8	Students match configurations of a variety of shapes (e.g., line segments in different arrangements, 3-6 tiled shapes, embedded shapes) to corresponding configurations, given only a brief view of the goal shapes.	5–12
Geometry: Shapes (Identifying)	Memory Geometry 1–5	Students match familiar geometric shapes (shapes in same or similar sizes, same orientation) within the framework of a "Concentration" card game.	3–5
Geometry: Shapes (Matching)	Mystery Pictures 1–4 and Mystery Pictures Free Explore	Students construct predefined pictures by selecting shapes that match a series of target shapes.	3–8
Geometry: Shapes (Parts)	Shape Parts 1–7	Students build or fix some real-world object, exploring shape and properties of shapes.	5–12
Geometry: Shapes (Properties)	Legends of the Lost Shape	Students identify target shapes using textual clues provided.	8–12
Geometry: Shapes (Properties)	Shape Shop 1–3	Students identify a wide range of shapes given their names, with more difficult distracters.	8–12
Measurement: Length	Comparisons	Students are shown pictures of two objects and are asked to click on the one that fits the prompt (longer, shorter, heavier, etc.).	4–8
Measurement: Length	Deep Sea Compare	Students compare the length of two objects by representing them with a third object.	5–7
Measurement: Length	Workin' on the Railroad	Students identify the length (in non-standard units) of railroad trestles they built to span a gully.	6–9
Measurement: Length	Reptile Ruler	Students learn about linear measurement by using a ruler to determine the length of various reptiles.	7–10
Multiplication/Division	Arrays in Area	Students build arrays and then determine the length of those arrays.	8–11
Multiplication/Division	Comic Book Shop	Students use skip counting to produce products that are multiples of 10s, 5s, 2s, and 3s. The task is to identify the product, given a number and bundles.	7–9
Multiplication/Division	Egg-stremely Equal	Students divide large sets of eggs into several equal parts.	4–8
Multiplication/Division	Field Trip	Students solve multidigit multiplication problems in a field trip environment (e.g., equal number of students on each bus; the number of tickets needed for all students.)	8–11
Multiplication/Division	Snack Time	Students use direct modeling to solve multiplication problems.	6–8
Multiplication/Division	Word Problems with Tools 5–6, 10	Students use number tools to solve single and multidigit multiplication and division problems.	8–11
Multiplication/Division	Clean the Plates	Students use skip counting to produce products that are multiples of 10s, 5s, 2s, and 3s.	7–9
Numbers: Adding and Subtracting	Barkley's Bones 1–10 and Barkley's Bones 1–20	Students determine the missing addend in $X + \underline{} = Z$ problems to feed bone treats to a dog. (Z = 10 or less)	5–8
Number: Adding and Subtracting	Double Compare 1–10 and Double Compare 1–20	Students compare sums of cards (to 10 or 20) to determine which sum is greater.	5–8

Building Blocks

Domain: Trajectory	Activity Name	Description	Age Range (in years)
Number: Adding and Subtracting	Word Problems with Tools 1–4, 7–9, 11–12	Students use number tools to solve single and multidigit addition and subtraction problems.	8–12
Number: Adding and Subtracting and Counting	Counting Activities (Road Race Counting Game, Numeral Train Game, etc.)	Students identify numerals or dot amounts (totals to 20) and move forward a corresponding number of spaces on a game board.	3–9
Number: Adding and Subtracting and Multiplication and Division	Function Machine 1–4	Students provide inputs to a function and examine the resulting outputs to determine the definition of that function. Functions include addition, subtraction, multiplication, or division.	6–12
Number: Comparing	Ordinal Construction Company	Students learn ordinal positions (1st through 10th) by moving objects between the floors of a building.	5–7
Number: Comparing	Rocket Blast 1–3	Given a number line with only initial and final endpoints labeled and a location on that line, students determine the number label for that location.	6–12
Number: Comparing and Counting	Party Time 1–3 and Party Time Free Explore	Students use party utensils to practice one-to-one correspondence, identify numerals that represent target amounts, and match object amounts to target numerals.	4–6
Number: Comparing and Multiplication and Division	Number Compare 1–5	Students compare two cards and choose the one with the greater value.	4–11
Number: Comparing, Counting, Adding and Subtracting	Pizza Pizzazz 1–5 and Pizza Pizzazz Free Explore	Students count items, match target amounts, and explore missing addends related to toppings on pizzas.	3–8
Number: Counting (Object)	Countdown Crazy	Students click digits in sequence to count down from 10 to 0.	5–7
Number: Counting (Object)	Memory Number 1–3	Students match displays containing both numerals and collections to matching displays within the framework of a "Concentration" card game.	4–6
Number: Counting (Object) and Adding and Subtracting	Dinosaur Shop 1–4 and Dinosaur Shop Free Explore	Students use toy dinosaurs to identify numerals representing target amounts, match object amounts to target numerals, add groups of objects, and find missing addends.	4–7
Number: Counting (Objects)	Book Stacks	Students "count on" (through at least one decade) from a given number as they load books onto a cart.	6–8
Number: Counting (Objects)	School Supply Shop	Students count school supplies, bundled in groups of ten to reach a target number up to 100.	6–8
Number: Counting (Objects)	Tire Recycling	Students use skip-counting by 2s and 5s to count tires as they are moved.	6–8
Number: Counting (Strategies)	Build Stairs 1–3 and Build Stairs Free Explore	Students practice counting, sequencing, and ordering by building staircases.	4–7
Number: Counting (Strategies)	Math-O-Scope	Students identify the numbers that surround a given number in the context of a 100s chart.	7–9
Number: Counting (Strategies)	Tidal Tally	Students identify missing addends (hidden objects) by counting forward from given addends (visible objects) to reach a numerical total.	6–9
Number: Counting (Verbal)	Count and Race	Students count up to 50 by adding cars to a racetrack one at a time.	3–6
Number: Counting (Verbal)	Before and After Math	Students identify and select numbers that come either just before or right after a target number.	4–7
Number: Counting (Verbal)	Kitchen Counter	Students click on objects one at a time while the numbers from one to ten are counted aloud.	3–6
Number: Subitizing	Number Snapshots 1–10	Students match numerals or dot collections to a corresponding numeral or collection given only a brief view of the goal collections.	3–12
Patterning	Marching Patterns 1–3	Students extend a linear pattern of marchers by one full repetition of an entire unit (AB, AAB, ABB, and ABC patterns).	5–7
Patterning	Pattern Planes 1–3	Students duplicate a linear pattern of flags based on an outline that serves as a guide (AB, AAB, ABB, and ABC patterns).	4–6
Patterning	Pattern Zoo 1–3 and Patterns Free Explore	Students identify a linear pattern of fruit that matches a target pattern to feed zoo animals (AB, AAB, ABB, and ABC patterns). Students explore patterning by creating rhythmic patterns of their own.	3–6

eMathTools

Number Worlds integrates *eMathTools* in appropriate lessons throughout the program. This component provides multimedia formats for demonstrating and exploring concepts and solving problems. The *eMathTools* are described below.

Data Organization and Display Tools

- **Spreadsheet Tool**—allows students to manage, display, sort, and calculate data. Links to the graphing tool for further data display
- **Graphing Tool**—displays data in pie charts or circle graphs, line graphs, bar graphs, or coordinate grids
- **Venn Diagram**—allows students to sort data visually

Measurement and Conversion Tools

- **Stopwatch**—measures in real time for development of counting and time concepts
- **Calendar**—an electronic calendar to develop concepts of time
- **Metric and Customary Conversion Tool**—converts metric and customary measurements in length, distance, mass and weight, time, temperature, and capacity
- **Estimating Proportion Tool**—allows visual representations of proportions to develop understanding of ratios, fractions, and decimals

Geometric Exploration Tools

- **Tessellations**—allows students to create tessellation patterns by rotating, coloring, and tiling shapes
- **Net Tool**—allows students to manipulate 2-D shapes and then print them to create 3-D shapes
- **Shape Tools**—explores and manipulates shapes to create designs
- **Geometry Sketch Tool**—allows drawing, manipulating, and measuring a wide variety of shapes
- **Pythagorean Theorem Tool**—launches right triangles to explore the Pythagorean Theorem

Calculation and Counting Tools

- **Calculator**—allows students to launch a calculator to perform mathematical operations
- **Function Machine**—an electronic version of a function machine that students use to solve missing variable problems
- **Multiplication and Division Table**—an interactive version of a table that highlights relationships between multipli-cation and division facts
- **Addition and Subtraction Table**—an interactive version of a table that highlights relationships between addition and subtraction facts
- **100s Chart**—an interactive version of a table that highlights patterns and relationships among numbers
- **Number Line**—an electronic number line that allows students to skip count and see the relationships among whole numbers, fractions, decimals, and percents
- **Number Stairs**—a tool to illustrate counting in units
- **Probability Tool**—uses number cubes, spinner, or tumble drum to test scenarios of probability
- **Set Tool**—allows students to visually represent and manipulate different sets of objects for a variety of counting activities
- **Base-Ten Blocks**—allows students to manipulate base-10 block for counting
- **Coins and Money**—uses visual representations of coins and money to represent counting
- **Fraction Tool**—represents fractional units for counting and understanding relationships
- **Array Tool**—presents arrays to represent multiplication and division patterns and relationships

English Learners

English learners enter schools at different grade levels and bring varying levels of English proficiency and academic preparation attained in their primary language. Some students have high levels of academic preparation that allow them to focus more on learning the terminology associated with the math skills they already know. These students have a great advantage as they already understand that mathematics has its own terminology and have a grasp of the basic skills and concepts needed to participate in advancing their math skills.

Other students may have had little schooling or interrupted school experiences, which means they need to catch up to their grade-level peers academically, and they must begin the task of English acquisition at the same time. Many schools use some form of primary language instruction or support to accelerate students' skill base while learning English. This approach contains the added advantage of being able to draw on the student's family as a source of support and inclusion.

Number Worlds and English Learners

Creating Context and Alternate Vocabulary

In addition to the specialized terminology of mathematics, academic language often includes the use of common English idiomatic expressions and content area words that have multiple meanings in English. Previewing these expressions with English learners and helping them keep track of their usage and meaning is excellent practice for extending knowledge of English for academic purposes.

In each **Number Worlds** lesson several words or expressions are noted along with a quick description of their meaning. Some of the words and expressions are math phrases, but many are examples of wider vocabulary used in word problems and direction lines in the lessons. Creating context and alternate vocabulary features highlight some of these key phrases and give teachers a brief explanation.

The inclusion of alternate vocabulary such as the words with double meanings that students encounter in the subject area brings to the attention of teachers those complexities that English speakers may already have mastered but that can make the difference between deep comprehension and confusion for English learners. Alternate vocabulary also highlights some potentially puzzling idiomatic expressions.

Spanish Cognates

One way to rapidly accelerate the acquisition of more sophisticated academic language is to take advantage of words with shared roots across languages. Words that are related by root or are borrowed from another language are called cognates. In English many math and science terms come from Greek, Latin, and Arabic roots. English learners who speak Romance languages such as Spanish have an advantage in these subjects because much of the important concept vocabulary is similar in both languages. By teaching students to look for some basic word parts in English, these students can take advantage of what they know in their primary language to augment their knowledge of English. Pronunciation of these words may be quite different, but the written words often look very similar and mean the same thing.

Learning Trajectories

Children follow natural developmental progressions in learning, developing mathematical ideas in their own way. Curriculum research has revealed sequences of activities that are effective in guiding children through these levels of thinking. These developmental paths are the basis for Building Blocks learning trajectories. Learning trajectories have three parts: a mathematical goal, a developmental path through which children develop to reach that goal, and a set of activities matched to each of those levels that help children develop the next level. Thus, each learning trajectory has levels of understanding, each more sophisticated than the last, with tasks that promote growth from one level to the next. The Building Blocks Learning Trajectories give simple labels, descriptions, and examples of each level. Complete learning trajectories describe the goals of learning, the thinking and learning processes of children at various levels, and the learning activities in which they might engage. This document provides only the developmental levels.

Learning Trajectories for Primary Grades Mathematics

Developmental Levels

Frequently Asked Questions (FAQ)

1. **Why use learning trajectories?** Learning trajectories allow teachers to build the *mathematics of children*— the thinking of children as it develops naturally. So, we know that all the goals and activities are within the developmental capacities of children. We know that each level provides a natural *developmental building block* to the next level. Finally, we know that the activities provide the *mathematical building blocks* for school success, because the research on which they are based typically involves higher-income children.

2. **When are children "at" a level?** Children are at a certain level when most of their behaviors reflect the thinking— ideas and skills—of that level. Often, they show a few behaviors from the next (and previous) levels as they learn.

3. **Can children work at more than one level at the same time?** Yes, although most children work mainly at one level or in transition between two levels (naturally, if they are tired or distracted, they may operate at a much lower level). Levels are not "absolute stages." They are "benchmarks" of complex growth that represent distinct ways of thinking. So, another way to think of them is as a sequence of different patterns of thinking. Children are continually learning, within levels and moving between them.

4. **Can children jump ahead?** Yes, especially if there are separate "sub-topics." For example, we have combined many counting competencies into one "Counting" sequence with sub-topics, such as verbal counting skills. Some children learn to count to 100 at age 6 after learning to count objects to 10 or more, some may learn that verbal skill earlier. The sub-topic of verbal counting skills would still be followed.

5. **How do these developmental levels support teaching and learning?** The levels help teachers, as well as curriculum developers, assess, teach, and sequence activities. *Teachers who understand learning trajectories and the developmental levels that are at their foundation are more effective and efficient.* Through planned teaching and also encouraging informal, incidental mathematics, teachers help children learn *at an appropriate and deep level*.

6. **Should I plan to help children develop just the levels that correspond to my children's ages?** No! The ages in the table are typical ages children develop these ideas. *But these are rough guides only—children differ widely.* Furthermore, the ages below are lower bounds of what children achieve *without instruction*. So, these are *"starting levels" not goals*. We have found that children who are provided high-quality mathematics experiences are capable of developing to levels one or more years beyond their peers.

Each column in the table below, such as "Counting," represents a main developmental progression that underlies the learning trajectory for that topic.

For some topics, there are "subtrajectories"—strands within the topic. In most cases, the names make this clear. For example, in Comparing and Ordering, some levels are about the "Comparer" levels, and others about building a "Mental Number Line." Similarly, the related subtrajectories of "Composition" and "Decomposition" are easy to distinguish. Sometimes, for clarification, subtrajectories are indicated with a note in italics after the title. For example, in Shapes, Parts and Representing are subtrajectories within the Shapes trajectory.

Clements, D. H., Sarama, J., & DiBiase, A.-M. (Eds.). (2004). *Engaging Young Children in Mathematics: Standards for Early Childhood Mathematics Education.* Mahwah, NJ: Lawrence Erlbaum Associates.

Clements, D. H., & Sarama, J. (in press). "Early Childhood Mathematics Learning." In F. K. Lester, Jr. (Ed.), *Second Handbook of Research on Mathematics Teaching and Learning.* New York: Information Age Publishing.

Learning Trajectories

Developmental Levels for Counting

The ability to count with confidence develops over the course of several years. Beginning in infancy, children show signs of understanding number. With instruction and number experience, most children can count fluently by age 8, with much progress in counting occurring in kindergarten and first grade. Most children follow a natural developmental progression in learning to count with recognizable stages or levels. This developmental path can be described as part of a learning trajectory.

Age Range	Level Name	Level	Description
1–2	Pre-Counter	1	A child at the earliest level of counting may name some numbers meaninglessly. The child may skip numbers and have no sequence.
1–2	Chanter	2	At this level a child may sing-song numbers, but without meaning.
2	Reciter	3	At this level the child verbally counts with separate words, but not necessarily in the correct order.
3	Reciter (10)	4	A child at this level can verbally count to 10 with some correspondence with objects. They may point to objects to count a few items but then lose track.
3	Corresponder	5	At this level a child can keep one-to-one correspondence between counting words and objects—at least for small groups of objects laid in a line. A corresponder may answer "how many" by recounting the objects starting over with one each time.
4	Counter (Small Numbers)	6	At around 4 years children begin to count meaningfully. They accurately count objects to 5 and answer the "how many" question with the last number counted. When objects are visible, and especially with small numbers, begins to understand cardinality. These children can count verbally to 10 and may write or draw to represent 1–5.
4	Producer—Counter To (Small Numbers)	7	The next level after counting small numbers is to count out objects up to 5 and produce a group of four objects. When asked to show four of something, for example, this child can give four objects.
4–5	Counter (10)	8	This child can count structured arrangements of objects to 10. He or she may be able to write or draw to represent 10 and can accurately count a line of nine blocks and says there are 9. A child at this level can also find the number just after or just before another number, but only by counting up from 1.
5–6	Counter and Producer—Counter to (10+)	9	Around 5 years of age children begin to count out objects accurately to 10 and then beyond to 30. They can keep track of objects that have and have not been counted, even in different arrangements. They can write or draw to represent 1 to 10 and then 20 and 30, and can give the next number to 20 or 30. These children can recognize errors in others' counting and are able to eliminate most errors in one's own counting.
5–6	Counter Backward from 10	10	Another milestone at about age 5 is being able to count backwards from 10.
6–7	Counter from N (N+1, N−1)	11	Around 6 years of age children begin to count on, counting verbally and with objects from numbers other than 1. Another noticeable accomplishment is that children can determine immediately the number just before or just after another number without having to start back at 1.
6–7	Skip-Counting by 10s to 100	12	A child at this level can count by tens to 100. They can count through decades knowing that 40 comes after 39, for example.
6–7	Counter to 100	13	A child at this level can count by ones through 100, including the decade transitions from 39 to 40, 49 to 50, and so on, starting at any number.
6–7	Counter On Using Patterns	14	At this level a child keeps track of counting acts by using numerical patterns such as tapping as he or she counts.
6–7	Skip Counter	15	The next level is when children can count by 5s and 2s with understanding.
6–7	Counter of Imagined Items	16	At this level a child can count mental images of hidden objects.
6–7	Counter On Keeping Track	17	A child at this level can keep track of counting acts numerically with the ability to count up one to four more from a given number.
6–7	Counter of Quantitative Units	18	At this level a child can count unusual units such as "wholes" when shown combinations of wholes and parts. For example when shown three whole plastic eggs and four halves, a child at this level will say there are five whole eggs.
6–7	Counter to 200	19	At this level a child counts accurately to 200 and beyond, recognizing the patterns of ones, tens, and hundreds.
7+	Number Conserver	20	A major milestone around age 7 is the ability to conserve number. A child who conserves number understands that a number is unchanged even if a group of objects is rearranged. For example, if there is a row of ten buttons, the child understands there are still ten without recounting, even if they are rearranged in a long row or a circle.
7+	Counter Forward and Back	21	A child at this level counts in either direction and recognizes that sequence of decades sequence mirrors single-digit sequence.

Developmental Levels for Comparing and Ordering Numbers

Comparing and ordering sets is a critical skill for children as they determine whether one set is larger than another to make sure sets are equal and "fair." Prekindergartners can learn to use matching to compare collections or to create equivalent collections. Finding out how many more or fewer in one collection is more demanding than simply comparing two collections. The ability to compare and order sets with fluency develops over the course of several years. With instruction and number experience, most children develop foundational understanding of number relationships and place value at ages 4 and 5. Most children follow a natural developmental progression in learning to compare and order numbers with recognizable stages or levels. This developmental path can be described as part of a learning trajectory.

Age Range	Level Name	Level	Description
2	Object Corresponder	1	At this early level a child puts objects into one-to-one correspondence, but with only intuitive understanding of resulting equivalence. For example, a child may know that each carton has a straw, but doesn't necessarily know there are the same numbers of straws and cartons.
2	Perceptual Comparer	2	At the next level a child can compare collections that are quite different in size (for example, one is at least twice the other) and know that one has more than the other. If the collections are similar, the child can compare very small collections.
2–3	First-Second Ordinal Counter	3	A child at this level can identify the first and often second objects in a sequence.
3	Nonverbal Comparer of Similar Items	4	At this level a child can identify that different organizations of the same number of small groups are equal and different from other sets. (1–4 items).
3	Nonverbal Comparer of Dissimilar Items	5	At the next level a child can match small, equal collections of dissimilar items, such as shells and dots, and show that they are the same number.
4	Matching Comparer	6	As children progress they begin to compare groups of 1–6 by matching. For example, a child gives one toy bone to every dog and says there are the same number of dogs and bones.
4	Knows-to-Count Comparer	7	A significant step occurs when the child begins to count collections to compare. At the early levels children are not always accurate when larger collection's objects are smaller in size than the objects in the smaller collection. For example, a child at this level may accurately count two equal collections, but when asked, says the collection of larger blocks has more.
4	Counting Comparer (Same Size)	8	At the next level children make accurate comparisons via counting, but only when objects are about the same size and groups are small (about 1–5).
5	Counting Comparer (5)	9	As children develop their ability to compare sets, they compare accurately by counting, even when larger collection's objects are smaller. A child at this level can figure out how many more or less.
5	Ordinal Counter	10	At the next level a child identifies and uses ordinal numbers from "first" to "tenth." For example, the child can identify who is "third in line."
5	Counting Comparer	11	At this level a child can compare by counting, even when the larger collection's objects are smaller. For example, a child can accurately count two collections and say they have the same number even if one has larger objects.
5	Mental Number Line to 10	12	At this level a child uses internal images and knowledge of number relationships to determine relative size and position. For example, the child can determine whether 4 or 9 is closer to 6.
5	Serial Orderer to 6+	13	Children demonstrate development in comparing when they begin to order lengths marked into units (1–6, then beyond). For example, given towers of cubes, this child can put them in order, 1 to 6. Later the child begins to order collections. For example, given cards with one to six dots on them, puts in order.
6	Counting Comparer (10)	14	The next level can be observed when the child compares sets by counting, even when larger collection's objects are smaller, up to 10. A child at this level can accurately count two collections of 9 each, and says they have the same number, even if one collection has larger blocks.
6	Mental Number Line to 10	15	As children move into the next level they begin to use mental rather than physical images and knowledge of number relationships to determine relative size and position. For example, a child at this level can answer which number is closer to 6, 4, or 9 without counting physical objects.
6	Serial Orderer to 6+	16	At this level a child can order lengths marked into units. For example, given towers of cubes the child can put them in order.
7	Place Value Comparer	17	Further development is made when a child begins to compare numbers with place value understandings. For example, a child at this level can explain that "63 is more than 59 because six tens is more than five tens even if there are more than three ones."

Learning Trajectories

Age Range	Level Name	Level	Description
7	Mental Number Line to 100	18	Children demonstrate the next level in comparing and ordering when they can use mental images and knowledge of number relationships, including ones embedded in tens, to determine relative size and position. For example, a child at this level when asked, "Which is closer to 45, 30 or 50?" says "45 is right next to 50, but 30 isn't."

Age Range	Level Name	Level	Description
8+	Mental Number Line to 1000s	19	About age 8 children begin to use mental images of numbers up to 1,000 and knowledge of number relationships, including place value, to determine relative size and position. For example, when asked, "Which is closer to 3,500—2,000 or 7,000?" a child at this level says "70 is double 35, but 20 is only fifteen from 35, so twenty hundreds, 2,000, is closer."

Developmental Levels for Recognizing Number and Subitizing (Instantly Recognizing)

The ability to recognize number values develops over the course of several years and is a foundational part of number sense. Beginning at about age 2, children begin to name groups of objects. The ability to instantly know how many are in a group, called *subitizing*, begins at about age 3. By age 8, with instruction and number experience, most children can identify groups of items and use place values and multiplication skills to count them. Most children follow a natural developmental progression in learning to count with recognizable stages or levels. This developmental path can be described as part of a learning trajectory.

Age Range	Level Name	Level	Description
2	Small Collection Namer	1	The first sign of a child's ability to subitize occurs when the child can name groups of one to two, sometimes three. For example, when shown a pair of shoes, this young child says, "Two shoes."
3	Nonverbal Subitizer	2	The next level occurs when shown a small collection (one to four) only briefly, the child can put out a matching group nonverbally, but cannot necessarily give the number name telling how many. For example, when four objects are shown for only two seconds, then hidden, child makes a set of four objects to "match."
3	Maker of Small Collections	3	At the next level a child can nonverbally make a small collection (no more than five, usually one to three) with the same number as another collection. For example, when shown a collection of three, makes another collection of three.
4	Perceptual Subitizer to 4	4	Progress is made when a child instantly recognizes collections up to four when briefly shown and verbally names the number of items. For example, when shown four objects briefly, says "four."
5	Perceptual Subitizer to 5	5	The next level is the ability to instantly recognize briefly shown collections up to five and verbally name the number of items. For example, when shown five objects briefly, says "five."

Age Range	Level Name	Level	Description
5	Conceptual Subitizer to 5+	6	At the next level the child can verbally label all arrangements to five shown only briefly. For example, a child at this level would say, "I saw 2 and 2 and so I saw 4."
5	Conceptual Subitizer to 10	7	The next step is when the child can verbally label most briefly shown arrangements to six, then up to ten, using groups. For example, a child at this level might say, "In my mind, I made two groups of 3 and one more, so 7."
6	Conceptual Subitizer to 20	8	Next, a child can verbally label structured arrangements up to twenty, shown only briefly, using groups. For example, the child may say, "I saw three 5s, so 5, 10, 15."
7	Conceptual Subitizer with Place Value and Skip Counting	9	At the next level a child is able to use skip counting and place value to verbally label structured arrangements shown only briefly. For example, the child may say, "I saw groups of tens and twos, so 10, 20, 30, 40, 42, 44, 46...46!"
8+	Conceptual Subitizer with Place Value and Multiplication	10	As children develop their ability to subitize, they use groups, multiplication, and place value to verbally label structured arrangements shown only briefly. At this level a child may say, "I saw groups of tens and threes, so I thought, five tens is 50 and four 3s is 12, so 62 in all."

Developmental Levels for Composing Number
(Knowing Combinations of Numbers)

Composing and decomposing are combining and separating operations that allow children to build concepts of "parts" and "wholes." Most prekindergartners can "see" that two items and one item make three items. Later, children learn to separate a group into parts in various ways and then to count to produce all of the number "partners" of a given number. Eventually children think of a number and know the different addition facts that make that number. Most children follow a natural developmental progression in learning to compose and decompose numbers with recognizable stages or levels. This developmental path can be described as part of a learning trajectory.

Age Range	Level Name	Level	Description
4	Pre-Part-Whole Recognizer	1	At the earliest levels of composing a child only nonverbally recognizes parts and wholes. For example, When shown four red blocks and two blue blocks, a young child may intuitively appreciate that "all the blocks" include the red and blue blocks, but when asked how many there are in all, may name a small number, such as 1.
5	Inexact Part-Whole Recognizer	2	A sign of development in composing is that the child knows that a whole is bigger than parts, but does not accurately quantify. For example, when shown four red blocks and two blue blocks and asked how many there are in all, names a "large number," such as 5 or 10.
5	Composer to 4, then 5	3	The next level is that a child begins to know number combinations. A child at this level quickly names parts of any whole, or the whole given the parts. For example, when shown four, then one is secretly hidden, and then is shown the three remaining, quickly says "1" is hidden.

Age Range	Level Name	Level	Description
6	Composer to 7	4	The next sign of development is when a child knows number combinations to totals of seven. A child at this level quickly names parts of any whole, or the whole given parts and can double numbers to 10. For example, when shown six, then four are secretly hidden, and shown the two remaining, quickly says "4" are hidden.
6	Composer to 10	5	The next level is when a child knows number combinations to totals of 10. A child at this level can quickly name parts of any whole, or the whole given parts and can double numbers to 20. For example, this child would be able to say "9 and 9 is 18."
7	Composer with Tens and Ones	6	At the next level the child understands two-digit numbers as tens and ones; can count with dimes and pennies; and can perform two-digit addition with regrouping. For example, a child at this level can explain, "17 and 36 is like 17 and 3, which is 20, and 33, which is 53."

Developmental Levels for Adding and Subtracting

Learning single-digit addition and subtraction is generally characterized as "learning math facts." It is assumed that children must memorize these facts, yet research has shown that addition and subtraction have their roots in counting, counting on, number sense, the ability to compose and decompose numbers, and place value. Research has shown that learning methods for adding and subtracting with understanding is much more effective than rote memorization of seemingly isolated facts. Most children follow an observable developmental progression in learning to add and subtract numbers with recognizable stages or levels. This developmental path can be described as part of a learning trajectory.

Age Range	Level Name	Level	Description
1	Pre +/−	1	At the earliest level a child shows no sign of being able to add or subtract.
3	Nonverbal +/−	2	The first inkling of development is when a child can add and subtract very small collections nonverbally. For example, when shown two objects, then one object going under a napkin, the child identifies or makes a set of three objects to "match."

Age Range	Level Name	Level	Description
4	Small Number +/−	3	The next level of development is when a child can find sums for joining problems up to 3 + 2 by counting all with objects. For example, when asked, "You have 2 balls and get 1 more. How many in all?" counts out 2, then counts out 1 more, then counts all 3: "1, 2, 3, 3!"

Learning Trajectories

Age Range	Level Name	Level	Description
5	Find Result +/−	4	**Addition** Evidence of the next level in addition is when a child can find sums for joining (you had 3 apples and get 3 more, how many do you have in all?) and part-part-whole (there are 6 girls and 5 boys on the playground, how many children were there in all?) problems by direct modeling, counting all, with objects. For example, when asked, "You have 2 red balls and 3 blue balls. How many in all?" the child counts out 2 red, then counts out 3 blue, then counts all 5. **Subtraction** In subtraction, a child at this level can also solve take-away problems by separating with objects. For example, when asked, "You have 5 balls and give 2 to Tom. How many do you have left?" the child counts out 5 balls, then takes away 2, and then counts the remaining 3.
5	Find Change +/−	5	**Addition** At the next level a child can find the missing addend ($5 + \_\_ = 7$) by adding on objects. For example, when asked, "You have 5 balls and then get some more. Now you have 7 in all. How many did you get?" the child counts out 5, then counts those 5 again starting at 1, then adds more, counting "6, 7," then counts the balls added to find the answer, 2. **Subtraction** Compares by matching in simple situations. For example, when asked, "Here are 6 dogs and 4 balls. If we give a ball to each dog, how many dogs won't get a ball?" a child at this level counts out 6 dogs, matches 4 balls to 4 of them, then counts the 2 dogs that have no ball.
5	Make It N +/−	6	A significant advancement in addition occurs when a child is able to count on. This child can add on objects to make one number into another, without counting from 1. For example, when asked, "This puppet has 4 balls but she should have 6. Make it 6," puts up 4 fingers on one hand, immediately counts up from 4 while putting up two fingers on the other hand, saying, "5, 6" and then counts or recognizes the two fingers.
6	Counting Strategies +/−	7	The next level occurs when a child can find sums for joining (you had 8 apples and get 3 more...) and part-part-whole (6 girls and 5 boys...) problems with finger patterns or by adding on objects or counting on. For example, when asked "How much is 4 and 3 more?" the child answers "4 . . . 5, 6, 7 [uses rhythmic or finger pattern]. 7!" Children at this level also can solve missing addend ($3 + \_\_ = 7$) or compare problems by counting on. When asked, for example, "You have 6 balls. How many more would you need to have 8?" the child says, "6, 7 [puts up first finger], 8 [puts up second finger]. 2!"
6	Part-Whole +/−	8	Further development has occurred when the child has part-whole understanding. This child can solve all problem types using flexible strategies and some derived facts (for example, "5 + 5 is 10, so 5 + 6 is 11"), sometimes can do start unknown ($\_\_ + 6 = 11$), but only by trial and error. This child when asked, "You had some balls. Then you get 6 more. Now you have 11 balls. How many did you start with?" lays out 6, then 3 more, counts and gets 9. Puts 1 more with the 3, says 10, then puts 1 more. Counts up from 6 to 11, then recounts the group added, and says, "5!"
6	Numbers-in-Numbers +/−	9	Evidence of the next level is when a child recognizes that a number is part of a whole and can solve problems when the start is unknown ($\_\_ + 4 = 9$) with counting strategies. For example, when asked, "You have some balls, then you get 4 more balls, now you have 9. How many did you have to start with?" this child counts, putting up fingers, "5, 6, 7, 8, 9." Looks at fingers, and says, "5!"
7	Deriver +/−	10	At the next level a child can use flexible strategies and derived combinations (for example, "7 + 7 is 14, so 7 + 8 is 15") to solve all types of problems. For example, when asked, "What's 7 plus 8?" this child thinks: $7 + 8 □ 7 + [7 + 1] □ [7 + 7] + 1 = 14 + 1 = 15$. A child at this level can also solve multidigit problems by incrementing or combining tens and ones. For example, when asked "What's 28 + 35?" this child thinks: 20 + 30 = 50; +8 = 58; 2 more is 60, 3 more is 63. Combining tens and ones: 20 + 30 = 50. 8 + 5 is like 8 plus 2 and 3 more, so, it's 13—50 and 13 is 63.
8+	Problem Solver +/−	11	As children develop their addition and subtraction abilities, they can solve all types of problems by using flexible strategies and many known combinations. For example, when asked, "If I have 13 and you have 9, how could we have the same number?" this child says, "9 and 1 is 10, then 3 more to make 13. 1 and 3 is 4. I need 4 more!"
8+	Multidigit +/−	12	Further development is evidenced when children can use composition of tens and all previous strategies to solve multidigit +/− problems. For example, when asked, "What's 37 − 18?" this child says, "I take 1 ten off the 3 tens; that's 2 tens. I take 7 off the 7. That's 2 tens and 0 . . . 20. I have one more to take off. That's 19." Another example would be when asked, "What's 28 + 35?" thinks, 30 + 35 would be 65. But it's 28, so it's 2 less . . . 63.

Developmental Levels for Multiplying and Dividing

Multiplication and division builds on addition and subtraction understandings and is dependent upon counting and place value concepts. As children begin to learn to multiply they make equal groups and count them all. They then learn skip counting and derive related products from products they know. Finding and using patterns aids in learning multiplication and division facts with understanding. Children typically follow an observable developmental progression in learning to multiply and divide numbers with recognizable stages or levels. This developmental path can be described as part of a learning trajectory.

Age Range	Level Name	Level	Description
2	Nonquantitive Sharer "Dumper"	1	Multiplication and division concepts begin very early with the problem of sharing. Early evidence of these concepts can be observed when a child dumps out blocks and gives some (not an equal number) to each person.
3	Beginning Grouper and Distributive Sharer	2	Progression to the next level can be observed when a child is able to make small groups (fewer than 5). This child can share by "dealing out," but often only between two people, although he or she may not appreciate the numerical result. For example, to share four blocks, this child gives each person a block, checks each person has one, and repeats this.
4	Grouper and Distributive Sharer	3	The next level occurs when a child makes small equal groups (fewer than 6). This child can deal out equally between two or more recipients, but may not understand that equal quantities are produced. For example, the child shares 6 blocks by dealing out blocks to herself and a friend 1 at a time.
5	Concrete Modeler ×/÷	4	As children develop, they are able to solve small-number multiplying problems by grouping—making each group and counting all. At this level a child can solve division/sharing problems with informal strategies, using concrete objects—up to twenty objects and two to five people—although the child may not understand equivalence of groups. For example, the child distributes twenty objects by dealing out two blocks to each of five people, then one to each, until blocks are gone.
6	Parts and Wholes ×/÷	5	A new level is evidenced when the child understands the inverse relation between divisor and quotient. For example, this child understands "If you share with more people, each person gets fewer."
7	Skip Counter ×/÷	6	As children develop understanding in multiplication and division they begin to use skip counting for multiplication and for measurement division (finding out how many groups). For example, given twenty blocks, four to each person, and asked how many people, the child skip counts by 4, holding up one finger for each count of 4. A child at this level also uses trial and error for partitive division (finding out how many in each group). For example, given twenty blocks, five people, and asked how many should each get, this child gives three to each, then one more, then one more.
8+	Deriver ×/÷	7	At the next level children use strategies and derived combinations and solve multidigit problems by operating on tens and ones separately. For example, a child at this level may explain "7 × 6, five 7s is 35, so 7 more is 42."
8+	Array Quantifier	8	Further development can be observed when a child begins to work with arrays. For example, given 7 × 4 with most of 5 × 4 covered, a child at this level may say, "There's eight in these two rows, and five rows of four is 20, so 28 in all."
8+	Partitive Divisor	9	The next level can be observed when a child is able to figure out how many are in each group. For example, given twenty blocks, five people, and asked how many should each get, a child at this level says "four, because 5 groups of 4 is 20."
8+	Multidigit ×/÷	10	As children progress they begin to use multiple strategies for multiplication and division, from compensating to paper-and-pencil procedures. For example, a child becoming fluent in multiplication might explain that "19 times 5 is 95, because twenty 5s is 100, and one less 5 is 95."

Learning Trajectories

Developmental Levels for Measuring

Measurement is one of the main real-world applications of mathematics. Counting is a type of measurement, determining how many items are in a collection. Measurement also involves assigning a number to attributes of length, area, and weight. Prekindergarten children know that mass, weight, and length exist, but they don't know how to reason about these or to accurately measure them.

As children develop their understanding of measurement, they begin to use tools to measure and understand the need for standard units of measure. Children typically follow an observable developmental progression in learning to measure with recognizable stages or levels. This developmental path can be described as part of a learning trajectory.

Age Range	Level Name	Level	Description
3	Length Quantity Recognizer	1	At the earliest level children can identify length as an attribute. For example, they might say, "I'm tall, see?"
4	Length Direct Comparer	2	In the next level children can physically align two objects to determine which is longer or if they are the same length. For example, they can stand two sticks up next to each other on a table and say, "This one's bigger."
5	Indirect Length Comparer	3	A sign of further development is when a child can compare the length of two objects by representing them with a third object. For example, a child might compare length of two objects with a piece of string. Additional evidence of this level is that when asked to measure, the child may assign a length by guessing or moving along a length while counting (without equal length units). The child may also move a finger along a line segment, saying 10, 20, 30, 31, 32.
5	Serial Orderer to 6+	4	At the next level a child can order lengths, marked in one to six units. For example, given towers of cubes, a child at this level puts in order, 1 to 6.
6	End-to-End Length Measurer	5	At the next level the child can lay units end-to-end, although he or she may not see the need for equal-length units. For example, a child might lay 9-inch cubes in a line beside a book to measure how long it is.

Age Range	Level Name	Level	Description
7	Length Unit Iterater	6	A significant change occurs when a child can use a ruler and see the need for identical units.
7	Length Unit Relater	7	At the next level a child can relate size and number of units. For example, the child may explain, "If you measure with centimeters instead of inches, you'll need more of them, because each one is smaller."
8	Length Measurer	8	As children develop measurement ability they begin to measure, knowing the need for identical units, the relationships between different units, partitions of unit, and zero point on rulers. At this level the child also begins to estimate. The child may explain, "I used a meter stick three times, then there was a little left over. So, I lined it up from 0 and found 14 centimeters. So, it's 3 meters, 14 centimeters in all."
8	Conceptual Ruler Measurer	9	Further development in measurement is evidenced when a child possesses an "internal" measurement tool. At this level the child mentally moves along an object, segmenting it, and counting the segments. This child also uses arithmetic to measure and estimates with accuracy. For example, a child at this level may explain, "I imagine one meterstick after another along the edge of the room. That's how I estimated the room's length is 9 meters."

C24 **Number Worlds** • Learning Trajectories

Developmental Levels for Recognizing Geometric Shapes

Geometric shapes can be used to represent and understand objects. Analyzing, comparing, and classifying shapes helps create new knowledge of shapes and their relationships. Shapes can be decomposed or composed into other shapes. Through their everyday activity, children build both intuitive and explicit knowledge of geometric figures. Most children can recognize and name basic two-dimensional shapes at 4 years of age. However, young children can learn richer concepts about shape if they have varied examples and nonexamples of shape, discussions about shapes and their characteristics, a wide variety of shape classes, and interesting tasks. Children typically follow an observable developmental progression in learning about shapes with recognizable stages or levels. This developmental path can be described as part of a learning trajectory.

Age Range	Level Name	Level	Description
2	Shape Matcher—	1	The earliest sign of understanding shape is when a child can match basic shapes (circle, square, typical triangle) with the same size and orientation. Example: Matches ☐ to ☐. A sign of development is when a child can match basic shapes with different sizes. Example: Matches ☐ to ☐. The next sign of development is when a child can match basic shapes with different orientations. Example: Matches ☐ to ◇.
3	Shape Prototype Recognizer and Identifier	2	A sign of development is when a child can recognize and name prototypical circle, square, and, less often, a typical triangle. For example, the child names this a square ☐. Some children may name different sizes, shapes, and orientations of rectangles, but also accept some shapes that look rectangular but are not rectangles. Children name these shapes "rectangles" (including the non-rectangular parallelogram).
3	Shape Matcher— More Shapes	3	As children develop understanding of shape, they can match a wider variety of shapes with the same size and orientation. —4 Matches wider variety of shapes with different sizes and orientations. Matches these shapes ∣ ╱. —5 Matches combinations of shapes to each other. Matches these shapes ⊛ ⊛.
4	Shape Recognizer— Circles, Squares, and Triangles	4	The next sign of development is when a child can recognize some nonprototypical squares and triangles and may recognize some rectangles, but usually not rhombi (diamonds). Often, the child doesn't differentiate sides/corners. The child at this level may name these as triangles △ △.
4	Constructor of Shapes from Parts – Looks Like	5	A significant sign of development is when a child represents a shape by making a shape "look like" a goal shape. For example, when asked to make a triangle with sticks, the child creates the following ☐.
5	Shape Recognizer— All Rectangles	6	As children develop understanding of shape, they recognize more rectangle sizes, shapes, and orientations of rectangles. For example, a child at this level correctly names these shapes "rectangles".
5	Side Recognizer	7	A sign of development is when a child recognizes parts of shapes and identifies sides as distinct geometric objects. For example, when asked what this shape is △, the child says it is a quadrilateral (or has four sides) after counting and running a finger along the length of each side.
5	Angle Recognizer	8	At the next level a child can recognize angles as separate geometric objects. For example, when asked, "Why is this a triangle," says, "It has three angles" and counts them, pointing clearly to each vertex (point at the corner).
5	Shape Recognizer	9	As children develop they are able to recognize most basic shapes and prototypical examples of other shapes, such as hexagon, rhombus (diamond), and trapezoid. For example, a child can correctly identify and name all the following shapes.
6	Shape Identifier	10	At the next level the child can name most common shapes, including rhombi, "ellipses-is-not-circle." A child at this level implicitly recognizes right angles, so distinguishes between a rectangle and a parallelogram without right angles. Correctly names all the following shapes:
6	Angle Matcher	11	A sign of development is when the child can match angles concretely. For example, given several triangles, finds two with the same angles by laying the angles on top of one another.

Learning Trajectories

Age Range	Level Name	Level	Description
7	Parts of Shapes Identifier	12	At the next level the child can identify shapes in terms of their components. For example, the child may say, "No matter how skinny it looks, that's a triangle because it has three sides and three angles."
7	Constructor of Shapes from Parts Exact	13	A significant step is when the child can represent a shape with completely correct construction, based on knowledge of components and relationships. For example, asked to make a triangle with sticks, creates the following:
8	Shape Class Identifier	14	As children develop, they begin to use class membership (for example, to sort), not explicitly based on properties. For example, a child at this level may say, "I put the triangles over here, and the quadrilaterals, including squares, rectangles, rhombi, and trapezoids, over there."
8	Shape Property Identifier	15	At the next level a child can use properties explicitly. For example, a child may say, "I put the shapes with opposite sides parallel over here, and those with four sides but not both pairs of sides parallel over there."

Age Range	Level Name	Level	Description
8	Angle Size Comparer	16	The next sign of development is when a child can separate and compare angle sizes. For example, the child may say, "I put all the shapes that have right angles here, and all the ones that have bigger or smaller angles over there."
8	Angle Measurer	17	A significant step in development is when a child can use a protractor to measure angles.
8	Property Class Identifier	18	The next sign of development is when a child can use class membership for shapes (for example, to sort or consider shapes "similar") explicitly based on properties, including angle measure. For example, the child may say, "I put the equilateral triangles over here, and the right triangles over here."
8	Angle Synthesizer	19	As children develop understanding of shape, they can combine various meanings of angle (turn, corner, slant). For example, a child at this level could explain, "This ramp is at a 45° angle to the ground."

Developmental Levels for Composing Geometric Shapes

Children move through levels in the composition and decomposition of two-dimensional figures. Very young children cannot compose shapes but then gain ability to combine shapes into pictures, synthesize combinations of shapes into new shapes, and eventually substitute and build different kinds of shapes. Children typically follow an observable developmental progression in learning to compose shapes with recognizable stages or levels. This developmental path can be described as part of a learning trajectory.

Age Range	Level Name	Level	Description
2	Pre-Composer	1	The earliest sign of development is when a child can manipulate shapes as individuals, but is unable to combine them to compose a larger shape. Make a Picture Outline Puzzle
3	Pre-DeComposer	2	At the next level a child can decompose shapes, but only by trial and error. For example, given only a hexagon, the child can break it apart to make this simple picture by trial and error:

Age Range	Level Name	Level	Description
4	Piece Assembler	3	Around age 4 a child can begin to make pictures in which each shape represents a unique role (for example, one shape for each body part) and shapes touch. A child at this level can fill simple outline puzzles using trial and error. Make a Picture Outline Puzzle

Age Range	Level Name	Level	Description
5	Picture Maker	4	As children develop they are able to put several shapes together to make one part of a picture (for example, two shapes for one arm). A child at this level uses trial and error and does not anticipate creation of the new geometric shape. The child can choose shapes using "general shape" or side length and fill "easy" outline puzzles that suggest the placement of each shape (but note below that the child is trying to put a square in the puzzle where its right angles will not fit). Make a Picture Outline Puzzle
5	Simple Decomposer	5	A significant step occurs when the child is able to decompose ("take apart" into smaller shapes) simple shapes that have obvious clues as to their decomposition.
5	Shape Composer	6	A sign of development is when a child composes shapes with anticipation ("I know what will fit!"). A child at this level chooses shapes using angles as well as side lengths. Rotation and flipping are used intentionally to select and place shapes. For example, in the outline puzzle below, all angles are correct, and patterning is evident. Make a Picture Outline Puzzle
6	Substitution Composer	7	A sign of development is when a child is able to make new shapes out of smaller shapes and uses trial and error to substitute groups of shapes for other shapes to create new shapes in different ways. For example, the child can substitute shapes to fill outline puzzles in different ways.

Age Range	Level Name	Level	Description
6	Shape Decomposer (with Help)	8	As children develop they can decompose shapes by using imagery that is suggested and supported by the task or environment. For example, given hexagons, the child at this level can break it apart to make this shape:
7	Shape Composite Repeater	9	The next level is demonstrated when the child can construct and duplicate units of units (shapes made from other shapes) intentionally, and understands each as being both multiple small shapes and one larger shape. For example, the child may continue a pattern of shapes that leads to tiling.
7	Shape Decomposer with Imagery	10	A significant sign of development is when a child is able to decompose shapes flexibly by using independently generated imagery. For example, given hexagons, the child can break it apart to make shapes such as these:
8	Shape Composer— Units of Units	11	Children demonstrate further understanding when they are able to build and apply units of units (shapes made from other shapes). For example, in constructing spatial patterns the child can extend patterning activity to create a tiling with a new unit shape—a unit of unit shapes that he or she recognizes and consciously constructs. For example, the child builds Ts out of four squares, uses four Ts to build squares, and uses squares to tile a rectangle.
8	Shape DeComposer with Units of Units	12	As children develop understanding of shape they can decompose shapes flexibly by using independently generated imagery and planned decompositions of shapes that themselves are decompositions. For example, given only squares, a child at this level can break them apart—and then break the resulting shapes apart again—to make shapes such as these:

Learning Trajectories

Developmental Levels for Comparing Geometric Shapes

As early as 4 years of age children can create and use strategies, such as moving shapes to compare their parts or to place one on top of the other for judging whether two figures are the same shape. From Pre-K to Grade 2 they can develop sophisticated and accurate mathematical procedures for comparing geometric shapes. Children typically follow an observable developmental progression in learning about how shapes are the same and different with recognizable stages or levels. This developmental path can be described as part of a learning trajectory.

Age Range	Level Name	Level	Description
3	"Same Thing" Comparer	1	The first sign of understanding is when the child can compare real-world objects. For example, the child says two pictures of houses are the same or different.
4	"Similar" Comparer	2	The next sign of development occurs when the child judges two shapes the same if they are more visually similar than different. For example, the child may say, "These are the same. They are pointy at the top."
4	Part Comparer	3	At the next level a child can say that two shapes are the same after matching one side on each. For example, "These are the same" (matching the two sides).
4	Some Attributes Comparer	4	As children develop they look for differences in attributes, but may examine only part of a shape. For example, a child at this level may say, "These are the same" (indicating the top halves of the shapes are similar by laying them on top of each other).

Age Range	Level Name	Level	Description
5	Most Attributes Comparer	5	At the next level the child looks for differences in attributes, examining full shapes, but may ignore some spatial relationships. For example, a child may say, "These are the same."
7	Congruence Determiner	6	A sign of development is when a child determines congruence by comparing all attributes and all spatial relationships. For example, a child at this level says that two shapes are the same shape and the same size after comparing every one of their sides and angles.
7	Congruence Superposer	7	As children develop understanding they can move and place objects on top of each other to determine congruence. For example, a child at this level says that two shapes are the same shape and the same size after laying them on top of each other.
8	Congruence Representer	8	Continued development is evidenced as children refer to geometric properties and explain transformations. For example, a child at this level may say, "These must be congruent, because they have equal sides, all square corners, and I can move them on top of each other exactly."

Developmental Levels for Spatial Sense and Motions

Infants and toddlers spend a great deal of time exploring space and learning about the properties and relations of objects in space. Very young children know and use the shape of their environment in navigation activities. With guidance they can learn to "mathematize" this knowledge. They can learn about direction, perspective, distance, symbolization, location, and coordinates. Children typically follow an observable developmental progression in developing spatial sense with recognizable stages or levels. This developmental path can be described as part of a learning trajectory.

Age Range	Level Name	Level	Description
4	Simple Turner	1	An early sign of spatial sense is when a child mentally turns an object to perform easy tasks. For example, given a shape with the top marked with color, correctly identifies which of three shapes it would look like if it were turned "like this" (90 degree turn demonstrated) before physically moving the shape.

Age Range	Level Name	Level	Description
5	Beginning Slider, Flipper, Turner	2	The next sign of development is when a child can use the correct motions, but is not always accurate in direction and amount. For example, a child at this level may know a shape has to be flipped to match another shape, but flips it in the wrong direction.

Age Range	Level Name	Level	Description
6	Slider, Flipper, Turner	3	As children develop spatial sense they can perform slides and flips, often only horizontal and vertical, by using manipulatives. For example, a child at this level can perform turns of 45, 90, and 180 degrees and knows a shape must be turned 90 degrees to the right to fit into a puzzle.
7	Diagonal Mover	4	A sign of development is when a child can perform diagonal slides and flips. For example, a child at this level knows a shape must be turned or flipped over an oblique line (45 degree orientation) to fit into a puzzle.
8	Mental Mover	5	Further signs of development occur when a child can predict results of moving shapes using mental images. A child at this level may say, "If you turned this 120 degrees, it would be just like this one."

Developmental Levels for Patterning and Early Algebra

Algebra begins with a search for patterns. Identifying patterns helps bring order, cohesion, and predictability to seemingly unorganized situations and allows one to make generalizations beyond the information directly available. The recognition and analysis of patterns are important components of the young child's intellectual development because they provide a foundation for the development of algebraic thinking. Although prekindergarten children engage in pattern-related activities and recognize patterns in their everyday environment, research has revealed that an abstract understanding of patterns develops gradually during the early childhood years. Children typically follow an observable developmental progression in learning about patterns with recognizable stages or levels. This developmental path can be described as part of a learning trajectory.

Age Range	Level Name	Level	Description
2	Pre-Patterner	1	A child at the earliest level does not recognize patterns. For example, a child may name a striped shirt with no repeating unit a "pattern."
3	Pattern Recognizer	2	At the next level the child can recognize a simple pattern. For example, a child at this level may say, "I'm wearing a pattern" about a shirt with black, white, black, white stripes.
3–4	Pattern Fixer	3	A sign of development is when the child fills in a missing element of a pattern. For example, given objects in a row with one missing, the child can identify and fill in the missing element.
4	Pattern Duplicator AB	3	A sign of development is when the child can duplicate an ABABAB pattern, although the child may have to work close to the model pattern. For example, given objects in a row, ABABAB, makes their own ABBABBABB row in a different location.

Age Range	Level Name	Level	Description
4	Pattern Extender AB	4	At the next level the child is able to extend AB repeating patterns.
4	Pattern Duplicator	4	At this level the child can duplicate simple patterns (not just alongside the model pattern). For example, given objects in a row, ABBABBABB, makes their own ABBABBABB row in a different location.
5	Pattern Extender	5	A sign of development is when the child can extend simple patterns. For example, given objects in a row, ABBABBABB, adds ABBABB to the end of the row.
7	Pattern Unit Recognizer	7	At this level a child can identify the smallest unit of a pattern. For example, given objects in a ABBAB_BABB patterns, identifies the core unit of the pattern as ABB.

Learning Trajectories

Developmental Levels for Classifying and Analyzing Data

Data analysis contains one big idea: classifying, organizing, representing, and using information to ask and answer questions. The developmental continuum for data analysis includes growth in classifying and counting to sort objects and quantify their groups.... Children eventually become capable of simultaneously classifying and counting, for example, counting the number of colors in a group of objects.

Children typically follow an observable developmental progression in learning about patterns with recognizable stages or levels. This developmental path can be described as part of a learning trajectory.

Age Range	Level Name	Level	Description
2	Similarity Recognizer	1	The first sign that a child can classify is when he or she recognizes, intuitively, two or more objects as "similar" in some way. For example, "that's another doggie."
2	Informal Sorter	2	A sign of development is when a child places objects that are alike on some attribute together, but switches criteria and may use functional relationships are the basis for sorting. A child at this level might stack blocks of the same shape or put a cup with its saucer.
3	Attribute Identifier	3	The next level is when the child names attributes of objects and places objects together with a given attribute, but cannot then move to sorting by a new rule. For example, the child may say, "These are both red."
4	Attribute Sorter	4	At the next level the child sorts objects according to a given attributes, forming categories, but may switch attributes during the sorting. A child at this stage can switch rules for sorting if guided. For example, the child might start putting red beads on a string, but switches to the spheres of different colors.
5	Consistent Sorter	5	A sign of development is when the child can sort consistently by a given attribute. For example, the child might put several identical blocks together.
6	Exhaustive Sorter	6	At the next level, the child can sort consistently and exhaustively by an attribute, given or created. This child can use terms "some" and "all" meaningfully. For example, a child at this stage would be able to find all the attribute blocks of a certain size and color.
6	Multiple Attribute Sorter	7	A sign of development is when the child can sort consistently and exhaustively by more than one attribute, sequentially. For example, a child at this level, can put all the attribute blocks together by color, then by shape.
7	Classifier and Counter	8	At the next level, the child is capable of simultaneously classifying and counting. For example, the child counts the number of colors in a group of objects.
7	List Grapher	9	In the early stage of graphing, the child graphs by simply listing all cases. For example, the child may list each child in the class and each child's response to a question.
8+	Multiple Attribute Classifier	10	A sign of development is when the child can intentionally sort according to multiple attributes, naming and relating the attributes. This child understands that objects could belong to more than one group. For example, the child can complete a two-dimensional classification matrix or forming subgroups within groups.
8+	Classifying Grapher	11	At the next level the child can graph by classifying data (e.g., responses) and represent it according to categories. For example, the child can take a survey, classify the responses, and graph the result.
8+	Classifier	12	At sign of development is when the child creates complete, conscious classifications logically connected to a specific property. For example, a child at this level gives definition of a class in terms of a more general class and one or more specific differences and begins to understand the inclusion relationship.
8+	Hierarchical Classifier	13	At the next level, the child can perform hierarchical classifications. For example, the child recognizes that all squares are rectangles, but not all rectangles are squares.
8+	Data Representer	14	Signs of development are when the child organizes and displays data through both simple numerical summaries such as counts, tables, and tallies, and graphical displays, including picture graphs, line plots, and bar graphs. At this level the child creates graphs and tables, compares parts of the data, makes statements about the data as a whole, and determines whether the graphs answer the questions posed initially.

Trajectory Progress Chart

Student's Name _____

Number

Age Range	Counting	Comparing and Ordering Number	Recognizing Number and Subitizing (instantly recognizing)	Composing Number (knowing combinations of numbers)	Adding and Subtracting	Multiplying and Dividing (sharing)
1 year	___ Pre-Counter ___ Chanter				___ Pre +/−	
2	___ Reciter	___ Object Corresponder ___ Perceptual Comparer	___ Small Collection Namer			___ Nonquantitative Sharer
3	___ Reciter (10) ___ Corresponder	___ First-Second Ordinal Counter ___ Nonverbal Comparer of Similar Items (1–4 items)	___ Nonverbal Subitizer ___ Maker of Small Collections		___ Nonverbal +/−	___ Beginning Grouper and Distributive Sharer
4	___ Counter (small numbers) ___ Producer (small numbers) ___ Counter (10)	___ Nonverbal Comparer of Dissimilar Items ___ Matching Comparer ___ Knows-to-Count Comparer ___ Counting Comparer (same size)	___ Perceptual Subitizer to 4	___ Pre-Part-Whole Recognizer	___ Small Number +/−	___ Grouper and Distributive Sharer
5	___ Counter and Producer (10+) ___ Counter Backward from 10	___ Counting Comparer (5) ___ Ordinal Counter	___ Perceptual Subitizer to 5 ___ Conceptual Subitizer to 5+ ___ Conceptual Subitizer to 10	___ Inexact Part-Whole Recognizer ___ Composer to 4, then 5	___ Find Result +/− ___ Find Change +/− ___ Make It N +/−	___ Concrete Modeler ×/÷
6	___ Counter from N (N+1, N−1) ___ Skip Counter by tens to 100 ___ Counter to 100 ___ Counter On Using Patterns ___ Skip Counter ___ Counter of Imagined Items ___ Counter On Keeping Track ___ Counter of Quantitative Units ___ Counter to 200	___ Counting Comparer (10) ___ Mental Number Line to 10 ___ Serial Orderer to 6+	___ Conceptual Subitizer to 20	___ Composer to 7 ___ Composer to 10	___ Counting Strategies +/− ___ Part-Whole +/−	___ Parts and Wholes ×/÷
7	___ Number Conserver ___ Counter Forward and Back	___ Place Value Comparer ___ Mental Number Line to 100	___ Conceptual Subitizer with Place Value and Skip Counting	___ Composer with Tens and Ones	___ Numbers-in-Numbers +/− ___ Deriver +/−	___ Skip Counter ×/÷
8+		___ Mental Number Line to 1,000s	___ Conceptual Subitizer with Place Value and Multiplication		___ Problem Solver +/− ___ Multidigit +/−	___ Deriver ×/÷ ___ Array Quantifier ___ Partitive Divisor ___ Multidigit ×/÷

Number Worlds • Trajectory Progress Chart   C31

Trajectory Progress Chart

Student's Name _____

Geometry

Age Range	Shapes	Composing Shapes	Comparing Shapes	Motions and Spatial Sense	Measuring	Patterning	Classifying and Analyzing Data
2 years	___ Shape Matcher—Identical ___ —Sizes ___ —Orientations					___ Pre-Patterner	___ Similarity Recognizer ___ Informal Sorter
3	___ Shape Recognizer—Typical ___ Shape Matcher—More Shapes ___ —Sizes and Orientations ___ —Combinations	___ Pre-Composer ___ Pre-Decomposer	___ "Same Thing" Comparer		___ Length Quantity Recognizer	___ Pattern Recognizer	___ Attribute Identifier
4	___ Shape Recognizer—Circles, Squares, and Triangles + ___ Constructor of Shapes from Parts—Looks Like Representing	___ Piece Assembler	___ "Similar" Comparer ___ Part Comparer ___ Some Attributes Comparer	___ Simple Turner	___ Length Direct Comparer	___ Pattern Fixer ___ Pattern Duplicator AB ___ Pattern Extender AB ___ Pattern Duplicator	___ Attribute Sorter
5	___ Shape Recognizer—All Rectangles ___ Side Recognizer ___ Angle Recognizer ___ Shape Recognizer—More Shapes	___ Picture Maker ___ Simple Decomposer ___ Shape Composer	___ Most Attributes Comparer	___ Beginning Slider, Flipper, Turner	___ Indirect Length Comparer	___ Pattern Extender	___ Consistent Sorter
6	___ Shape Identifier ___ Angle Matcher Parts	___ Substitution Composer ___ Shape Decomposer (with help)		___ Slider, Flipper, Turner	___ Serial Orderer to 6+ ___ End-to-End Length Measurer		___ Exhaustive Sorter ___ Multiple Attribute Sorter
7	___ Parts of Shapes Identifier ___ Constructor of Shapes from Parts—Exact Representing	___ Shape Composite Repeater ___ Shape Decomposer with Imagery	___ Congruence Determiner ___ Congruence Superposer	___ Diagonal Mover	___ Length Unit Iterater ___ Length Unit Relater	___ Pattern Unit Recognizer	___ Classifier and Counter ___ List Grapher
8+	___ Shape Class Identifier ___ Shape Property Identifier ___ Angle Size Comparer ___ Angle Measurer ___ Property Class Identifier ___ Angle Synthesizer	___ Shape Composer—Units of Units ___ Shape Decomposer with Units of Units	___ Congruence Representer	___ Mental Mover	___ Length Measurer ___ Conceptual Ruler Measurer		___ Multiple Attribute Classifier ___ Classifying Grapher ___ Classifier ___ Hierarchical Classifier ___ Data Representer

C32 Number Worlds • Trajectory Progress Chart

Glossary

A

acute angle An angle with a measure greater than 0 degrees and less than 90 degrees.

addend One of the numbers being added in an addition sentence. In the sentence 41 + 27 = 68, the numbers 41 and 27 are addends.

addition A mathematical operation based on "putting things together." Numbers being added are called *addends*. The result of addition is called a *sum*. In the number sentence 15 + 63 = 78, the numbers 15 and 63 are addends.

additive inverses Two numbers whose sum is 0. For example, 9 + −9 = 0. The additive inverse of 9 is −9, and the additive inverse of −9 is 9.

adjacent angles Two angles with a common side that do not otherwise overlap. In the diagram, angles 1 and 2 are adjacent angles; so are angles 2 and 3, angles 3 and 4, and angles 4 and 1.

algorithm A step-by-step procedure for carrying out a computation or solving a problem.

angle Two rays with a common endpoint. The common endpoint is called the vertex of the angle.

area A measure of the surface inside a closed boundary. The formula for the area of a rectangle or parallelogram is $A = b \times h$, where A represents the area, b represents the length of the base, and h is the height of the figure.

array A rectangular arrangement of objects in rows and columns in which each row has the same number of elements, and each column has the same number of elements.

attribute A feature such as size, shape, or color.

average See **mean**. The **median** and **mode** are also sometimes called the *average*.

axis (plural **axes**) A number line used in a coordinate grid.

B

bar graph A graph in which the lengths of horizontal or vertical bars represent the magnitude of the data represented.

base ten The commonly used numeration system, in which the ten digits 0, 1, 2,..., 9 have values that depend on the place in which they appear in a numeral (ones, tens, hundreds, and so on, to the left of the decimal point; tenths, hundredths, and so on, to the right of the decimal point).

bisect To divide a segment, angle, or figure into two parts of equal measure.

C

capacity A measure of how much liquid or substance a container can hold. See also **volume**.

centi- A prefix for units in the metric system meaning one hundredth.

centimeter (cm) In the metric system, a unit of length defined as 1/100 of a meter; equal to 10 millimeters or 1/10 of a decimeter.

circle The set of all points in a plane that are a given distance (the radius) from a given point (the center of the circle).

circle graph A graph in which a circular region is divided into sectors to represent the categories in a set of data. The circle represents the whole set of data.

circumference The distance around a circle or sphere.

closed figure A figure that divides the plane into two regions, inside and outside the figure. A closed space figure divides space into two regions in the same way.

common denominator Any nonzero number that is a multiple of the denominators of two or more fractions.

common factor Any number that is a factor of two or more numbers.

complementary angles Two angles whose measures total 90 degrees.

composite function A function with two or more operations. For example, this function multiplies the input number by 5 then adds 3.

composite number A whole number that has 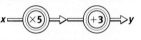 more than two whole number factors. For example, 14 is a composite number because it has more than two whole number factors.

cone A space figure having a circular base, curved surface, and one vertex.

congruent Having identical sizes and shapes. Congruent figures are said to be congruent to each other.

coordinate One of two numbers used to locate a point on a coordinate grid. See also **ordered pair**.

coordinate grid A device for locating points in a plane by means of ordered pairs or coordinates. A coordinate grid is formed by two number lines that intersect at their 0-points.

corresponding angles Two angles in the same relative position in two figures, or in similar locations in relation to a transversal intersecting two lines. In the diagram above, angles 1 and 5, 3 and 7, 2 and 6, and 4 and 8 are corresponding angles. If the lines are parallel, then the corresponding angles are congruent.

corresponding sides Two sides in the same relative position in two figures. In the diagram AB and A'B', BC and B'C', and AC and A'C' are corresponding sides.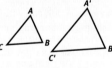

cube A space figure whose six faces are congruent squares that meet at right angles.

cubic centimeter (cm³) A metric unit of volume; the volume of a cube 1 centimeter on an edge. 1 cubic centimeter is equal to 1 milliliter.

cubic unit A unit used in a volume and capacity measurement.

customary system of measurement The measuring system used most often in the United States. Units for linear measure (length, distance) include inch, foot, yard, and mile; units for weight include ounce and pound; units for capacity (amount of liquid or other substance a container can hold) include fluid ounce, cup, pint, quart, and gallon.

cylinder A space figure having a curved surface and parallel circular or elliptical bases that are congruent.

D

decimal A number written in standard notation, usually one containing a decimal point, as in 3.78.

decimal approximation A decimal that is close to the value of a rational number. By extending the decimal approximation to additional digits, it is possible to come as close as desired to the value of the rational number. For example, decimal approximations of $\frac{1}{12}$ are 0.083, 0.0833, 0.08333, and so on.

decimal equivalent A decimal that names the same number as a fraction. For example, the decimal equivalent of $\frac{3}{4}$ is 0.75. The only rational numbers with decimal equivalents are those that can be written as fractions whose denominators have prime factors only of 2 and 5. For example, $\frac{1}{2}, \frac{1}{4},$ and $\frac{1}{20}$ have decimal equivalents, but $\frac{1}{6}, \frac{1}{7},$ and $\frac{1}{9}$ have only decimal approximations.

degree (°) A unit of measure for angles; based on dividing a circle into 360 equal parts. Also, a unit of measure for temperature.

degree Celsius (°C) In the metric system, the unit for measuring temperature. Water freezes at 0°C and boils at 100°C.

degree Fahrenheit (°F) In the U.S. customary system, the unit for measuring temperature. Water freezes at 32°F and boils at 212°F.

denominator The number of equal parts into which a whole is divided. In the fraction $\frac{a}{b}$, b is the denominator. See also **numerator**.

diameter A line segment, going through the center of a circle, that starts at one point on the circle and ends at the opposite point on the circle; also, the length of such a line segment. The diameter of a circle is twice its radius. AB is a diameter of this circle. See also **circle**.

difference The result of subtraction. In the subtraction sentence 40 − 10 = 30, the difference is 30.

digit In the base-ten numeration system, one of the symbols 0, 1, 2, 3, 4, 5, 6, 7, 8, 9. Digits can be used to write a numeral for any whole number in the base-ten numbering system. For example, the numeral 145 is made of the digits 1, 4, and 5.

distributive law A law that relates two operations on numbers, usually multiplication and addition, or multiplication and subtraction. Distributive law of multiplication over addition: $a \times (b + c) = (a \times b) + (a \times c)$

dividend See **division**.

division A mathematical operation based on "equal sharing" or "separating into equal parts." The *dividend* is the total before sharing. The divisor is the number of equal parts or the number in each equal part. The *quotient* is the result of division. For example, in 35 ÷ 5 = 7, 35 is the dividend, 5 is the divisor, and 7 is the quotient. If 35 objects are separated into 5 equal parts, there are 7 objects in each part. If 35 objects are separated into parts with 5 in each part, there are 7 equal parts. The number left over when a set of objects is shared equally or separated into equal groups is called the *remainder*. For 35 ÷ 5, the quotient is 7 and the remainder is 0. For 36 ÷ 5, the quotient is 7 and the remainder is 1.

Glossary

divisor See **division**.

E

edge The line segment where two faces of a polyhedron meet.

endpoint The point at either end of a line segment; also, the point at the end of a ray. Line segments are named after their endpoints; a line segment between and including points A and B is called segment AB or segment BA.

equation A mathematical sentence that states the equality of two expressions. For example, $3 + 7 = 10$, $y = x + 7$, and $4 + 7 = 8 + 3$ are equations.

equilateral polygon A polygon in which all sides are the same length.

equivalent Equal in value, but in a different form. For example, $\frac{1}{2}$, $\frac{2}{4}$, 0.5, and 50% are equivalent forms of the same number.

equivalent fractions Fractions that have different numerators and denominators but name the same number. For example, $\frac{2}{3}$ and $\frac{6}{9}$ are equivalent fractions.

estimate A judgment of time, measurement, number, or other quantity that may not be exactly right.

evaluate an algebraic expression To replace each variable in an algebraic expression with a particular number and then to calculate the value of the expression.

evaluate a numerical expression To carry out the operations in a numerical expression to find the value of the expression.

even number A whole number such as 0, 2, 4, 6, and so on, that can be divided by 2 with no remainder. See also **odd number**.

event A happening or occurrence. The tossing of a coin is an event.

expression A group of mathematical symbols (numbers, operation signs, variables, grouping symbols) that represents a number (or can represent a number if values are assigned to any variables it contains).

F

face A flat surface on a space figure.

fact family A group of addition or multiplication facts grouped together with the related subtraction or division facts. For example, $4 + 8 = 12$, $8 + 4 = 12$, $12 - 4 = 8$, and $12 - 8 = 4$ form an addition fact family. The facts $4 \times 3 = 12$, $3 \times 4 = 12$, $12 \div 3 = 4$, and $12 \div 4 = 3$ form a multiplication fact family.

factor (noun) One of the numbers that is multiplied in a multiplication expression. For example, in $4 \times 1.5 = 6$, the factors are 4 and 1.5. See also **multiplication**.

factor (verb) To represent a quantity as a product of factors. For example, 20 factors to 4×5, 2×10, or $2 \times 2 \times 5$.

factor of a whole number n A whole number, which, when multiplied by another whole number, results in the number n. The whole number n is divisible by its factors. For example, 3 and 5 are factors of 15 because $3 \times 5 = 15$, and 15 is divisible by 3 and 5.

factor tree A method used to obtain the prime factorization of a number. The original number is represented as a product of factors, and each of those factors is represented as a product of factors, and so on, until the factor string consists of prime numbers.

formula A general rule for finding the value of something. A formula is usually written as an equation with variables representing unknown quantities. For example, a formula for distance traveled at a constant rate of speed is $d = r \times t$, where d stands for distance, r is for rate, and t is for time.

fraction A number in the form $\frac{a}{b}$, where a and b are integers and b is not 0. Fractions are used to name part of a whole object or part of a whole collection of objects, or to compare two quantities. A fraction can represent division; for example, $\frac{2}{5}$ can be thought of as 2 divided by 5.

frequency The number of times an event or value occurs in a set of data.

function machine An imaginary machine that processes numbers according to a certain rule. A number (input) is put into the machine and is transformed into a second number (output) by application of the rule.

G

greatest common factor The largest factor that two or more numbers have in common. For example, the common factors of 24 and 30 are 1, 2, 3, and 6. The greatest common factor of 24 and 30 is 6.

H

height (of a parallelogram) The length of the line segment between the base of the parallelogram and the opposite side (or an extension of the opposite side), running perpendicular to the base.

height (of a polyhedron) The perpendicular distance between the bases of the polyhedron or between a base and the opposite vertex.

height (of a rectangle) The length of the side perpendicular to the side considered base of the rectangle. (Base and height of a rectangle are interchangeable.)

height (of a triangle) The length of the line segment perpendicular to the base of the triangle (or an extension of the base) from the opposite vertex.

hexagon A polygon with six sides.

histogram A bar graph in which the labels for the bars are numerical intervals.

hypotenuse In a right triangle, the side opposite the right angle.

I

improper fraction A fraction that names a number greater than or equal to 1; a fraction whose numerator is equal to or greater than its denominator. Examples of improper fractions are $\frac{4}{3}$, $\frac{10}{8}$, and $\frac{4}{4}$.

inch (in.) In the U. S. customary system, a unit of length equal to $\frac{1}{12}$ of a foot.

indirect measurement Methods for determining heights, distances, and other quantities that cannot be measured or are not measured directly.

inequality A number sentence stating that two quantities are not equal. Relation symbols for inequalities include < (is less than), > (is greater than), and ≠ (is not equal to).

integers The set of integers is {..., −4, −3, −2, −1, 0, 1, 2, 3, 4, ...}. The set of integers consists of whole numbers and their opposites.

intersect To meet (at a point, a line, and so on), sharing a common point or points.

interior The set of all points in a plane "inside" a closed plane figure, such as a polygon or circle. Also, the set of all points in space "inside" a closed space figure, such as a polyhedron or sphere.

isosceles Having two sides of the same length; commonly used to refer to triangles and trapezoids.

K

kilo- A prefix for units in the metric system meaning one thousand.

L

least common denominator The least common multiple of the denominators of every fraction in a given set of fractions. For example, 12 is the least common denominator of $\frac{2}{3}$, $\frac{1}{4}$, and $\frac{5}{6}$. See also **least common multiple**.

least common multiple The smallest number that is a multiple of two or more numbers. For example, some common multiples of 6 and 8 are 24, 48, and 72. 24 is the least common multiple of 6 and 8.

leg of a right triangle A side of a right triangle that is not the hypotenuse.

line A straight path that extends infinitely in opposite directions.

line graph (broken-line graph) A graph in which points are connected by line segments to represent data.

line of symmetry A line that separates a figure into halves. The figure can be folded along this line into two parts which exactly fit on top of each other.

line segment A straight path joining two points, called *endpoints* of the line segment. A straight path can be described as the shortest distance between two points.

line symmetry A figure has line symmetry (also called *bilateral symmetry*) if a line of symmetry can be drawn through the figure.

liter (L) A metric unit of capacity, equal to the volume of a cube 10 centimeters on an edge. $1 L = 1,000 mL = 1,000 cm^3$. A liter is slightly larger than a quart. See also **milliliter (mL)**.

M

map scale A ratio that compares the distance between two locations shown on a map with the actual distance between them.

mean A typical or central value that may be used to describe a set of numbers. It can be found by adding the numbers in the set and dividing the sum by the number of numbers. The mean is often referred to as the *average*.

median The middle value in a set of data when the data are listed in order from least to greatest (or greatest to least). If the number of values in the set is even (so that there is no "middle" value), the median is the mean of the two middle values.

meter (m) The basic unit of length in the metric system, equal to 10 decimeters, 100 centimeters, and 1,000 millimeters.

metric system of measurement A measurement system based on the base-ten numeration system and used in most countries in the world. Units for linear measure (length, distance) include millimeter, centimeter, meter, kilometer; units for mass (weight) include gram and kilogram; units for capacity (amount of liquid or other substance a container can hold) include milliliter and liter.

Glossary

midpoint A point halfway between two points.

milli- A prefix for units in the metric system meaning one thousandth.

milliliter (mL) A metric unit of capacity, equal to 1/1,000 of a liter and 1 cubic centimeter.

millimeter (mm) In the metric system, a unit of length equal to 1/10 of a centimeter and 1/1,000 of a meter.

minuend See **subtraction**.

mixed number A number greater than 1, written as a whole number and a fraction less than 1. For example, $5\frac{1}{2}$ is equal to $5 + \frac{1}{2}$.

mode The value or values that occur most often in a set of data.

multiple of a number n The product of a whole number and the number n. For example, the numbers 0, 4, 8, 12, and 16 are all multiples of 4 because $4 \times 0 = 0$, $4 \times 1 = 4$, $4 \times 2 = 8$, $4 \times 3 = 12$, and $4 \times 4 = 16$.

multiplication A mathematical operation used to find the total number of things in several equal groups, or to find a quantity that is a certain number of times as much or as many as another number. Numbers being multiplied are called *factors*. The result of multiplication is called the *product*. In $8 \times 12 = 96$, 8 and 12 are the factors and 96 is the product.

multiplicative inverses Two numbers whose product is 1. For example, the multiplicative inverse of $\frac{2}{5}$ is $\frac{5}{2}$, and the multiplicative inverse of 8 is $\frac{1}{8}$. Multiplicative inverses are also called *reciprocals* of each other.

N

negative number A number less than 0; a number to the left of 0 on a horizontal number line.

number line A line on which equidistant points correspond to integers in order.

number sentence A sentence that is made up of numerals and a relation symbol ($<$, $>$, or $=$). Most number sentences also contain at least one operation symbol. Number sentences may also have grouping symbols, such as parentheses.

numeral The written name of a number.

numerator In a whole divided into a number of equal parts, the number of equal parts being considered. In the fraction $\frac{a}{b}$, a is the numerator.

O

obtuse angle An angle with a measure greater than 90 degrees and less than 180 degrees.

octagon An eight-sided polygon.

odd number A whole number that is not divisible by 2, such as 1, 3, 5, and so on. When an odd number is divided by 2, the remainder is 1. A whole number is either an odd number or an even number.

opposite of a number A number that is the same distance from 0 on the number line as the given number, but on the opposite side of 0. If a is a negative number, the opposite of a will be a positive number. For example, if $a = -5$, then $-a$ is 5. See also **additive inverses**.

ordered pair Two numbers or objects for which order is important. Often, two numbers in a specific order used to locate a point on a coordinate grid. They are usually written inside parentheses; for example, (2, 3). See also **coordinate**.

ordinal number A number used to express position or order in a series, such as first, third, tenth. People generally use ordinal numbers to name dates; for example, "May fifth" rather than "May five."

origin The point where the x- and y-axes intersect on a coordinate grid. The coordinates of the origin are (0, 0).

outcome The result of an event. Heads and tails are the two outcomes of the event of tossing a coin.

P

parallel lines (segments, rays) Lines (segments, rays) going in the same direction that are the same distance apart and never meet.

parallelogram A quadrilateral that has two pairs of parallel sides. Pairs of opposite sides and opposite angles of a parallelogram are congruent.

parentheses A pair of symbols, (and), used to show in which order operations should be done. For example, the expression $(3 \times 5) + 7$ says to multiply 5 by 3 then add 7. The expression $3 \times (5 + 7)$ says to add 5 and 7 and then multiply by 3.

pattern A model, plan, or rule that uses words or variables to describe a set of shapes or numbers that repeat in a predictable way.

pentagon A polygon with five sides.

percent A rational number that can be written as a fraction with a denominator of 100. The symbol % is used to represent percent. 1% means 1/100 or 0.01. For example, "53% of the students in the school are girls" means that of every 100 students in the school, 53 are girls.

perimeter The distance along a path around a plane figure. A formula for the perimeter of a rectangle is $P = 2 \times (B + H)$, where B represents the base and H is the height of the rectangle. Perimeter may also refer to the path itself.

perpendicular Two rays, lines, line segments, or other figures that form right angles are said to be perpendicular to each other.

pi The ratio of the circumference of a circle to its diameter. Pi is the same for every circle, approximately 3.14 or $\frac{22}{7}$. Also written as the Greek letter π.

pictograph A graph constructed with pictures or icons, in which each picture stands for a certain number. Pictographs make it easier to visually compare quantities.

place value A way of determining the value of a digit in a numeral, written in standard notation, according to its position, or place, in the numeral. In base-ten numbers, each place has a value ten times that of the place to its right and one-tenth the value of the place to its left.

plane A flat surface that extends forever.

plane figure A figure that can be contained in a plane (that is, having length and width but no height).

point A basic concept of geometry; usually thought of as a location in space, without size.

polygon A closed plane figure consisting of line segments (sides) connected endpoint to endpoint. The interior of a polygon consists of all the points of the plane "inside" the polygon. An n-gon is a polygon with n sides; for example, an 8-gon has 8 sides.

polyhedron A closed space figure, all of whose surfaces (faces) are flat. Each face consists of a polygon and the interior of the polygon.

power A product of factors that are all the same. For example, $6 \times 6 \times 6$ (or 216) is called 6 to the third power, or the third power of 6, because 6 is a factor three times. The expression $6 \times 6 \times 6$ can also be written as 6^3.

power of 10 A whole number that can be written as a product using only 10 as a factor. For example, 100 is equal to 10×10 or 10^2, so 100 is called 10 squared, the second power of 10, or 10 to the second power. Other powers of 10 include 10^1, or 10, and 10^3, or 1,000.

prime factorization A whole number expressed as a product of prime factors. For example, the prime factorization of 18 is $2 \times 3 \times 3$. A number has only one prime factorization (except for the order in which the factors are written).

prime number A whole number greater than 1 that has exactly two whole number factors, 1 and itself. For example, 13 is a prime number because its only factors are 1 and 13. A prime number is divisible only by 1 and itself. The first five prime numbers are 2, 3, 5, 7, and 11. See also **composite number**.

prism A polyhedron with two parallel faces (bases) that are the same size and shape. Prisms are classified according to the shape of the two parallel bases. The bases of a prism are connected by parallelograms that are often rectangular.

probability A number between 0 and 1 that indicates the likelihood that something (an event) will happen. The closer a probability is to 1, the more likely it is that an event will happen.

product See **multiplication**.

protractor A tool for measuring or drawing angles. When measuring an angle, the vertex of the angle should be at the center of the protractor and one side should be aligned with the 0 mark.

pyramid A polyhedron in which one face (the base) is a polygon and the other faces are formed by triangles with a common vertex (the apex). A pyramid is classified according to the shape of its base, as a triangular pyramid, square pyramid, pentagonal pyramid, and so on.

Pythagorean Theorem A mathematical theorem, proven by the Greek mathematician Pythagoras and known to many others before and since, that states that if the legs of a right triangle have lengths a and b, and the hypotenuse has length c, then $a^2 + b^2 = c^2$.

Q

quadrilateral A polygon with four sides.

quotient See **division**.

Glossary

R

radius A line segment that goes from the center of a circle to any point on the circle; also, the length of such a line segment.

random sample A sample taken from a population in a way that gives all members of the population the same chance of being selected.

range The difference between the maximum and minimum values in a set of data.

rate A ratio comparing two quantities with unlike units. For example, a measure such as 23 miles per gallon of gas compares mileage with gas usage.

ratio A comparison of two quantities using division. Ratios can be expressed with fractions, decimals, percents, or words. For example, if a team wins 4 games out of 5 games played, the ratio of wins to total games is $\frac{4}{5}$, 0.8, or 80%.

rational number Any number that can be represented in the form $a \div b$ or $\frac{a}{b}$, where a and b are integers and b is positive. Some, but not all, rational numbers have exact decimal equivalents.

ray A straight path that extends infinitely in one direction from a point, which is called its *endpoint*.

reciprocal See **multiplicative inverses**.

rectangle A parallelogram with four right angles.

reduced form A fraction in which the numerator and denominator have no common factors except 1.

reflection A transformation in which a figure "flips" so that its image is the reverse of the original.

regular polygon A convex polygon in which all the sides are the same length and all the angles have the same measure.

relation symbol A symbol used to express the relationship between two numbers or expressions. Among the symbols used in number sentences are = for "is equal to," < for "is less than," > for "is greater than," and ≠ for "is not equal to."

remainder See **division**.

rhombus A parallelogram whose sides are all the same length.

right angle An angle with a measure of 90 degrees, representing a quarter of a full turn.

right triangle A triangle that has a right angle.

rotation A transformation in which a figure "turns" around a center point or axis.

rotational symmetry Property of a figure that can be rotated around a point (less than a full, 360-degree turn) in such a way that the resulting figure exactly matches the original figure. If a figure has rotational symmetry, its order of rotational symmetry is the number of different ways it can be rotated to match itself exactly. "No rotation" is counted as one of the ways.

rounding Changing a number to another number that is easier to work with and is close enough for the purpose. For example, 12,924 rounded to the nearest thousand is 13,000 and rounded to the nearest hundred is 12,900.

S

sample A subset of a group used to represent the whole group.

scale The ratio of the distance on a map or drawing to the actual distance.

scalene triangle A triangle in which all three sides have different lengths.

scale drawing An accurate picture of an object in which all parts are drawn to the same scale. If an actual object measures 32 by 48 meters, a scale drawing of it might measure 32 by 48 millimeters.

scale model A model that represents an object or display in proportions based on a determined scale.

similar figures Figures that are exactly the same shape but not necessarily the same size.

space figure A figure which cannot be contained in a plane. Common space figures include the rectangular prism, square pyramid, cylinder, cone, and sphere.

sphere The set of all points in space that are a given distance (the radius) from a given point (the center). A ball is shaped like a sphere.

square number A number that is the product of a whole number and itself. The number 36 is a square number, because 36 = 6 × 6.

square of a number The product of a number multiplied by itself. For example, 2.5 squared is $(2.5)^2$.

square root The square root of a number n is a number which, when multiplied by itself, results in the number n. For example, 8 is a square root of 64, because 8 × 8 = 64.

square unit A unit used to measure area—usually a square that is 1 inch, 1 centimeter, 1 yard, or other standard unit of length on each side.

standard notation The most familiar way of representing whole numbers, integers, and decimals by writing digits in specified places; the way numbers are usually written in everyday situations.

statistics The science of collecting, classifying, and interpreting numerical data as it is related to a particular subject.

stem-and-leaf plot A display of data in which digits with larger place values are named as stems, and digits with smaller place values are named as leaves.

straight angle An angle of 180 degrees; a line with one point identified as the vertex of the angle.

subtraction A mathematical operation based on "taking away" or comparing ("How much more?"). The number being subtracted is called the *subtrahend;* the number it is subtracted from is called the *minuend;* the result of subtraction is called the *difference.* In the number sentence 63 − 45 = 18, 63 is the minuend, 45 is the subtrahend, and 18 is the difference.

subtrahend See **subtraction**.

supplementary angles Two angles whose measures total 180 degrees.

surface area The sum of the areas of the faces of a space figure.

symmetrical Having the same size and shape across a dividing line or around a point.

T

tessellation An arrangement of closed shapes that covers a surface completely without overlaps or gaps.

tetrahedron A space figure with four faces, each formed by an equilateral triangle.

transformation An operation that moves or changes a geometric figure in a specified way. Rotations, reflections, and translations are types of transformations.

translation A transformation in which a figure "slides" along a line.

transversal A line which intersects two or more other lines.

trapezoid A quadrilateral with exactly one pair of parallel sides.

tree diagram A tool used to solve probability problems in which there is a series of events. This tree diagram represents a situation where the first event has three possible outcomes and the second event has two possible outcomes.

triangle A polygon with three sides. An *equilateral* triangle has three sides of the same length. An *isosceles* triangle has two sides of the same length. A *scalene* triangle has no sides of the same length.

U

unit (of measure) An agreed-upon standard with which measurements are compared.

unit fraction A fraction whose numerator is 1. For example, $\frac{1}{2}$, $\frac{1}{3}$, and $\frac{1}{10}$ are unit fractions.

unit cost The cost of one item or one specified amount of an item. If 20 pencils cost 60¢, then the unit cost is 3¢ per pencil.

unlike denominators Unequal denominators, as in $\frac{3}{4}$ and $\frac{5}{6}$.

V

variable A letter or other symbol that represents a number, one specific number, or many different values.

Venn diagram A picture that uses circles to show relationships between sets. Elements that belong to more than one set are placed in the overlap between the circles.

vertex The point at which the rays of an angle, two sides of a polygon, or the edges of a polyhedron meet.

vertical angles Two intersecting lines form four adjacent angles. In the diagram, angles 2 and 4 are vertical angles. They have no sides in common. Their measures are equal. Similarly, angles 1 and 3 are vertical angles.

volume A measure of the amount of space occupied.

W

whole number Any of the numbers 0, 1, 2, 3, 4, and so on. Whole numbers are the numbers used for counting and zero.

Scope and Sequence

The topics addressed at each grade level were determined after extensive analysis of national and state mathematics teaching expectations, standardized assessments, and topics covered in mathematics basal programs. Across all levels, the program follows the optimal sequence outlined by the learning trajectories of primary mathematics.

	A	B	C	D	E	F	G	H
Addition (whole numbers)								
Basic facts	•	•	•	•	•	•	•	•
Three or more addends				•	•	•	•	•
Two-digit numbers				•	•	•	•	•
Three-digit numbers					•	•	•	•
Greater numbers						•	•	•
Estimating sums						•	•	•
Algebra								
Properties of whole numbers	•	•	•	•	•	•	•	•
Integers (negative numbers)								
Operations with integers							•	•
Make and solve number sentences and equations		•	•	•	•	•	•	•
Variables							•	•
Order of operations							•	•
Writing variable expressions							•	•
Evaluating expressions							•	•
Solving one-step equations	•	•	•	•	•	•	•	•
Solving two-step equations				•	•	•	•	•
Combining like terms							•	•
Solving inequalities							•	•
Function machines/tables						•	•	•
Coordinate graphing						•	•	•
Graphing linear functions						•	•	•
Using formulas						•	•	•
Decimals and Money								
Place value						•	•	•
Adding						•	•	•
Subtracting						•	•	•
Multiplying by a whole number						•	•	•
Multiplying by a decimal								•
Multiplying by powers of 10						•	•	•

Scope and Sequence

	A	B	C	D	E	F	G	H
Decimals and Money								
Dividing by a whole number							•	•
Dividing by a decimal								•
Identifying and counting currency				•	•	•	•	•
Exchanging money				•	•	•	•	•
Computing with money				•	•	•	•	•
Division								
Basic facts							•	•
Remainders							•	•
One-digit divisors							•	•
Two-digit divisors							•	•
Greater divisors							•	•
Estimating quotients							•	•
Fractions								
Fractions of a whole				•	•	•	•	•
Comparing/ordering				•	•	•	•	•
Equivalent fractions							•	•
Reduced form						•	•	•
Mixed numbers/improper fractions							•	•
Adding–like denominators								•
Adding–unlike denominators								•
Geometry								
Plane figures	•	•	•	•	•	•	•	•
Classifying figures	•	•	•	•	•	•	•	•
Solid figures	•	•	•	•	•	•	•	
Congruence				•	•	•	•	•
Symmetry				•	•	•	•	•
Symmetry (rotational)					•	•	•	•
Slides/Flips/Turns					•	•	•	•
Angles					•	•	•	•
Classifying triangles				•	•	•	•	•
Classifying quadrilaterals						•	•	•
Parallel and perpendicular lines						•	•	•
Perimeter					•	•	•	•
Radius and diameter							•	•
Circumference							•	•
Surface Area				•	•	•	•	•
Volume				•	•	•	•	•

Scope and Sequence

Measurement

	A	B	C	D	E	F	G	H
Length								
Use customary units				•	•	•	•	•
Use metric units				•	•	•	•	•
Mass/Weight								
Use customary units				•	•	•	•	•
Use metric units				•	•	•	•	•
Capacity								
Use customary units				•	•	•	•	•
Use metric units				•	•	•	•	•
Temperature								
Use degrees Fahrenheit				•	•	•	•	•
Use degrees Celsius				•	•	•	•	•
Converting within customary system						•	•	•
Converting within metric system						•	•	•
Telling Time								
to the hour	•	•			•			
to the half hour					•			
to the quarter hour					•			
to the minute					•			
Converting units of time					•	•		
Reading a calendar					•			

Multiplication

	A	B	C	D	E	F	G	H
Basic facts					•	•	•	•
One-digit multipliers						•	•	•
Two-digit multipliers						•	•	•
Greater multipliers							•	•
Estimating products							•	•

Number and Numeration

	A	B	C	D	E	F	G	H
Reading and writing numbers	•	•	•	•	•	•		
Counting	•	•	•	•	•	•		
Skip counting				•	•	•	•	
Ordinal numbers				•	•	•		
Place value				•	•	•	•	
Even/odd numbers				•	•			
Comparing and ordering numbers	•	•	•	•	•	•	•	•
Rounding							•	•
Estimation/Approximation				•	•	•	•	•

Scope and Sequence

	A	B	C	D	E	F	G	H
Comparing and ordering integers							•	•
Integers (negative numbers)							•	•
Prime and composite numbers							•	•
Factors and prime factorization							•	•
Common factors							•	•
Common multiples						•	•	•

Subtraction (whole numbers)

	A	B	C	D	E	F	G	H
Basic facts	•	•	•	•	•	•	•	•
Two-digit numbers			•	•	•	•	•	•
Three-digit numbers					•	•	•	•
Greater numbers						•	•	•
Estimating differences	•	•	•	•	•	•	•	•

Patterns, Relations, and Functions

	A	B	C	D	E	F	G	H
Number patterns	•	•	•	•	•	•	•	•
Geometric patterns	•	•	•	•	•	•	•	•
Inequalities	•	•	•				•	•

Ratio and Proportion

	A	B	C	D	E	F	G	H
Meaning/Use							•	•
Similar Figures							•	•
Meaning of Percent							•	•
Percent of a Number							•	•

Statistics and Graphing

	A	B	C	D	E	F	G	H
Real and picture graphs	•	•	•	•	•	•	•	•
Bar graphs	•	•	•	•	•	•	•	•
Line graphs				•	•	•	•	•
Circle graphs							•	•
Analyzing graphs	•	•	•	•	•	•	•	•
Finding the mean							•	•
Finding the median							•	•
Finding the mode							•	•
Finding the range							•	•